Climate and Agriculture

Climate and Agriculture
An Ecological Survey

Jen-Hu Chang

Taylor & Francis Group
LONDON AND NEW YORK

First published 1968 by Transaction Publishers

Published 2017 by Routledge
2 Park Square, Milton Park, Abingdon, Oxon OX14 4RN
711 Third Avenue, New York, NY 10017, USA

Routledge is an imprint of the Taylor & Francis Group, an informa business

Copyright © 1968 by Jen-hu Chang

All rights reserved. No part of this book may be reprinted or reproduced or utilised in any form or by any electronic, mechanical, or other means, now known or hereafter invented, including photocopying and recording, or in any information storage or retrieval system, without permission in writing from the publishers.

Notice:
Product or corporate names may be trademarks or registered trademarks, and are used only for identification and explanation without intent to infringe.

Library of Congress Catalog Number: 2008055478

Library of Congress Cataloging-in-Publication Data

Chang, Jen-hu.
   Climate and agriculture : an ecological survey / Jen-hu Chang.
      p. cm.
   Originally published: Chicago : Aldine, 1968.
   Includes bibliographical references and index.
   ISBN 978-0-202-36249-6 (alk. paper)
     1. Crops and climate. I. Title.

S600.C43 2009
338.1'4--dc22

2008055478

ISBN 13: 978-0-202-36249-6(pbk)

This book is dedicated to my father,
Dr. Chi-yun Chang,
a geographer as well as an historian,
who sees in nature's bounty not chaos but a
challenge to man to utilize its resources.

CONTENTS

List of Symbols		xiii
1.	INTRODUCTION	1
2.	RADIATION BALANCE	4
	Solar constant	4
	Solar spectrum	4
	Depletion by the atmosphere	6
	Measurements of global radiation	7
	Radiation and duration of sunshine	8
	Radiation and cloudiness	9
	Distribution of solar radiation	9
	Reflectivity	11
	Outgoing radiation	14
	Net radiation	16
	Radiation balance in the greenhouse	20
	Biosphere as the locale of energy exchange	22
3.	PHOTOSYNTHESIS	23
	General effects of radiation on plant growth	23
	Basic process of photosynthesis	23
	Saturation light intensity and the efficiency of light utilization	24
	Photosynthesis in relation to temperature	28
	Photosynthesis in relation to CO_2 concentration	28
	Respiration and net photosynthesis	32
	Other factors affecting the rate of net photosynthesis	34
4.	RADIATION DISTRIBUTION WITHIN THE PLANT COMMUNITY	36
	Leaf transmissibility	36
	Leaf arrangement	36
	Radiation and light distribution within the canopy	39
	Net radiation profile	42

5.	**LEAF AREA INDEX**	46
	Basic concept	46
	Net photosynthesis as a function of leaf area index and extinction coefficient	47
	Dry matter production as a function of radiation and leaf area index	49
	Variation of leaf area index throughout the crop cycle	51
	Leaf area index as a guide to cultural practices	53
6.	**RADIATION UTILIZATION BY FIELD CROPS**	57
	Efficiency of radiation utilization by field crops	57
	Computation of potential photosynthesis	58
	Comparison between the tropics and temperate regions	59
	Radiation utilization during successive stages of crop development	60
	Empirical relationship between radiation and crop yield	62
	Shade experiments and artificial light	65
7.	**PHOTOPERIODISM**	70
	Historical background	70
	Classification	70
	Photoperiodic induction	71
	Relation to temperature	71
	Photoperiodism as a factor in plant distribution	73
	Response of tropical plants	73
	Practical applications	74
8.	**AIR AND LEAF TEMPERATURE**	75
	Cardinal temperatures	75
	The Van't Hoff law	76
	Degree-day	77
	Thermoperiodicity	79
	Thermal regime in the tropics	82
	Temperature records	82
	Leaf temperature	83
9.	**SOIL TEMPERATURE**	87
	Significance of soil temperature	87
	Methods of measurements and records	87
	Thermal properties of soils	88

Physical laws governing the change of soil temperature	89
Soil texture	90
Aspect and slope	94
The effect of tilth	95
Soil temperature and crop yield	95
Methods of modifying soil temperature	97

10. FROST PROTECTION — 100
The damaging effects of freezing temperature	1.00
Frost weather	100
Heater	102
Wind machine	104
Flooding and sprinkling	104
Brushing	107
Sanding	108
Windbreaks	108

11. WIND PROFILE NEAR THE GROUND — 109
The logarithmic equation	109
Deacon's equation and the Richardson number	111
Wind profile over tall crops	112
Roughness and zero plane displacement	113
Significance of roughness	117

12. WATER IN RELATION TO PLANT GROWTH — 118
General effects	118
Effect of water on photosynthesis	119
Transpiration	121
Transpiration and dry matter production	123

13. EVAPOTRANSPIRATION — 129
Definition	129
Meteorological factors determining potential evapotranspiration	131
Evaporation and transpiration differences between	133
Actual evapotranspiration	133
Evaporation from bare soil	140
Advection	140
The effect of plant height on the rate of evapotranspiration	143

14. LYSIMETERS	145
Determining potential evapotranspiration	145
Installation of the lysimeter	145
Drainage lysimeter	146
Weighing lysimeter	147
15. EMPIRICAL FORMULAE	149
Thornthwaite's method	149
The Blaney-Criddle formula	151
Makkink's formula	153
Turc's formula	154
Other empirical formulae	156
16. THE AERODYNAMIC APPROACH	157
Evaporation as a process of diffusion	157
Dalton equation	157
Principle of similarity	158
The Thornthwaite-Holzman equation	159
Eddy correlation technique	160
17. THE ENERGY BUDGET APPROACH	163
The energy budget equation	163
The Bowen ratio	165
The Penman equation	166
The fraction of radiation used in evapotranspiration	170
Diurnal variation of the energy budget	171
Energy budget throughout the crop cycle	172
Relative magnitude of evaporation and transpiration	174
18. EVAPORIMETERS	178
Limitations and advantages	178
Design and installation	179
Ratios between evapotranspiration and pan evaporation	185
Atmometers	190
19. WATER BALANCE	194
Limitations of soil moisture measurements	194
The water balance equation	195
Moisture storage capacity of the soil	198
Precipitation	199

Evapotranspiration	201
Deficit	203
Surplus	205

20. WATER AND YIELD RELATIONSHIP — 209
Irrigation practice for maximum yield — 209
Relationship between actual evapotranspiration and yield — 211
   Irrigation experiment — 215
   Efficiency of water use in dry matter production — 218
   Effect of irrigation on crop quality — 222

21. DEW, FOG, AND HUMIDITY — 225
Dew — 225
Fog — 230
Humidity — 231

22. WIND — 233
Effects of wind on plant growth — 233
Shelterbelts — 238

23. CONCLUSION — 243

Bibliography — 247

Glossary — 291

Index — 297

LIST OF SYMBOLS

I. Capital Letters
- A = Heat flux to air; *also* Effective pan ratio
- C = Cloudiness; *also* Control pan ratio
- C_p = Specific heat of air
- D = Difference between black and white atmometers; *also* Rate of dew formation
- E = Evapotranspiration
- E_a = Aerodynamic term in Penman's equation
- E_o = Evaporation from free-water surface
- F = Leaf area index
- I = Light intensity at given height within plant community; *also* Annual heat index in Thornthwaite formula
- I_o = Light intensity at top of plant community
- K = Crop coefficient
- K_c = Eddy diffusivity for carbon dioxide
- K_h = Eddy diffusivity for heat
- K_m = Eddy diffusivity for momentum
- K_w = Eddy diffusivity for water vapor
- Kh = Thermal diffusivity of soil
- M = Dry matter production; *also* Soil moisture storage capacity
- N = Maximum possible duration of sunshine; *also* Number of irrigation rounds
- P = Oscillation period; *also* Precipitation
- Q = Global radiation; *also* Carbon dioxide flux
- Q_A = Angot's value, or theoretical amount of radiation that would reach earth's surface in absence of atmosphere
- Q_b = Effective outgoing radiation
- Q_{bo} = Net long-wave radiation when Q is zero
- Q_n = Net radiation
- Q_t = Terrestrial outgoing radiation
- Q_T = Radiation intercepted by isolated tall vegetation
- R_i = Richardson number
- R_e = Effective rainfall
- R_o = Temperature range at surface
- R_z = Temperature range at depth z
- S = Heat flux to the soil

T = Temperature
T_a = Air temperature
T_s = Surface temperature
U = Consumptive use
V = Additional soil moisture available for evaporation (through vegetation) from cultivated soil
W = Transpiration
Z = Length of growing season

II. Lower Case Letters
a = Soil moisture available for evaporation from bare soil
b = Heating coefficient
c = Carbon dioxide concentration
d = Zero plane displacement
e = Vapor pressure of air
e_d = Saturation vapor pressure at dew point temperature
e_s = Vapor pressure at evaporating surface
g = Acceleration of gravity
h = Relative humidity; *also* Height of vegetation
i = Monthly heat index in Thornthwaite formula
k = Extinction coefficient; *also* Von Karman's constant
l = Evaporation power of air in Turc's equation
n = Actual duration of sunshine received
p = Daytime hours of any month as a percentage of the yearly total daytime hours.
q = Specific humidity
r = Reflection coefficient
s = Fraction of light passing through unit leaf layer without interception
t = Time
u = Wind speed
w = Vertical wind velocity; *also* Width of the vegetation
z = Depth; *also* Height; *also* Solar zenith angle
z_o = Roughness parameter

III. Greek Letters
α = Constant in Deacon's wind profile equation
β = Bowen ratio; *also* Stability index in Deacon's wind profile equation
Γ = Adiabatic lapse rate
γ = Psychrometric constant, 0.49 for °C and mm of mercury, or 0.27 for °F and mm of mercury
Δ = Soil moisture deficit from a reference value; *also* Slope of the saturated vapor pressure-temperature curve at the mean air temperature
ϵ = Emmisivity
ρ = Air density
σ = The Stefan-Bolzman constant, 8.17×10^{-11} langley/T^4/minute, or 5.70×10^{-5} erg/cm^2/T^4/sec.
τ = Leaf transmission coefficient; *also* Momentum density flux; *also* Shearing stress
ϕ = Degree of latitude

PREFACE

Studies of the relationships between plants and climate have made great strides in the last 20 years. On the one hand, the rapid progress in the study of micrometeorology, particularly the energy budget and water balance, permits a precise quantitative characterization of the climatic environment. On the other hand, the elaborate experimentation of biological and agricultural scientists has brought to light an array of new findings about the varied plant responses to the environment. However, these studies are widely scattered in journals dealing with plant physiology, botany, ecology, agronomy, soil, forestry, agricultural engineering, hydrology, meteorology, climatology, and geography, and, as such, may not be utilized readily by research workers except in their respective fields. This book is compiled upon the premise that a broad survey of recent advances will not only show the inadequacy of the traditional concepts but also will stimulate further research.

This presentation is intended both as a textbook in agricultural climatology and as a reference manual for research workers in several fields. The emphasis is on general principles rather than on experimental and technical details that require very long treatment and are of interest only to a restricted audience. However, the comprehensive references in the text should enable the reader to pursue further subjects of special interest.

Doubtless, no two scientists will agree upon what material should be included in such a broad interdisciplinary survey and how it should be organized. I have found the present order of presentation quite satisfactory in a course offered to students from a multitude of disciplines. The main body of the text, with the exception of Chapter 22, may be divided into two parts, namely, the energy budget of radiant and sensible heat (Chapters 2-10), and water balance (Chapters 11-21). Chapter 2 deals with the physical processes of radiation flux from the top of the atmosphere to the ground surface. The general discussion of the process of photosynthesis in Chapter 3 is essential for an understanding of radiation utilization by field crops as affected by energy distribution within the canopy and the leaf area index (Chapters 4-6). Photoperiodism, in Chapter 7, differs from photosynthesis in that it is a photostimulus process requiring only very weak light intensity. Chapters 8 through 10 are concerned with various facets of temperature effect on plant growth.

The second part of the book, the water balance section, starts with a study of wind profile, which may seem out of place at first glance. However, it is necessary for an understanding of the physics of ground-level vapor flux and hence the plant and water relationship. The discussion of the effects of water on plant photosynthesis and transpiration in Chapter 12 is followed by a treatment of the broadened concept of evapotranspiration (Chapter 13) and the various methods of measuring or estimating it (Chapters 14-18). The determination of evapotranspiration makes it possible to develop the meteorological approach of water balance technique (Chapter 19), which has immediate applications in irrigation practice and water resources management (Chapter 20). Chapter 21 deals with minor hydrometeors that are usually ignored in the water balance computation. Chapter 22, on the effect of wind on plant growth, forms an isolated topic in the book and is a difficult one to handle adequately.

The order presented in this book doesn't need to be followed in classroom presentation. The organization and the emphasis could vary according to the background and interest of the students.

I was introduced into the realm of micrometeorology and topoclimatology by the late Dr. C. W. Thornthwaite. Although his approach suffers from an occasional lapse in scientific rigor, his many fresh ideas have dominated the progress of microclimatology for the last twenty years. I am indebted to him for his stimulation and inspiration in my early years. Subsequently, I studied under the late Professor C. F. Brooks of Harvard University and Dr. V. E. Suomi and Dr. R. A. Bryson of the University of Wisconsin. Their searching instruction further acquainted me with the intricacy of micrometeorological instruments and sophisticated meteorological concepts. From 1958 to 1964, I had the opportunity to gain first-hand experience on the application of climatological research to sugar cane culture in Hawaii. I soon came to realize that the bulging file of neatly written climatological records of conventional measurements was scarcely used and then only in the crudest manner. Equally discouraging was the fact that some agriculturists were dubious of basic micrometeorological doctrine. The need for a ready reference book to bridge the gap between theory and application and between physical and agricultural sciences was apparent. Realizing the broad scope of the subject and the inadequacy of my training, I undertook this task with full expectation that some ideas presented here would soon be superseded or refined as the world of scholarship rapidly moves on.

This manuscript was written as part of a research project sponsored by the Department of Geography and the Water Resources Research Center of the University of Hawaii. Professors P. C. Ekern and John M. Street kindly read the first draft and offered many helpful criticisms. I am thankful to Mrs. Rose T. Pfund for her patience in editing the manuscript.

JEN-HU CHANG

CHAPTER 1

INTRODUCTION

Plants depend for growth and development on their genetic constitution and on the environmental conditions of soil and climate. As an ecological factor in agriculture, soil has been thoroughly studied and is better understood than climate. In general, farmers know more about soil management than they do about fully exploiting climatic resources. One reason for the slow progress of agricultural meteorology is the common misconception that studies of plant and climate relationships have only limited practical value. Although man is not yet able to change the weather, except on a very limited scale, he is capable of adjusting agricultural practices to fit the climate. Apart from the many applications of weather forecasting to current problems, meteorological studies may benefit agriculture in at least three other ways.

The first relates to the problem of selecting the production site for a given crop or the crop for a given site. Although the locations of many agricultural regions—for example, the Corn and Cotton Belts—were selected by farmers long before the development of the modern science of climatology, the lack of detailed knowledge of plant and climate relationships has hampered intelligent planning of land use on a wider scale. Until the interaction of the climatic complex with the physiological processes of the crop is understood, the production of a crop suitable for local climatic conditions must remain a matter of pure empiricism. The common practice of defining the so-called "climatic analogues" primarily in terms of monthly means of temperature and precipitation (Nuttonson, 1947) has proved to be inadequate as a guide to plant introduction or land-use planning. Radiation, evapotranspiration, diurnal temperature range, water balance, and other meteorological parameters must be fully analyzed before one can blueprint a plan that will realize the maximum economic return for a given climatic regime.

Secondly, climatological measurements are needed in agronomic experiments. Unless climatic variations are taken into account, it is difficult to interpret crop variety, fertilizer, and other agronomic experiments consisting of repetitious year-by-year field trials. Collis-George and Davey (1961) presented the problem:

> The interpretation of many field trials involving biological parameters would appear, after a century, to have come to a stalemate for lack of recorded "control" or environmental data. . . . Until we use a comprehensive style of experimentation and recording to determine the principal soil, site, and micrometeorological parameters controlling biological responses, we cannot elucidate the importance of, and the interrelationship between these factors.

Consequently, they urged comprehensive climatological documentation and expressed the hope that, with such information, a limited number of field trials would yield more applicable information than the large number of uninstrumentated trials now in existence.

The third application of climatological research concerns cultural practices. Problems such as irrigation, row spacing, timing of fertilizer application, variety selection, and transplanting can be best solved when viewed in the light of climatic environment. Another promising field of research is the artificial modification of microclimate by means of shelter belt, shading and mulching, heating, evaporation suppression, and the like. It is with regard to these problems that climatological studies promise to pay the most handsome dividends in the years to come; for even in the most technologically advanced countries, managerial control of cultural practice is more often a matter of personal judgment than of science.

Thus, meteorology and agriculture, whether they be considered as science or basic techniques of daily life, are closely linked. Climate assumes significance in nearly every phase of agricultural activity: from the selection of sites to agronomic experiments and from long-range planning to daily operation. As an interdisciplinary subject, agricultural meteorology is indeed very broad and complex. Research workers in the field must have an intimate understanding of the problems involved in both agriculture and meteorology. Dale (1950) pointed out the difficulties:

> The agriculturist often used the grossest weather statistics or, at times, adopted the responsibility for setting up standards of meteorological observations. On the other hand, the meteorologist assumed certain agronomic limits without really understanding the limitations of his assumptions, and made forecasts or climatic studies in terms of these limits. Little attempt was made by either science to match the factors involved in plant growth to the weather statistics.

Not surprisingly, a majority of the textbooks on agricultural meteorology, written for the most part by meteorologists, have labored on the rudiments of weather sciences, but have neglected many important studies by plant scientists that have profound implications in improving cultural practices and crop productivity.

The influence exercised by climate on living organisms is exceedingly complex, not only because of the many variables, but also because of the constant interaction between them. The usual experimental setup is to vary only one factor at a time—not an easy task under field conditions, or even in the ordinary greenhouse because of the nonhomogeneity of the environment during growth. Until the completion of the Clark Greenhouse in 1939 and the Earhart Plant Research Laboratory in 1949 (both at the California Institute of Technology), no large-scale growing facilities for plants under completely controlled conditions existed. Such a controlled laboratory, known as the "phototron," has greatly improved the reproducibility of biological experiments. Went (1956) cited an instance where the coefficient of variability of experimental results for tomato plants was about 20 per cent in an ordinary greenhouse, as against 5 per cent in the phototron.

In spite of the valuable information furnished by the phototron and other laboratory work, complete understanding of the quantitative relationship between plant and climate cannot be expected for a very long time. Each advance in the knowledge of plant physiology seems only to add to the complexity. Furthermore, the application of laboratory results to field problems often involves uncertainties in extrapolation. For many practical problems, there still is no better solution than to conduct *ad hoc* field experiments. The results of field trials, though less reliable than controlled experiments, and sometimes applicable only to local conditions, can nevertheless be used to derive general agronomic principles when carefully culled and interpreted. This book endeavors to review and summarize the important studies of the characterization of micrometeorological environment and the relationship between plant and climate. To simplify the discussion, the important climatic variables will be examined one at a time. However, some consideration will be given to their interaction whenever possible.

CHAPTER 2

RADIATION BALANCE

The ultimate source of practically all of the energy for all physical and biological processes occuring on Earth is solar radiation. The importance of radiation to agronomy is best stated in a definition by Monteith (1958) that agriculture is an exploitation of solar energy, made possible by an adequate supply of water and nutrients to maintain plant growth.

SOLAR CONSTANT

The rate at which radiation is received outside the Earth's atmosphere on a surface normal to the incident radiation, and at the Earth's mean distance from the sun, is known as the solar constant. For many years, the solar constant had been estimated to be 1.94 langleys per minute,[1] based on the measurements by the Smithsonian Institution. However, a recent study by Johnson (1954) leads to a mean value of 2.00 langleys per minute. This higher value is due mainly to an upward revision in the estimation of the ultraviolet part of the spectrum. The solar constant is not a true constant but fluctuates by as much as 1.5 per cent about its mean value. Much of the fluctuation is in the ultraviolet portion of the spectrum (Rense, 1961).

SOLAR SPECTRUM

Brooks (1956) plotted the spectral distribution of Johnson's solar constant in Figure 1. Since more than 99 per cent of the energy is contained in the wavelength region of 0.3 to 4 microns, solar radiation is referred to as short-wave radiation. In Figure 1, the equivalent wave number is shown across the bottom. A wave number is the reciprocal of the wavelength and is proportional to the frequency of radiation.

---

1. Langley is a unit of energy equivalent to one gram-calorie per square centimeter. One langley per minute equals to 220 BTU per square foot per hour, or 69.7 milliwatts per square centimeter.

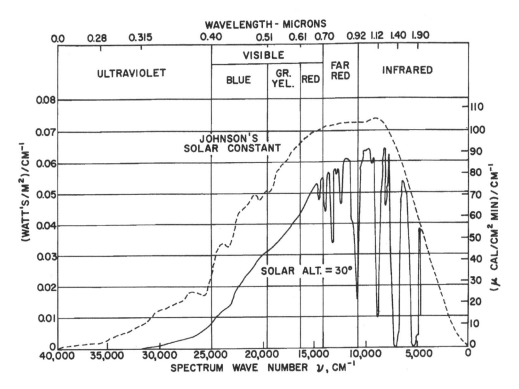

FIGURE 1. Spectral distribution of solar radiation at top of the atmosphere and at surface for optical air path of two atmospheres (after Brooks).

In Figure 1, the spectral distribution of solar radiation at the surface for a solar altitude of 30° or an optical air path of two air masses is also shown. The energy in the ultraviolet below the wavelength 0.4 micron comprises about 9 per cent of the total incident energy, while the energy in the visible region comprises 41 per cent, and that in the infrared, beyond 0.72 micron, contains about 50 per cent. The spectral distribution of solar radiation varies with the altitude. When the sun is directly overhead (air mass 1), visible radiation constitutes about 49 per cent of the direct radiation. At a solar altitude of 20° (air mass 3) only 29 per cent of the radiation is visible.

Ultraviolet and infrared radiation is reduced much more on a cloudy day than on a sunny day. This also holds true for high-latitude stations as opposed to the tropics. High mountains are noted for the enrichment of ultraviolet radiation.

The Dutch Committee on Plant Irradiation (1953) further discussed the various bands of the solar spectrum and their significance to plant growth:

1st band: Radiation with a wavelength longer than 1.0 micron. No specific effects are known of this radiation upon plants. This radiation, as it is absorbed by the plant, is transformed into heat without interfering with the biochemical processes.

2nd band: Radiation between 1.0 and 0.72 micron. This is the region of specific elongating effect upon plants. Although the spectral region of the elongating effect does not coincide precisely with the limits of this band, it may be provisionally accepted as an adequate measure of the elongating activity of the radiation.

The far red region is also important for photoperiodism, seed germination, control of flowering, and coloration of fruit.

3rd band: Radiation between 0.72 and 0.61 micron. This spectral region is strongly absorbed by the chlorophyll. It generates strong photosynthetic activity, in many cases also showing intense photoperiodic activity.

4th band: Radiation between 0.61 and 0.51 micron. This is a spectral region of low photosynthetic effectiveness in the green and of weak formative activity.

5th band: Radiation between 0.51 and 0.40 micron. This is essentially the region of strongest chlorophyll and yellow pigment absorption. It is also a region of strong photosynthetic activity in the blue-violet, and of strong, formative effects.

6th band: Radiation between 0.40 and 0.315 micron. This band produces formative effects; plants become shorter and leaves thicker.

7th band: Radiation between 0.315 and 0.28 micorn. Radiation in this zone is detrimental to most plants.

8th band: Radiation with a wavelength shorter than 0.28 micron. These wavelengths rapidly kill plants.

The ultraviolet radiation also produces significant germicidal action.

DEPLETION BY THE ATMOSPHERE

In its passage downward through the atmosphere, the depletion of solar energy results chiefly from selective absorption by the constituent gases and water vapor, from the scattering by molecules of air and small solid and liquid particles, and from reflection outward to space by larger particles and cloud surfaces.

The constituents of the atmosphere that take a significant part in the absorption of solar radiation are: (1) oxygen atoms in the upper air absorbing the extreme ultraviolet (0.12-0.18 micron), (2) ozone absorption mainly in the ultraviolet region, consisting of the Hartley band, 0.20-0.33 micron, and to a much smaller extent, the visible (Chappuis) band, 0.44-0.76 micron, (3) water vapor absorption in the near infrared bands centered at 0.93, 1.13, 1.42, and 1.47 microns, and (4) carbon dioxide in the near infrared bands at 2.7 microns.

Figure 1 presents a typical, depleted solar spectrum reaching the ground surface from a solar altitude of 30°, or an optical air path of two atmospheres. Practically all the radiation in the ultraviolet region at wavelengths smaller than 0.33 micron is cut off by the oxygen atoms and ozone in the upper atmosphere. This is of immense significance to life on Earth. We can tolerate this radiation in only minute doses; any excess is fatal.

MEASUREMENTS OF GLOBAL RADIATION

The solar radiation received at the Earth's surface consists of two parts: direct solar radiation and diffuse sky radiation. The sum of these two radiation components is known as the global radiation. The intensity of diffuse radiation is dependent upon the latitude, the altitude, the sun's angle, cloudiness, and atmosphere turbidity. The diffuse radiation, though usually less intense than the direct beam, may be significant in high latitudes. In South Africa, for example, diffuse radiation accounts for about 30 per cent of the total short-wave energy received at the ground surface (Drummond, 1958). In polar regions of large solar zenith distance, the diffuse radiation may comprise almost the entire global radiation during the winter season.

Most radiation stations record only global radiation, which is adequate for general agricultural purposes, although a knowledge of the relative magnitude of direct and diffuse radiation may be helpful for some physiological studies. The Eppley pyrheliometer, which measures radiation of wavelengths less than three microns, has been adopted by the U. S. Weather Bureau as their standard instrument for measuring global radiation. The instrument uses a thermopile formed of alloys of platinum-rhodium and gold-palladium wires as its sensor. It is fairly accurate but subject to several possible sources of mechanical errors if not properly checked and handled (Fuquay and Buettner, 1957). The high cost of installation, together with the need for electrical power and expert care have, however, limited its wide use as a field instrument.

Among the low-cost field instruments now in use, the photochemical tubes (Brodie, 1964, and Heinicke, 1963) and the wig-wag (Brodie, 1965) deserve attention. The photochemical method uses oxalic acid as the reagent and uranyl sulfate as the catalyst. The chemicals, exposed to the sunshine, are titrated at intervals of several days to determine the radiation intensity for that period. The wig-wag utilizes the principle that the gaseous phase of a volatile liquid expands as a result of the conversion of radiant energy to sensible heat. It requires only a reading of the count at whatever interval the observer chooses.

The field instruments are usually less accurate than the Eppley, and they do not give continuous records. For most agricultural problems, we can use to advantage any method that gives integrated values of solar radiation for a day or even a week, with an accuracy of within 5 per cent.

RADIATION AND DURATION OF SUNSHINE

Many meteorological stations record only the duration of sunshine. From the few meteorological stations where both radiation and duration of sunshine are recorded, it is possible to derive a relationship between these two variables in the form first proposed by Ångström (1924):

$$Q/Q_A = a + b\, n/N$$

where Q is the radiation actually received; Q_A is Angot's value, or the theoretical amount of radiation that would reach the Earth's surface in the

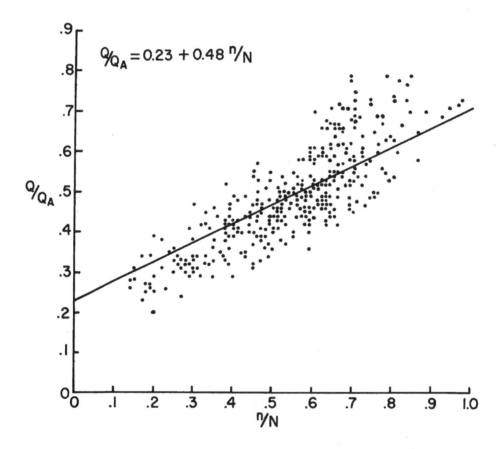

FIGURE 2. Relationship between monthly solar radiation and percentage of possible sunshine for 32 stations (after Black, Bonython, and Prescott).

absence of an atmosphere (List, 1966), n is the actual duration of sunshine received, N is the maximum possible duration of sunshine, and a and b are constants.

Black, Bonython, and Prescott (1954) analyzed the monthly values of radiation and duration of sunshine of 32 stations from the tropics to the polar regions. In the overall regression equation, a equals 0.23 and b equals 0.48 (Figure 2). They have also noted that the regression coefficient b, is more or less a constant, whereas the value of a shows a marked variation. Glover and McCulloch (1958) explained that the constant a in the Ångström equation is dependent upon the optical air mass, and hence, the latitude of an area. Using the data of seven stations covering the range of 0° to 60° latitude, they established that $b = 0.52$, and $a = 0.29 \cos \phi$, where ϕ is the latitude of the area.

The Ångström type of equation can be expected to give only an approximate estimation of solar radiation because the regional variation of atmospheric turbidity is not considered. Hounam (1958) cited an example in Australia: "There might be a large difference in water content on clear and cloudy days at Townsville, but not at Melbourne; therefore, the ratio Q/Q_A would be lower at Townsville." It is advisable to determine the constants from the local data whenever possible. For the United States, the extensive study by Fritz and MacDonald (1949) established: $a = 0.35$ and $b = 0.61$. For Canada, the values for the summer months are: $a = 0.355$ and $b = 0.68$. (Mateer, 1955)

RADIATION AND CLOUDINESS

For areas where the records of sunshine duration are not available, solar radiation may be estimated from cloudiness. Black (1956) has derived a general relationship from the records of 150 stations throughout the world:

$$Q/Q_A = 0.803 - 0.340\,C - 0.458\,C^2$$

where C is the mean monthly cloudiness in tenths.

The relationship between radiation and cloudiness is not linear, because on overcast days, with $C = 1.0$, average clouds appear to be relatively denser than on days with intermediate values of C. The estimates are approximations because the formula does not discriminate as to the cloud form and times of occurrence.

DISTRIBUTION OF SOLAR RADIATION

Several attempts have been made to map the world solar radiation

FIGURE 3. Generalized isolines of global radiation, surface (kcal/cm²/yr).

distribution. The latest one by Landsberg (1961) incorporated the data of over 300 stations (Figure 3). His map was highly generalized and he made no attempt to depict monthly variations, since the existing data were inadequate for that purpose. For several parts of the world, detailed radiation maps are, however, available.[2] Radiation maps are useful guides for assessing the agricultural potentialites of different regions. De Vries (1963) is of the opinion that, from a biological standpoint, it would be preferable to utilize the solar radiation as a fundamental climatic factor instead of air temperature in the classification of macroenvironments.

REFLECTIVITY

Not all the incoming global radiation is retained by the ground surface, as part of it is reflected and part reradiated. The reflectivity of the total short-wave radiation is usually measured by an inverted pyrheliometer, and that in the visible range of the spectrum by photometers. The term "albedo" has generally been used to denote the reflectivity of either the total short-wave or the visible spectrum. To eliminate this confusion, Monteith (1959) suggested that albedo be used exclusively for the visible light, and the term "reflection coefficient" for total short-wave radiation.

The reflectivity of leaves varies with wavelength, showing a minimum in the middle of the visible spectrum, and a maximum in the near infrared. In the far infrared, leaves appear to have an almost zero reflectivity. In general, the visible albedo of vegetation is smaller than its total reflection coefficient. For example, the reflectivity of a corn crop may vary from 40 per cent in the near infrared to 7 per cent in the visible spectrum (Allen and Brown, 1965). The reflectivity of single leaves is slightly higher than that of the crop surface, because when leaves form the canopy, a part of the radiation is trapped between them by multiple reflection.

The reflection coefficient of a crop surface is dependent upon its color, the moisture condition, the density of crop cover, leaf arrangement, and the angle of the sun. Reflectivity generally increases with the visual brightness of the surface. Desert vegetation, for example, with white or very hairy leaves has relatively high reflectivity.

The darkening effect caused by the moistening of a light-colored surface also lowers the reflection coefficient. As an example, Ångström (1925) observed that after a rain the albedo of light-colored grass decreased from 0.32 to 0.22. Bowers and Hanks (1965) found an

2. See Fritz (1957), Ramdas and Yegnanarayanan (1956), Xiao (1959), Sekihara and Kano (1957), Kawabata and Fujito (1955), Bleksley (1956), Drummond and Vowinckel (1957), Thompson (1965), Verle and Svinukhov (1960), Berliand and Efimova (1955), Gabites (1951), Black (1960), Day (1961), and Wallen (1966).

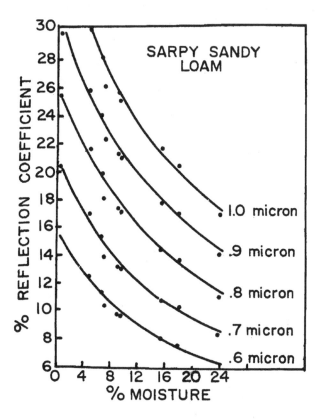

FIGURE 4. Percentage of reflectivity vs. moisture content for various wavelengths (after Bowers and Hanks).

excellent relationship between the moisture content of a sandy loam soil and its reflectivity for various wavelengths (Figure 4).

During the early stage of crop development, when the ground cover is incomplete, the reflection coefficient is usually reduced because of the low reflectivity of the bare ground. The reflection coefficient of sugar cane in Hawaii has been observed to increase from 0.06 to 0.08 for a newly planted field, to 0.16 to 0.18 after the full development of the canopy. (Chang, 1961)

The reflection coefficient of a surface changes with the sun's angle during the course of a day. On June 14, 1959, Monteith and Szeicz (1961) observed in England that reflection by grass and bare soil was least at midday, and increased almost linearly with decreasing solar radiation, more rapidly over grass than over soil (Figure 5). The trend was similar on August 18. The high values on that morning were caused by the increased reflection over a dewy night. Cloud cover reduces the diurnal variations of the reflection coefficient by proportionately increasing the diffuse radiation, which is almost independent of the solar elevation.

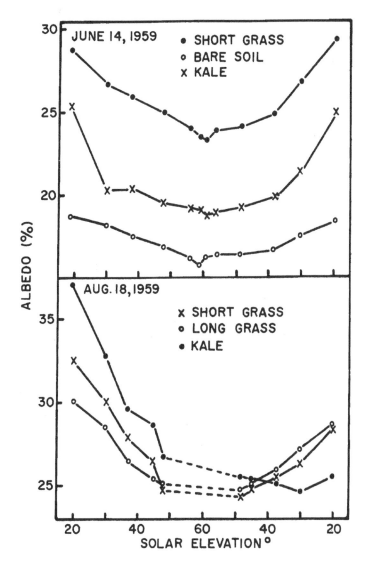

FIGURE 5. Diurnal variation of reflection coefficient on June 14, and August 18, 1959, at Rothamstead Experiment Station, England (after Monteith and Szeicz).

Table 1 summarizes a number of observations on the reflection coefficients of field crops. Note that in middle and high latitudes, the maximum reflection coefficients for mature crops that completely cover the ground are approximately 0.25; whereas in the tropics, the values are much lower. The low reflectivity in the tropics is in part caused by the greater solar elevations.

Posey and Clapp (1965) have constructed surface albedo maps of the world using previously existing data. Detailed seasonal albedo maps for

TABLE 1
REFLECTION COEFFICIENTS OF FIELD CROPS

Locality	Crops	Reflection coefficient	Reference
England	Grass, lucerne, potatoes, sugar beets, spring wheat	0.25-0.27	Monteith (1959)
USSR	Rye	0.10-0.25	Budyko (1958)
	Potatoes	0.15-0.25	
	Meadow grass	0.15-0.25	
	Cotton	0.20-0.25	
New Jersey	Spinach	0.24-0.28	Thornthwaite (1954)
Canada	Corn	0.12-0.21	Graham and King (1961b)
New York	Corn	0.235	Allen and Brown (1965)
Australia	Irrigated cotton	0.171-0.196	Fitzpatrick and Stern (1966)
England	Short grass	0.25-0.27	Monteith and Szeicz (1961)
	Long grass	0.26	
	Kale	0.19-0.28	
Hawaii	Sugar cane	0.05-0.18	Chang (1961)
Hawaii	Pineapple	0.05-0.08	Ekern (1965b)

North America were later compiled by Kung, Bryson, and Lenschow (1964). Barry and Chambers (1966) presented a summer albedo map for England and Wales; whereas Larsson (1963) made a detailed analysis of albedo in the Arctic.

OUTGOING RADIATION

Earth's surface emits long-wave radiation to the sky. More than 99 per cent of the emitted energy is contained in the wavelength region of 4 to 100 microns, with a peak at about 10 microns. The intensity of the terrestrial radiation may be expressed by the formula:

$$Q_t = \epsilon \sigma T^4$$

where Q_t is the terrestrial outgoing radiation; ϵ, the emissivity of the surface; σ, the Stefan-Boltzman constant, 8.17×10^{-11} langley $/T^4/$ minute, or 5.70×10^{-5} $erg/cm^2/T^4$ sec; T, the absolute temperature of the surface.

An object that at every wavelength emits the maximum radiation possible at its temperature is termed a "black body," and has an emissivity of 1.0. The surface of the Earth is a good approximation of a black body. Most vegetative surfaces have an emissivity of 0.90 to 0.95 (Brooks, 1959). However, the emissivity varies with the wavelength of the spectrum, reaching a primary minimum of 0.1 to 0.5 in the infrared between 0.74 and 1.10 microns and a secondary minimum of 0.6 and 0.8 at 0.55 micron in the visible portion of the spectrum. (Gates, Keegan, Schleter, and Weidner, 1965).

About 90 per cent of the outgoing radiation from the Earth's surface is absorbed by water vapor (5.3 to 7.7 microns and beyond 20 microns), ozone (9.4 to 9.8 microns), carbon dioxide (13.1 to 16.9 microns), and clouds (all wavelengths). The terrestrial radiation escapes freely to the outer space only in the "atmospheric window" between 8.5 and 11.0 microns.

A large portion of the radiation absorbed by the atmosphere is reradiated back to the surface, known as counterradiation or back radiation. The atmospheric counterradiation effectively prevents the Earth's surface from excessive cooling at night. The intensity of the counterradiation varies with the air temperature, the water vapor content of the air, and the cloud cover. The difference between the outgoing terrestrial radiation and the atmospheric counterradiation is known as the effective outgoing radiation.

The only available instrument for measuring outgoing radiation is the pyrgeometer, which is used rarely because it is suitable only for spot reading on clear nights. Effective outgoing radiation can also be estimated from a radiation chart or by empirical equations. The radiation chart most commonly used in the United States was developed by Elsasser (1942) and improved by Elsasser and Culbertson (1960). Other similar or simplified charts have been devised by Robinson (1950). Möller (1951), and Yamamoto (1952). The use of radiation charts requires radiosonde observation of temperature and humidity to a height of several miles. Because of the complexity of the charts and the necessity for aerological observation, radiation charts are not suitable for most applications to agriculture.

In the absence of aerological data, empirical formulae, such as the one derived by Brunt (1934), have been used. The Brunt formula reads:

$$Q_b = \sigma T^4 (0.56 - 0.08\sqrt{e})(1 - aC)$$

where Q_b is the effective outgoing radiation; σ is the Stefan-Boltzman constant, 8.17 x 10^{-11} langley/T^4/ minute, or 5.70 x 10^{-5} $erg/cm^2/T^4/$ sec; T is the absolute temperature of the air near the surface; e is the vapor pressure of the air expressed in millibars; C is the cloudiness; and a is a constant depending upon cloud type (0.025, 0.06, and 0.09 for high, medium, and low clouds respectively).

If data for the cloud type are not available, a less exact estimate is available by substituting $(0.10 + 0.09C)$ for $(1 - aC)$.

The constants in Brunt's formula may vary according to the climatic condition. Goss and Brooks (1956) found it necessary to revise the equation to the following form for Davis, California:

$$Q_b = \sigma T^4 (0.36 - 0.05\sqrt{e})(0.1 + 0.9C)$$

An empirical formula, established for a particular region, cannot be used elsewhere with the same confidence as a theoretical method based on fundamental principles. However, it can provide an estimate over the region for which it is valid with much less effort than any chart method and with comparable accuracy. Lönnqvist (1954), after comparing the various procedures for estimating long-wave radiation at Kew Observatory, concluded that the Brunt formula was as accurate as the chart method. Picha and Villanueva (1962) also obtained good results in applying Brunt's formula at Atlanta, Georgia. On the other hand, the formula may seriously underestimate back radiation during occasions of strong heating because it uses air temperature instead of the ground temperature. Calculations show that in the case of a grass surface that is 20° F. warmer than the air, the net outgoing radiation may be 50 per cent more than that estimated from the Brunt formula (Deacon, Priestley, and Swinbank, 1958). Anderson (1952) is of the opinion that the formula is accurate only to within 15 to 20 per cent for estimates over a long period of time.

Similar equations for estimating outgoing radiation have been derived by Ångström (1916), Swinbank (1963), and others.

NET RADIATION

Incoming radiation that is not reflected or back-radiated is known as the net radiation.

$$Q_n = (1 - r)Q + Q_b$$

where Q_n is the net radiation, r is the reflection coefficient, Q is incoming short-wave radiation, and Q_b is the net long-wave radiation.

Thus the net radiation is the difference between total upward and downward radiation flux, and is a measure of the energy available at the ground surface.

Instruments made to measure net radiation have been devised by Gier and Dunkle (1951), Soumi (1958), Funk (1962), Fritschen (1963, 1965a), and others. These early designs are far from perfect. Rain or dew on the black body-sensing surface disturbs the reading of the Gier and Dunkle net radiometer, which measure radiation thermoelectrically. High wind speeds and even wind direction with respect to the sensing elements may affect the reading (Portman and Dias, 1959). At low sun's angle, the cosine response of the Soumi net radiometer is poor. For example, at zenith angle of 25°, it measures about 90 per cent of the net flux, at 65° zenith angle, the value is 83 per cent, and at 80°, it is 73 per cent. There is still an urgent need for developing a precise net radiometer.

At present, almost all the published data on net radiation are from sporadic "fair weather" observations in conjunction with specific experimental work. Table 2 summarizes a number of net radiation measurements

TABLE 2
THE MEASURED FRACTION OF GLOBAL RADIATION RETAINED AS NET RADIATION IN 24-HOUR PERIODS

Location	Cover	Global radiation (langleys/day)	Ratio: Net radiation/sunlight	References
Copenhagen, Denmark 55° 40′ N, 28 m	Grass sod	463 (July) 373 (Apr.-Sept.)	0.51 0.44	Aslyng and Nielsen (1960)
Wageningen, Netherlands 51° 58′ N, 37 m	Grass sod	550 (Aug.)	0.56	Scholte-Ubing (1959)
Rothamstead, England 51° 48′ N, 158 m	Grass sod Tall crops	550 550	0.41 0.46	Monteith and Szeicz (1961)
Davis California 38° 30′ N, 12 m	Grass sod	750 (June) 175 (Dec.)	0.56 0.33	Pruitt and Angus (1961)
Aspendale, Australia 38° 02′ S, 3 m	Grass	181 (July) 689 (Jan.)	0.18 0.69	Funk (1963)
Tempe, Arizona 33° 30′ N, 350 m	Grass sod	675 (July)	0.57	Van Bavel, Fritschen, and Reeves (1963)
Honolulu, Hawaii 21° 18′ N, 15 m	Sugar cane	725 (summer) 400 (winter)	0.69 0.65	Chang (1961)
Wahiawa, Hawaii 21° 18′ N, 216 m	Pine-apple	710 May-June, clear days 500 (Dec.)	0.66 0.53	Ekern (1965b)
Mauna Loa, Hawaii 19° 32′ N, 3400 m	Lava	815 (June clear day) 551 (Dec. clear day)	0.50 0.41	Ekern (1965b)

expressed as a fraction of global radiation in 24-hour periods (Ekern, 1965b). In general, the ratio between net radiation and global radiation decreases with latitude. For the same locality, it also decreases from summer to winter. In the tropical mountains (e.g., Mauna Loa), the rapid loss of long-wave radiation through the dry air, however, reduces the 24-hour fraction of net radiation to values comparable to those of high latitudes in the summer.

A world map of annual net radiation (Figure 6), based on ingenious methods of estimation, has been prepared by Budyko, Yefimova, Zubenok, and Strokhina (1962). The highest values are found in the tropical oceans, especially the northern Arabian Sea. As a result of the high reflectivity and high ground temperature, the annual net radiation over the subtropical deserts is only about half that over the tropical oceans. The differences diminish with latitude. The annual net radiation is about the same over land and water beyond 50° in both the northern and southern hemispheres.

FIGURE 6. The radiation balance (in kilocalories per square centimeter per year).

In the winter, net radiation is negative in middle and high latitudes, reaching a minimum value of around −30 langleys per day in continental interiors and −40 to −50 langleys/day in coastal regions. According to a detailed study by Orvig (1961) at Knob Lake, Canada (54° 48′ N, 66° 49′ W), the net radiation is negative during the greater part of the cold season, from September to March, when the daily incoming radiation is less than 284 langleys per day. In general, the net radiation is negative over snow-cover because of its high albedo.

In the tropics and during the summer in middle and high latitudes, a good linear relationship usually exists between net and incoming radiation. Since a few representative readings over a short period would permit the establishment of this relationship, such studies should be given high priority in the research program. In addition to the several studies cited in Table 2, which deals with 24-hour periods, the daytime hour relationships alone have been investigated on sodgrass at Ames, Iowa, by Shaw, (1956) and over paddy rice fields in Japan by Nakagawa (1963). Shaw has found is advisable to separate clear and cloudy days to obtain more accurate equations of the form:

$$Q_n = 0.87Q - 82.0 \text{ (clear days)}$$
$$Q_n = 0.75Q - 21.4 \text{ (cloudy days)}$$

where Q_n is the total net radiation from half an hour after sunrise to half an hour before sunset, and Q is the total daily global radiation.

In Japan, the relationship obtained with the paddy fields at the most active tillering stage during the summer can be expressed either in values of an hour or a minute:

$$Q_n = 0.648Q - 3.63 \text{ (langleys per hour)}$$
$$Q_n = 0.648Q - 0.0605 \text{ (langleys per minute)}$$

The net radiation is negative at night. It becomes positive during the day from about an hour after sunrise until an hour before sunset. The net radiation is usually less than the incident radiation, except on rare occasions when the sky is partially cloudy but the clouds are not directly overhead. During such periods, the incoming radiation is augmented by the reflected radiation, while the outgoing radiation is checked by the clouds. For equal short-wave radiation intensities, the net radiation is smaller in the afternoon than in the morning because of the higher afternoon surface temperature.

In an attempt to explain the varying relationships between incoming and net radiation, Monteith and Szeicz (1961) proposed the following formula:

$$Q_n = \frac{(1-r)Q}{(1-b)} + Q_{bo}$$

where Q_n is the net radiation, r is the reflection coefficient, Q is incoming short-wave radiation, b is heating coefficient, and Q_{bo} is the net long-wave radiation when Q is zero.

Monteith and Szeicz explained that the heating coefficient is the increase in net long-wave radiation per unit increase in total net radiation and as such it is dependent upon the roughness of the surface. They obtained heating coefficient values of 0.3 for soil, 0.15 to 0.3 for short grass, and 0.09 to 0.12 for tall grass during the summer in England. However, Ekern (1965b) gave a much lower estimated value of 0.03 for pineapple fields in Hawaii. He attributed this low value of heating coefficient to the extremely rough surfaces of the pineapple field and its rapid turbulent heat exchange with the air. On the rare occasions when the daytime increase in downward, long-wave radiation from the atmosphere is greater than that of the upward flux from the surface, the value of the heating coefficient may be negative. Such negative values have been observed by Stanhill, Hofstede, and Kalma (1966) over irrigated cotton fields in Israel.

Since tall vegetation usually has a rough surface and a greater capacity for heat absorption, it should retain a greater portion of net radiation than short crops. For instance, Decker (1959) found that the net radiation over bluegrass is 12 per cent less than over corn, and 8 per cent less than over alfalfa. Bahrani and Taylor (1961) reported a marked decrease in net radiation from about 400 to 240 langleys per day when alfalfa was harvested.

Hanks, Bowers, and Bark (1961) measured net radiation on clear summer days over bare soil and over soil covered with different mulches. Net radiation was highest on black-painted gravel, followed by plastic-covered surfaces, bare soil, straw-covered surfaces, and aluminum-painted gravel. Assuming a value of 100 per cent for the bare soil, the relative values are 116, 105, 100, 94, and 93 per cents respectively. In general, net radiation increased with the darkening of the surface.

RADIATION BALANCE IN THE GREENHOUSE

In a study in Madison, Wisconsin, Hasselkus and Beck (1963) reported that the percentage of light transmission through a greenhouse varied from about 40 per cent in December to 80 per cent in June, with an annual average of 64 per cent. The differences in the radiation balance in the open, and under a glass cover, have been investigated by Scholte-Ubing (1961a), who summarized his findings qualitatively:

The results . . . show a considerable decrease in the flux densities of short-

Radiation Balance

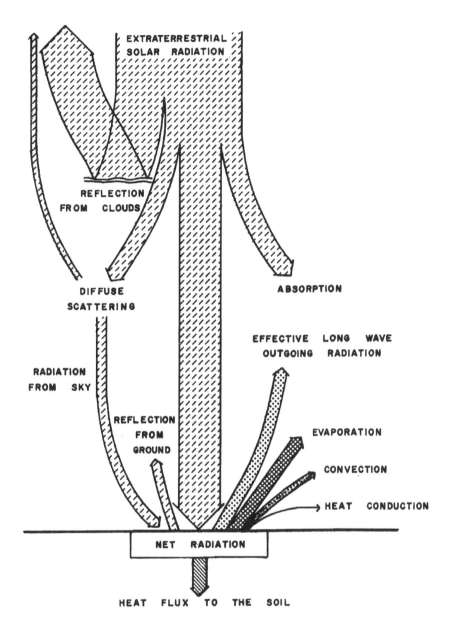

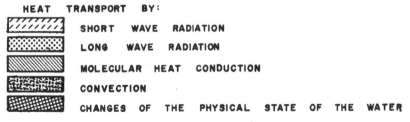

FIGURE 7. Schematic diagram showing heat exchange at noon for summer day. Width of arrows corresponds to transferred heat amounts (modified after Geiger).

wave and net radiation under glass over 24 hours as compared with the similar radiant flux densities in the open. The net long-wave (back) radiation under glass was fairly constant and low, even under clear skies, as compared with the net back radiation in the open. The nocturnal net radiation under glass was higher than it was in the open. However, during the daytime and also over 24 hours, the net radiation under glass still was always lower than it was in the open. The net radiation under glass did not differ much from the net short-wave radiation under glass.

These differences must be borne in mind when one interprets experimental results from the greenhouse.

BIOSPHERE AS THE LOCALE OF ENERGY EXCHANGE

In the preceding sections, we have discussed radiation flux from outer space to the ground surface. The various processes involved have been schematically shown by Geiger (1965) for a summer day (Figure 7). As the solar radiation is absorbed by the Earth and the atmosphere, it is converted into internal and geopotential energy. Whereas the former manifests itself as heat, the latter exists by virtue of gravity. In the process of evaporation, the internal energy may be converted into latent energy. However, part of the sensible heat from the ground is transferred to the atmosphere by conduction and convection. Convection generates kinetic energy (wind), which will again be transferred to the Earth as heat through friction near the ground.

The complicated processes of energy exchanges have engaged the attention of atmospheric scientists for some time. However, their significance to biological activities has been explored only recently. Plants are constantly immersed in radiation and their growth is profoundly affected by energy exchanges. Near the Earth's surfaces, geopotential and kinetic energy are of only secondary importance. The two main domains of biological significance are: (1) the energy budget of radiant and sensible heat and, (2) the water balance or flux of latent energy. One of the central problems in agricultural meteorology is to understand the mechanism of energy exchange between plants and the physical environment and to seek means for improving the efficiency of energy and water use by plants.

CHAPTER 3

PHOSYNTHESIS

GENERAL EFFECTS OF RADIATION ON PLANT GROWTH

Best (1962) has classified the effects of radiation on a green plant in the following manner:

I. Photo-energy processes: photosynthesis
II. Photo-stimulus processes:
   A. Movement processes: 1. nastic movement, 2. movements of orientation, 3. tropism, and 4. tactic movements
   B. Formative processes: 1. stem elongation, 2. leaf expansion, 3. pigment formation, 4. pubescence, 5. flowering in photoperiodically sensitive plants, 6. formation of photochlorophyll, and 7. anthocyanin formation.

In general, photoenergy processes require a higher radiation intensity than photostimulus processes. Formative processes are often determined by the relative lengths of light and dark periods to which plants are exposed, a phenomenon known as photoperiodism.

BASIC PROCESS OF PHOTOSYNTHESIS

Practically all the dry matter of higher plants originates from photosynthesis, a process by which plants, with the aid of the chlorophyll pigment, utilize the energy of solar radiation to produce carbohydrates out of water and carbon dioxide. In its simplified form, the chemical equation of photosynthesis is:

$$CO_2 + H_2O + \text{energy} \rightarrow (CH_2O) + O_2 - 112{,}000 \text{ calories}$$

where CH_2O is in a bracket as a reminder that the carbohydrates found in the plant (starch, sugars, celluloses, etc.) are much more complex.

Photosynthesis consists of several processes, and for the understanding of plant and climate relationships, the following three have been distinguished by Gaastra (1962):

(1) A diffusion process for the transportation of CO_2 from the external air toward the reaction center in the chloroplasts. The rate of this process depends mainly on the CO_2 concentration in the atmosphere, and only slightly on temperature. Light can affect the diffusion rate only through an indirect influence on temperature.

(2) A photochemical process resulting in the conversion of light energy into chemical energy that can be used for the reduction of CO_2 to carbohydrate. The photochemical process is influenced only by light.

(3) Biochemical processes, in which the energy produced by light conversion is used for the reduction of CO_2. The biochemical processes are strongly affected by temperature, though not by light.

Photosynthesis is inherently an inefficient process in the utilization of solar energy, partly because only the visible portion of the spectrum is active and partly because the quantum requirement for photosynthesis is much higher than its theoretical minimum value. Light is absorbed by atoms or molecules in the form of quanta of definite energy content. The work of reducing one mole of carbon dioxide to the level of carbohydrates is in the neighborhood of 112,000 calories. Light is made available for this process in quanta with an energy content of about 41,000 calories per mole. Thus the theoretical quantum requirement for photosynthesis is about three. However, experiments have demonstrated that more quanta are needed, and when photosynthesis is complete, the energy of surplus quanta is released as heat. The exact quantum requirement for photosynthesis has been a subject of controversy (Rabinowitch, 1948, and Warburg, 1958). The early work by Warburg (1919) gave a very low value of four. Later experiments seem to indicate that eight to twelve quanta are needed for the reduction of one CO_2 molecule. Thus, Bonner (1962) explained:

> Ten moles of quanta (10 einsteins) in the middle of the wavelength range usefully absorbed by chlorophyll supply about 520 kilogram calories. The reduction of 1 mole of CO_2 to the level of plant material captures and stores only 105 kilogram calories. The efficiency of the basic photosynthesis act is, therefore, $105/520 = 20$ per cent.

Assuming the incident radiation lost by reflection and transmission is 15 per cent and assuming the energy in the visible spectrum is 41 per cent, the maximum efficiency is $0.20 \times (1 - 0.15) \times 0.41$ or about 7 per cent of total incident radiation.

Saturation light intensity and the efficiency of light utilization. The photosynthetic rates of most leaves increase with light intensity almost linearly over a narrow range. At same light intensities, however, the photosynthetic rate becomes independent of the light intensity; the leaf becomes light-saturated. Böhning and Burnside (1956) classified plants into two groups according to their saturation light intensity (i.e., sun and shade species). According to their experimental results, sun species, to which most field crops belong, reach light saturation at about 2,500 foot-candles (Figure 8). For the shade species, light saturation is reached at a maximum of 1,000 foot-candles (Figure 9).

However, the saturation light intensities of 2,500 foot-candles given by Böhning and Burnside for several field crops are lower than those found by other investigators as summarized in Table 3. The saturation light intensity of sugar cane given by Hartt (1965) and that of rice by Yamada, Murata, Osada, and Iyama (1955) are particularly high. But even the highest value of 6,000 foot-candles is only about half of the full sunlight of the zenith sun, which usually exceeds 10,000 foot-candles and may reach 14,000 foot-candles in some areas. Thus, unshaded leaves are usually light-saturated from about 10 A.M. to 4 P.M.

TABLE 3
SATURATION LIGHT INTENSITIES OF SEVERAL CROPS

Crop	Saturation light intensity (foot-candle)[a]	Reference
Sugar beets	4,400 (Aug.-Oct.)	Thomas and Hill (1949)
Wheat	5,300 (June-July)	Thomas and Hill (1949)
Alfalfa	4,700 (summer) 3,400 (winter)	Thomas and Hill (1949)
Rice	5,000-6,000	Yamada, Murata, Osada, and Iyama (1955)
Rice	3,800	Matushima, Yamaguchi, and Okabe (1955)
Corn	2,500-3,000	Verduin and Loomis (1944)
Apples	4,050-4,400	Heinicke and Childers (1937)
Potatoes	3,000	Chapman and Loomis (1953)
Sugar cane	6,000	Hartt (1965)

[a] A foot-candle is the amount of illumination by a standard candle at a distance of one foot. One foot-candle aquals 10.76 lux. It expresses only the intensity of visible light. The conversion factor from foot-candle into langley varies with cloudiness and the angle of the sun. In 1924, Kimball established the following conversions at Washington, D.C.: for overcast skies, a mean value of 7,440 foot-candles per minute is equivalent to one langley per minute; for cloudy skies, it is 7,000; and for cloudless skies, 6,700. As an example of the effect of sun's angle, at 25° of the sun's zenith distance, 1 langley per minute is equivalent to 7,000 foot-candles per minute; at 78.7°, 6,200 foot-candles for a cloudless sky. In the tropics this value may exceed 10,000 foot-candles at noon.

For a few plants, the photosynthetic rates may even decline slightly as light intensity increases beyond the saturation point. In Figure 9, the photosynthetic rate of the fern *Nephrolephis* decreases at a light intensity above 2,000 foot-candles. Nutman (1937) measured the photosynthetic

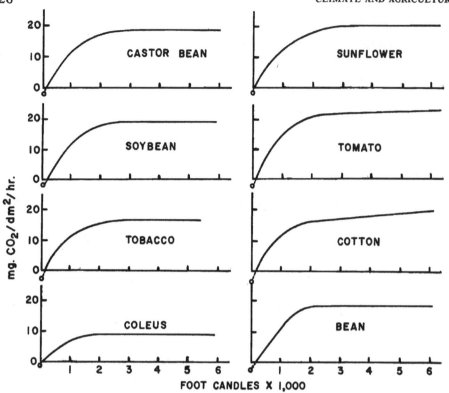

FIGURE 8. Light saturation curves of apparent photosynthesis for sun species. Circle represents dark respiration (after Böhning and Burnside).

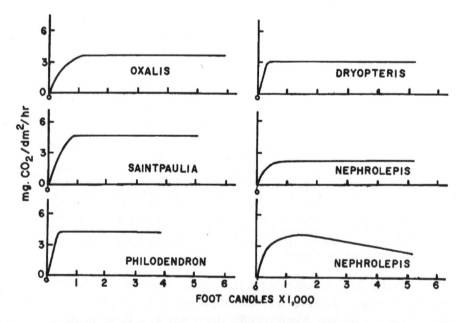

FIGURE 9. Light saturation curves of apparent photosynthesis for shade species. Circle represents dark respiration (after Böhning and Burnside).

rates of coffee leaves under natural conditions and found that they were reduced by the high intensity of midday sunlight through stomatal closure. The daily total assimilation of leaves in a coffee tree growing in the shade was greater than in the sun.

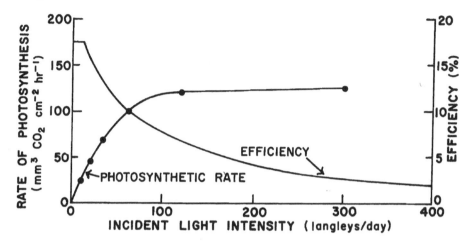

FIGURE 10. Rate of photosynthesis in sugar beet leaf as a function of incident radiation together with efficiency of radiation utilization in photosynthesis (after Bonner).

A leaf exposed to full sunlight cannot be expected to be completely efficient in the utilization of light energy for photosynthesis. Bonner (1962) has taken the photosynthetic curve for the sugar beet leaf, given by Gaastra (1958) to calculate the efficiency of radiation utilization (Figure 10). At extremely low radiation intensity, the efficiency may reach 17 per cent; it drops sharply to about 8 per cent at an intensity of 100 langleys per day, and to 3 per cent at 300 langleys per day, a value usually equaled or exceeded in the early spring of middle latitudes. The decrease in the efficiency of radiation utilization with increasing light intensity is caused by the finite resistance to the diffusion of carbon dioxide through the leaf to the chloroplasts.

The variation of the photosynthetic efficiency of higher plants is in general agreement with the prediction of Figure 10. This has been corroborated by a number of experiments. Bonner further explained:

> Thus Went [1957b] obtained an efficiency of light utilization of approximately 10% with tomatoes grown in light of an intensity approximately one-tenth that of full sunlight, while Gaastra [1958] has reported efficiencies with sugar beet of 12 to 19% under similar low, or even lower intensities. The work of Thomas and Hill [1937] has shown that the photosynthetic efficiency of alfalfa plots increases from approximately 2% at the intensity of full sunlight to 3.7 to 4% at intensities from one-third to one-half the intensity of full sunlight. The photosynthetic efficiency of sugar

cane at varying intensities reported by Burr [1961] increases with decreasing intensity and is in numerical agreement with the expectation of Figure 10.

The saturation light intensity is not a constant. Marked genotypic differences in photosynthetic rates have been noted for a number of crops (Watson, 1947; Chatterjee, 1961). Pieters (1960), working with *Acer Pseudoplatanus* L., demonstrated that the maximum rate of photosynthesis increases linearly with the thickness of the leaf. Sun species with thick leaves invariably have a higher saturation light intensity than shade species with thin leaves. Plant breeders can attain a higher level of saturation light intensity, at least for some crops, by reducing the chlorophyll-to-enzyme ratio in the photosynthetic unit, and by increasing the capacity for CO_2 absorption. Recently, there have been reports of extremely high light-saturation intensity values for some new varieties. Hesketh and Musgrave (1962) found a saturation light intensity of 10,000 foot-candles for corn, which was four times the values reported by previous workers. Similar high values have been noted for a few varieties of sugar cane, sunflowers, and soybeans. However, these high values were obtained in controlled assimilation chambers where the air flow rates, temperature, and moisture supply are nearly optimum.

Photosynthesis in relation to temperature. At saturation light intensities and normal CO_2 concentration, photosynthesis is affected by temperature because biochemical processes are limiting. Molga (1962) presented data showing the photosynthesis in potato, tomato, and cucumber leaves at different leaf temperatures (Figure 11). The photosynthetic rates increase with temperature reaching a maximum between 30° and 37° C. and then drop sharply at high temperatures. For most plants in temperate and tropical regions the optimum temperature exceeds 25° C. However, for some arctic and alpine plants the optimum temperature may be as low as 15° C. (Mooney and Billings, 1961). In general, plants in arctic regions have much lower photosynthetic rates than plants in temperate climates (Warren Wilson, 1960a). Some plants have the ability of adapting their optimum temperatures to the existing climate of an area. The optimum temperature for photosynthesis may also increase slightly with light intensity (Bolas, 1933).

Plants in the field are usually below the optimum temperature for most of the season. But the temperature of leaves in the sunlight is often higher than that of the air, sometimes by as much as 10° C., and during high-sun periods, the leaf temperature may very well exceed the optimum. Gates (1965a) has suggested that the midday depression in photosynthesis is often caused by the adverse effect of very high temperatures.

Photosynthesis in relation to CO_2 concentration. At very low light intensities, the photosynthetic rates in many plants are not severely af-

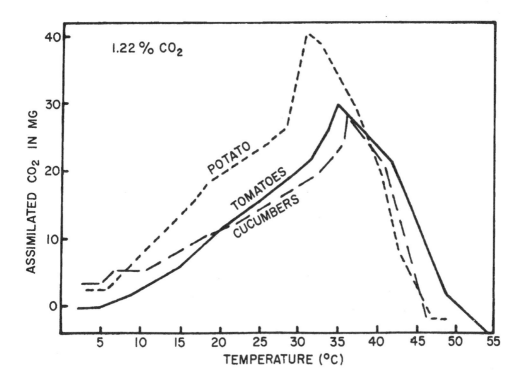

FIGURE 11. Relation between the photosynthesis in potato, tomato, and cucumber leaves, and leaf temperature (after Molga).

fected by CO_2 concentration in the air. Under high intensities, the photosynthetic rate of a cucumber leaf has been shown to increase with the enrichment of CO_2 from the normal concentration of 0.03 per cent, to the saturation value of 0.13 per cent (Gaastra, 1962) (Figure 12). Similar photosynthetic responses to carbon dioxide have been reported for potatoes (Chapman and Loomis, 1953), wheat (Hoover, Johnson, and Brackett, 1933), sugar beets (Gaastra, 1959), sugar cane (Waggoner, Moss, and Hesketh, 1963), corn (Moss, Musgrave, and Lemon, 1961), and many other crops. After examining the results of several workers on different plants, Thimann (1951) concluded that the photosynthetic rates of leaves can be raised at least three times by providing an adequate supply of CO_2 in the air.

With the enrichment of CO_2, both the saturation light intensity and the efficiency of light utilization also could be raised. For example, the saturation light intensity of a strawberry leaf increased from one-tenth sunlight at low CO_2 concentration to a half at high concentration (Thimann, 1956). According to a theoretical computation by Bonner (1962), the efficiency of light utilization would be doubled if the limitation imposed by CO_2 deficiency could be removed (Figure 13). The present-day world average of 0.33 per cent by volume, is not more than one-fourth

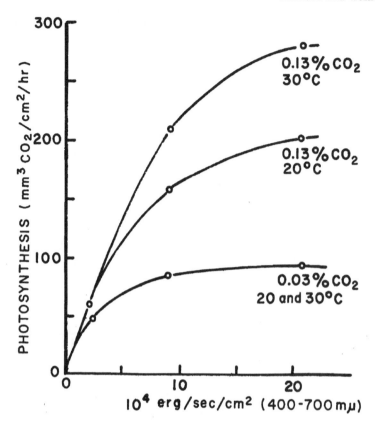

FIGURE 12. Photosynthesis of a cucumber leaf in relation to light intensity and temperature at a limiting (0.03%) and at a saturating (0.13%) CO_2 concentration (after Gaastra).

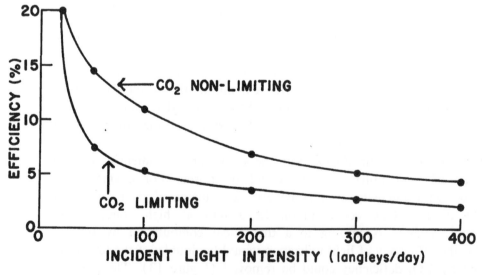

FIGURE 13. Expected efficiency of utilization of energy of incident light in photosynthesis when supply of CO_2 is nonlimiting (after Bonner).

the saturation level of CO_2 concentration for optimum photosynthesis. But the atmospheric CO_2 content has increased nearly 10 per cent during the last 50 years as a result of man's increased burning of fossil fuels and his greater agricultural activities.

There are local fluctuations in CO_2 content within the plant community. Figure 14 presents idealized profiles of CO_2 content in the air in an actively photosynthesizing corn field for different periods of a sunny day (Lemon, 1960). The daytime decrease in the CO_2 content of the air among plants carrying on active photosynthesis may extend, with some lag, to a height of 500 feet (Chapman, Gleason, and Loomis, 1954). During the daytime, the crop is a *sink* for CO_2 received from the ground and the atmosphere. Penman (1962) cited one instance when the summer flux from the soil accounted for one-third to one-half of the crop's assimilation. However, the normal contribution from the soil is much less. The CO_2 concentration within the canopy is especially low during a calm daytime period. At night, both soil and crop are sources of CO_2 for the atmosphere.

Tamm and Krysch (1961) found that minimum CO_2 concentrations within the crop canopy are usually between 0.025 to 0.029 per cent. However, absolute minimum values may reach 0.020 per cent (Chapman, Gleason, and Loomis, 1954). Thus, the local deficit of carbon dioxide will

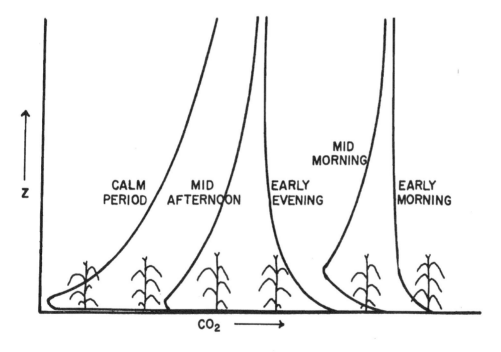

FIGURE 14. Idealized profiles of carbon dioxide content of the air in actively photosynthesizing corn field as function of height, z, above ground for various periods during sunny day (after Lemon).

cause a decrease of photosynthetic rate in the order of 10 to 20 per cent.

Fuller (1948) observed that in forests, on grasslands, and in river bottoms, carbon dioxide concentrations at or near the soil level may be two or three times above normal. The high concentration of CO_2 in the forest floor may be responsible for the growth and survival of plants in very low light intensity (Voight, 1962). The high CO_2 content over muck soils may account, in part, for their high productivity (Hopen and Ries, 1962).

Although it has not been possible to increase the CO_2 concentration on the ordinary cropland, carbon dioxide fertilization in greenhouses has been adopted as a commercial practice in Denmark, the Netherlands, England, and the United States in the growth of high-priced vegetables and flowers such as lettuce, tomatoes, cucumbers, roses, carnations, chrysanthemums, and snapdragons (Anonymous, 1963; Lindstrom, 1965, 1966b; Wittwer and Robb, 1964). Plants grown in CO_2-enriched greenhouses not only have a higher yield and better quality but also mature earlier. For most crops, the optimal concentration appears to be about 0.1 per cent. Younger plants have slightly higher optimal requirements for carbon dioxide than more mature plants. There are four commonly used sources of carbon dioxide: pure carbon dioxide in liquid form, carbon dioxide in the form of dry ice, propane (a liquid petroleum fuel), and paraffin (kerosene).

Respiration and net photosynthesis. Plant growth depends on the excess of dry matter increase by photosynthesis over the loss by respiration. The net gain is known as net photosynthesis, or net assimilation rate. Whereas photosynthesis take place mainly in the leaves during the day, respiration proceeds throughout the plants for the entire 24-hour period. In the process of respiration, plants burn up material to do work:

$$CH_2O + O_2 \rightarrow CO_2 + H_2O + \text{heat of combustion}$$

The heat of combustion of carbohydrate is usually between 4,000 and 5,000 calories per gram of dry matter.

Respiration rates increase with temperature up to a varying maximum for different plants. Thomas and Hill (1937) measured the respiration rates of alfalfa for 26 days from September 4 to October 1, 1936, and observed a fourfold increase in respiration rate, with a temperature rise from 0° to 20° C. (Figure 15). The respiration rates of plants also vary according to the light intensity (Rabinowitch, 1956), but the effect is much less significant than that of temperature and is often ignored.

The respiration rate of leaves is usually about 5 to 10 per cent of the gross photosynthesis at saturation light intensity and normal air. The respiration rate of whole plants is invariably higher than that of single leaves. The experiments by Thomas and Hill showed that total respiration of alfalfa was between 35 and 49 per cent of its photosynthetic rate, and

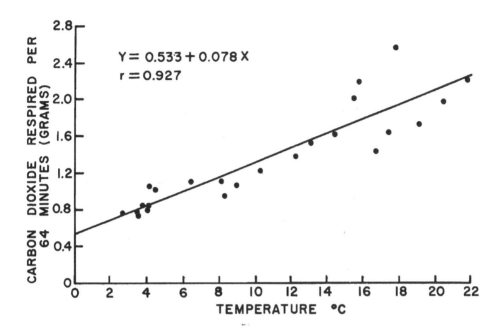

FIGURE 15. Influence of temperature on respiration of alfalfa, September 11, 1936, Logan, Utah (after Thomas and Hill).

for sugar beets the fraction was between 29 and 33 per cent. In temperate climates, the average value for most plant species is probably between 20 and 30 per cent.

At an extremely low light intensity, as at the forest floor, the net photosynthesis may be very small or even negative. The point at which the respiration rate equals the photosynthetic rate is known as the compensation point, which is a function of light intensity and temperature. For the rice plant, the compensation point is about 150 foot-candles at 40° F., about 400 foot-candles at 60° F., and about 1,400 foot-candles at 80° F. (Ormrod, 1961). Most plants, however, have a lower compensation point than rice. At normal room temperature, the compensation points are about 100 to 150 foot-candles for the sun species, and 50 foot-candles for the shade species (Figures 8 and 9).

Day and night temperatures exert opposite effects on net photosynthesis. A high night temperature increases the respiration loss and so reduces the net photosynthesis; whereas a high day temperature, up to about 30° C., may increase net photosynthesis. Gregory (1940) has found that the net photosynthesis of barley was positively correlated with daily maximum temperature but negatively correlated with night temperature. Under certain combinations of radiation and temperature, respiration may play such a dominant role that a negative relationship may exist between mean daily temperature and dry matter production. This has been observed by Stanhill (1962) for alfalfa at Gilat, Israel.

Since the ideal condition for net photosynthesis is a combination of relatively high daytime temperature and low nighttime temperature, a large diurnal temperature range is often desirable. Black (1955) constructed a graph showing the interaction between radiation and the range in diurnal temperature in determining the net photosynthesis of subterranean clover (Figure 16). As the diurnal temperature range increased from 6° to 24° F. in the experiment, the net photosynthesis was raised by 50 to 100 per cent depending on the prevailing light intensity.

Other factors affecting the rate of net photosynthesis. The rates of photosynthesis and respiration are affected by a number of other factors. The effect of water on photosynthesis will be discussed in Chapter 12.

Plant leaves infected with virus have a lower net assimilation

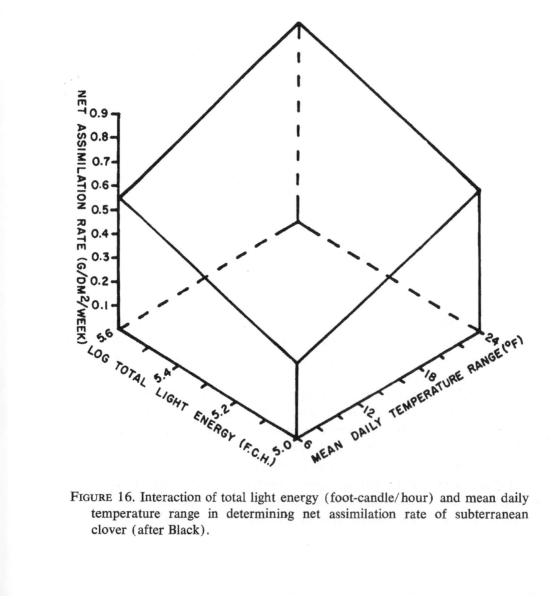

FIGURE 16. Interaction of total light energy (foot-candle/hour) and mean daily temperature range in determining net assimilation rate of subterranean clover (after Black).

rate. For instance, Owen (1958) reported that tobacco leaves infected with the potato virus X had a 30 per cent higher respiration rate and 20 per cent lower photosynthetic rate.

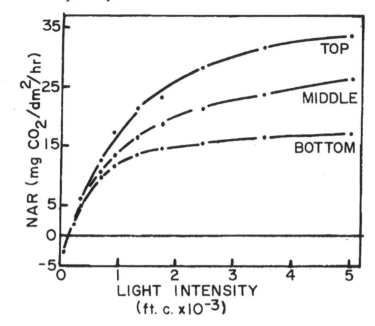

FIGURE 17. Net assimilation rates of top, middle, and bottom alfalfa leaves at several light intensities (after Brown, Cooper, and Blaser).

The photosynthetic rates of plant leaves usually decline with age (Singh and Lal, 1935). A year-old sugar cane leaf has only about half the photosynthetic capacity of a young leaf (Hawaiian Sugar Planters' Association, 1963). Brown, Cooper, and Blaser (1966) have measured the net assimilation rates of top, middle, and bottom alfalfa leaves at several light intensities. As the leaf ages decrease from the bottom to the top of the plants, the net assimilation rates increase in that order (Figure 17). In a controlled indoor experiment, Thorne (1960) found that the net assimilation rates of sugar beets, potatoes, and barley decreased approximately linearly with time.

CHAPTER 4

RADIATION DISTRIBUTION WITHIN THE PLANT COMMUNITY

Although the light saturation for a single leaf occurs at a radiation far short of full sunlight, the arrangement of leaf blades and stems in the field is such that a considerable part of the inner portion of the plant community is short of light. Understanding the relationship between radiation and crop production requires a knowledge of radiation distribution within the canopy based on the transmissibility of the leaf, leaf arrangement and inclination, plant density and height, and the sun's angle.

LEAF TRANSMISSIBILITY

Kasanaga and Monsi (1954) measured the light transmissibility of the mature leaves of 80 species. They found that leaves of deciduous trees and those of herbs and grasses, including cereals, have a transmissibility in the range of 5 to 10 per cent. The broad leaves of evergreen plants, however, have a somewhat lower value of 2 to 8 per cent. The transmissibility of the floating leaves of aquatic plants ranged from 4 to 8 per cent.

Transmissibility changes slightly with the age of a leaf. In spring and early summer the transmissibility of a young leaf is relatively high. With the maturing of the leaves, the transmissibility declines in summer. It rises again as the leaves turn yellow in autumn.

The transmissibility of a leaf is directly related to its chlorophyll content. In general, the logarithm of transmissibility decreases linearly with an increase in chlorophyll content, although the measured values are somewhat scattered as shown in Figure 18.

LEAF ARRANGEMENT

If leaves that transmit 10 per cent radiation were horizontally displayed in continuous layers, then only 1 per cent of the light, mostly in the green region, could penetrate the second layer. However, leaves are rarely

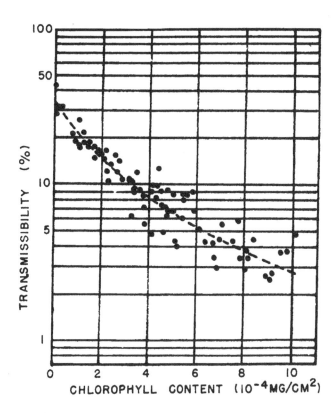

FIGURE 18. Relationship between transmissibility and the chlorophyll content per unit leaf area (after Kasanaga and Monsi).

displayed horizontally. Monsi and Saeki (1953) calculated the relative light interception by horizontal and erect foliage at 1:0.44. Therefore, the actual light gradient within the canopy is much less steep than the low transmissibility would suggest. For example, Brougham (1960) found that at the time when the total leaf area equaled the area of the ground, the mean transmissibility was 74 per cent for the more upright ryegrass, as against 50 per cent for the more horizontally disposed clover.

Using a curve similar to the one in Figure 10, and assuming that the most efficient light utilization is attained at 800 foot-candles, Warren Wilson (1960) computed the optimum leaf inclination angle from the horizontal as a function of light intensity (Figure 19a). In weak light, any departure from the horizontal position reduces net photosynthesis; in full sunlight, the optimum inclination for efficient light use is 81°, a nearly erect position. In Figure 19b, net photosynthesis is plotted against the intensity of vertical light for a horizontally displayed leaf and a leaf placed at the optimum inclination. The net photosynthesis of the latter increases linearly up to full sunlight; whereas the net photosynthesis of the former reaches a plateau at relatively weak light. At full sunlight, a leaf placed at

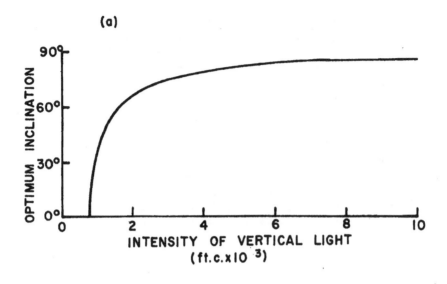

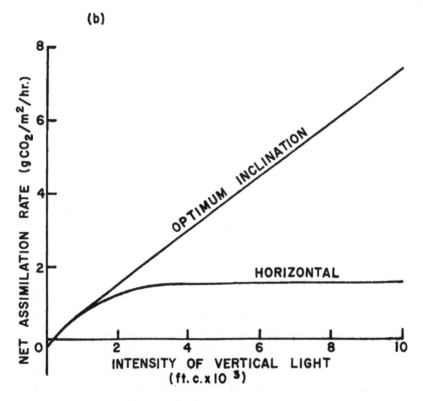

FIGURE 19. (a) Variation, according to intensity of vertical illumination, in foliage inclination giving most efficient utilization of light in net assimilation. (b) Effect of light intensity on net assimilation of leaves placed (1) horizontally and (2) at optimum inclination, assuming uniform projected areas of foilage (i.e., uniform area of light interception) (after Warren Wilson).

the optimum inclination is 4.5 times as efficient in using light as a horizontal leaf.

For the most efficient use of light, the upper leaves in a plant should have a nearly vertical orientation; whereas the lower foliage should be horizontal. Nichiprovich (1962) considered the ideal arrangement to be such that the lowest 13 per cent of the leaves lay at angles between 0° and 30° to the horizontal, that the adjoining 37 per cent of the leaves lay at 30° to 60°, and the upper 50 per cent of the leaves lay at 60° to 90°.

The genetic variability of leaf arrangement has only recently been exploited by plant breeders. Watson and Witts (1959) have reported that the development of the sugar beet from its wild ancestor by selection and breeding has not affected the intrinsic photosynthetic efficiency of the leaves. The higher yield of cultivated *Kleinwanzleben* sugar beets as contrasted with the wild varieties was attributed to the more upright leaf orientation of the former.

Light absorption of a field crop could also be improved by planting the seed with particular attention to the row direction. Corn, for example, displays its leaves in two ranks from opposite sides of the stem. The seed should be oriented so that the leaves are displayed at right angles to the row.

RADIATION AND LIGHT DISTRIBUTION WITHIN THE CANOPY

The distribution of light within a plant community can be expressed by Beer's law:

$$I = I_o\, e^{-kF}$$

where I is the light intensity at a given height within the plant community; I_0 is the light intensity at the top of the plant community; e is the base of natural logarithm; k is the extinction coefficient; F is leaf area index from the top to the height in question.

The extinction coefficient is primarily determined by the inclination and arrangement of leaves, and only secondarily by leaf transmissibility. In an herbaceous community, the extinction coefficients are usually 0.3 to 0.5 in stands with upright leaves, and 0.7 to 1.0 in stands with more or less horizontal leaves (Saeki, 1960).

The linear relationship between the logarithm of relative light intensity and the leaf area index is illustrated in Figure 20 from the work by Takeda (1961) on rice plants. With a leaf area index of 7, 95 per cent of the light is intercepted by the leaf assemblage. The leaf area index required to intercept 95 per cent of the light varies with crops. Brougham (1958b) found the following values: ryegrass, 7.1; timothy, 6.5, and white clover, 3.5. Other things being equal, the higher the leaf area index required to

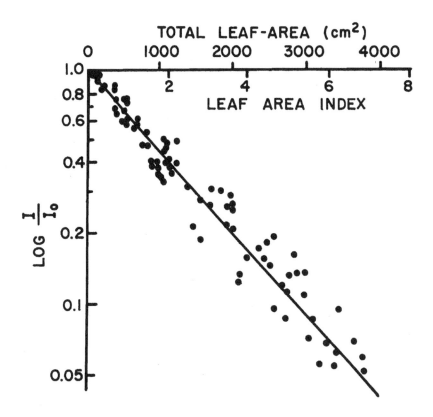

FIGURE 20. Relationship between the amount of leaf area and the relative light intensity in the community (after Takeda).

intercept the same fraction of radiation, the more efficient the crop is in utilizing radiation energy.

Recently, Monteith (1965a) proposed a somewhat different relationship to characterize the light distribution within the canopy. The equation is a binomial expansion of the form:

$$I = [s + (1-s)\tau]^F I_o$$

where I is the light intensity at a given height within the plant community; I_o is the light intensity at the top of the plant community; s is the fraction of light passing through unit leaf layer without interception; τ is leaf transmission coefficient; and F is the leaf area index.

Monteith gave values of s ranging from 0.4 for crops with predominantly horizontal leaves (e.g., kale, clover), to 0.8 for crops with nearly vertical leaves (e.g., cereals, grasses). He further explained:

> Because τ is a small fraction and $s \geqslant 0.4$, most of the radiation that penetrates a crop canopy when the sun is shining appears in the form of

sunflecks covering a fraction of s^F of the soil surface. Below a crop with $s = 0.4$, the relative area of sunflecks is less than 3 percent when the leaf area exceeds 4, but for a cereal with $s = 0.8$, the sunfleck area is 41 percent at $F = 4$ and 17 percent at $F = 8$. The light transmitted by cereals allows weeds to flourish but is sometimes exploited to undersow a second crop that develops when the cereal is harvested.

Although both Beer's law and Monteith's equation are very accurate in describing the radiation distribution within the canopy, they are seldom used because of the difficulty in determining the leaf area indices for the successive strata in the plant community. A simplified, though less accurate procedure, is to derive an empirical relationship between the plant height and the fraction of light received at the ground. Figure 21 illustrates such work by Stanhill (1962c) for alfalfa.

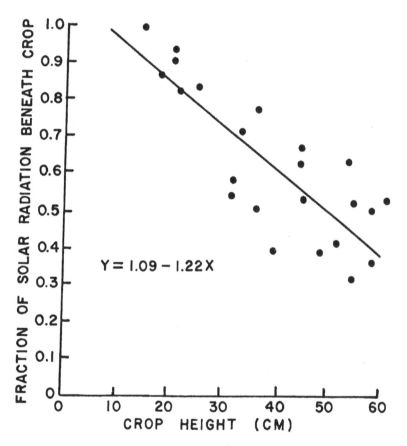

FIGURE 21. Relationship between crop height and fraction of solar radiation at ground level beneath crop canopy (after Stanhill).

Another interesting study was made by Baker and Musgrave (1964) who related the percentage of light interception in a corn stand with the age of the crop and the time of the day (Figure 22). The fit of the data was extremely good, with a multiple correlation coefficient of 0.944.

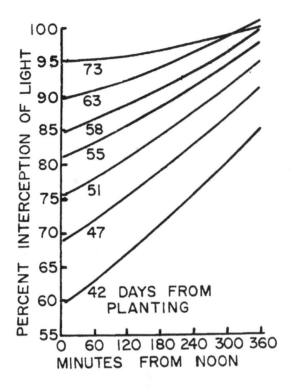

FIGURE 22. Light interception vs. time from solar noon and age of corn stand (after Baker and Musgrave).

Solar radiation has a changed spectral composition after transmission through the plant community. Radiation that is transmitted downward by leaves is mainly infrared, with a small amount of green light. The actual change depends on the proportion of radiation transmitted through the leaves and on that which reaches the ground through gaps in the crop canopy as unaltered sunflecks. Stanhill (1962c) found that for a tall alfalfa crop, the fraction of solar radiation reaching the ground was 30 per cent, as against 20 per cent for light. Yocum, Allen, and Lemon (1964) reported that for a tall corn crop the average percentage of transmission at the ground level was of the order of 5 to 10 per cent in the visible spectrum and 30 to 40 per cent in the near infrared.

The percentage of incident radiation penetrating the canopy changes markedly with the angle of the sun. The highest values are usually recorded at noon, and relatively high values also are recorded soon after sunrise and immediately before sunset. The high early morning and late afternoon values are attributed to a high proportion of diffuse light.

NET RADIATION PROFILE

The net radiation profile within a crop of bulrush millet has been recorded by Begg, Bierhuizen, Lemon, Misra, Slatyer, and Stern (1964) in tropical Australia on a clear day (Figure 23). The profile changes from a sharp increase of net radiation during high sun periods to small gradients during early morning and sunset. At night, the slope of the profile is negative.

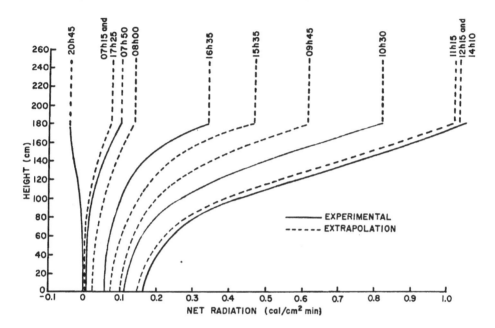

FIGURE 23. Net radiation profiles within crop of bulrush millet in tropical Australia March 29, 1963 (after Begg, Bierhuizen, Lemon, Misra, Slatyer, and Stern).

Throughout most of the day, net radiation decreases exponentially with height. The exponential relationship between net radiation and the height of a mature corn crop has been illustrated by Allen, Yocum, and Lemon (1964) in Figure 24. The fit is good except at the lowest layers. Although both the incident and net radiation decrease downward with the canopy exponentially, their extinction coefficients are different. Whereas the short wave is successively depleted as it goes downward through the foliage, the long wave may be augumented by radiation emitted by foliage. Therefore, the fraction of net radiation transmitted through the canopy exceeds that of visible light.

The net radiation profile of a clear day is markedly different from that of a cloudy day. The contrast is clearly illustrated by the observations of Denmead, Fritschen, and Shaw (1962) for a mature corn crop in rows spaced 40 inches apart (Figure 25). They plotted the average net radiation at various levels within the canopy against the leaf area index cumulated from the ground upward. The high retention of energy by the upper leaves on clear days and the even distribution of energy on cloudy days is evident.

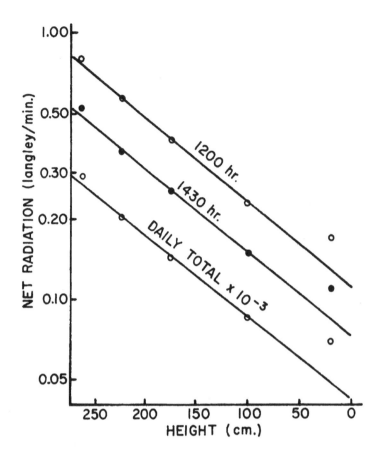

FIGURE 24. Net radiation vs. height in mature Ellis Hollow corn, September 10, 1961 (after Allen, Yocum, and Lemon).

The fraction of net radiation reaching the ground changes with the stage of crop development. Shaw (1959), working with corn, reported that early in the season, net radiation at the ground surface was close to 100 per cent. It decreased to 60 to 65 per cent when the crop reached a height of 60 inches, and to 14 per cent when the corn reached a height of 90 inches. Tanner, Peterson, and Love (1960), however, reported that for a typical mature corn crop the net radiation at ground level was about 40 per cent of that above the crop and that about 20 per cent of the net radiation at ground level was from long-wave exchange with the warmer crop. Since row spacing and differences in plant population both affect the net radiation profile, it is not surprising that widely different values have been quoted by different workers. For example, on clear summer days in Iowa, the fraction of net radiation reaching the ground was estimated to increase from 75 per cent to 90-95 per cent when the row spacing was reduced from 40 to 20 inches (Denmead, Fritschen, and Shaw, 1962).

Radiation Distribution within the Plant Community

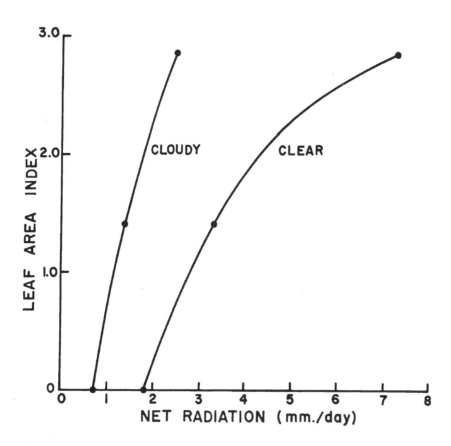

FIGURE 25. Average distribution of net radiation in the crop canopy for clear and cloudy days after maximum leaf area development. Leaf area index is cumulated from the ground (after Denmead, Fritschen, and Shaw).

CHAPTER 5

LEAF AREA INDEX

BASIC CONCEPT

A quantitative evaluation of dry matter production requires a knowledge of the net photosynthesis of individual leaves, as well as that of the total leaf area. Leaf foliage density is commonly characterized by the "leaf area index" developed by Watson (1947). The leaf area index is the leaf area subtended per unit area of land. This index usually includes only leaf blades, but the photosynthetically active leaf sheath should also be included (Thorne, 1959).

A number of methods have been devised for measuring the leaf area of plants. The more accurate methods are tedious and usually involve the destruction of the plants. There is still a need for a simple field method. The more commonly used methods are as follows:

(1) The outline of each leaf is reproduced by tracing, blueprinting, or imprinting on light-sensitive paper, and the area is measured by a planimeter.

(2) A correlation between leaf area and leaf weight is established from a large random sample of plants. The leaf area can be estimated from the leaf area/leaf weight ratio.

(3) The photoelectric method makes use of the principle that when the leaf is placed between a light source and a photocell, the reduction in photocell output gives a measure of the leaf area. Such photoelectric devices have been designed by Frear (1935) and Miller, Shadbolt, and Holm (1956). The accuracy of this method depends on leaves being placed flat and at right angles to the column of light. The possible error of this method has been discussed by Hurd and Rees (1966).

(4) Jenkins (1959) has devised an airflow planimeter for measuring leaf area. The apparatus consists of two identically perforated plates mounted on an airtight drum that is connected to a constant-speed rotary pump. The rate of air flow is a measure of the leaf area. This method is quite accurate, but the apparatus is expensive.

(5) The leaf area of a plant is usually closely related to the leaf width

and leaf length. Thus, the leaf area equals the leaf width multiplied by the leaf length multiplied by a factor. Once the factor is known, the leaf area can be estimated from linear measurements. This method has been adopted for determining leaf area in cotton (Ashley, Doss, and Bennett, 1963), wheat (Hopkins, 1939), sugar beets (Owen, 1957), and grass (Lal and Subba Rao, 1951, and Kemp, 1960).

(6) Another method is the use of series of standard leaf pictures for determining the area of individual leaves on an intact plant by visual comparison. Bald (1943) suggested that the mind judges size on the basis of proportionality or by the addition or subtraction of equal relative increments rather than of absolute increments, so that standards should form a logarithmic rather than an arithmetic series. He set up a twelve-point scale for potato leaves, varying in area from 21 square centimeters to 299 square centimeters. The same procedure has been adopted by Williams (1954) for tomatoes, by Humphries and French (1964) for potatoes and sugar beets, and by Williams, Evans, and Ludwig (1964) for clover and lucerne. Although the method is subject to errors of judgment as well as errors arising from the grouping of leaf areas into classes, it has the advantage of not destroying the leaves in the field, and reasonably accurate measurements may be obtained by experienced workers.

Most crop plants have a leaf area index ranging from 2 to 6. Pasture, pineapple, sugar cane, and rice, however, may have a leaf area index of 9 or even 12. In general, crops with vertical foliage have a higher maximum leaf area index. Donald (1961) pointed out that a high-yielding pasture may have a yield of four tons of dry matter per acre, mainly as erect leaves within a height of fifteen inches, as compared to a crop of maize with the same yield but less erect foliage distributed over a vertical span of six feet or more.

The optimum leaf area index is not necessarily the maximum leaf area index ever recorded. Instead, it should be such that the lowest leaf blades are barely maintained above the compensation point. If the lower leaves are below the compensation point, they will either lose weight or have to be supported by the translocation of metabolites from the upper leaves. Leaves below the compensation point will remain alive for some time until they have lost one-third to one-half their dry weight. On the other hand, if the leaf area index is less than the optimum, part of the radiation will be wasted, and the yield will fall short of the potential.

Net photosynthesis as a function of leaf area index and extinction coefficient. The light saturation of a single leaf occurs at a relatively low light intensity. As the leaf area index increases, the saturation light intensity of the plant stand also increases. In Figure 26, the relative net photosynthetic rates of subterranean clover are expressed as a function of light intensity for an individual leaf, and for whole plants with leaf area indices of 1, 5, and 9 (Davidson and Philip, 1958). At a high leaf area index, the

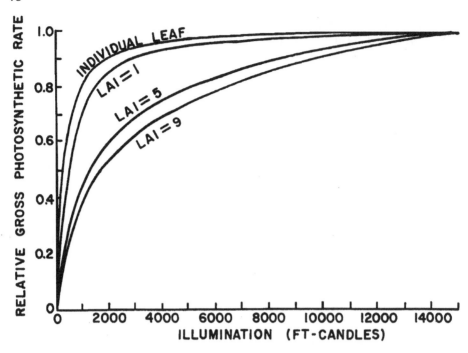

FIGURE 26. Relative gross photosynthetic rates expressed as functions of illumination for individual leaf and for whole plants with leaf area index of 1, 5, and 9 (after Davidson and Philip).

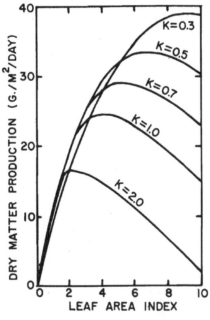

FIGURE 27. Daily production of *Celosia* stand under full sunlight as a function of leaf area index and K, extinction coefficient (after Saeki).

saturation light intensity is not reached even at 14,000 foot-candles. Also with greater leaf area index, the *total stand's photosynthesis* is more sensitive to light intensity changes at the higher end of the scale.

Under the same light intensity, the relationship between net photosynthesis and the leaf area index varies with the extinction coefficient of the plant stand. This changing relationship has been brought out by Saeki (1960) for a Celosia stand under full sunlight (Figure 27). At a low leaf area index, there is not much difference in the daily production of dry matter among plant species having different extinction coefficients. But with an increase in leaf area index, this difference becomes very marked.

Chapter 4 noted that the cultivated *Kleinwanzleben* sugar beet has more erect leaves and, hence, a lower extinction coefficient than the wild varieties. Therefore, in the early stages of crop development, net photosynthesis is almost the same in both the cultivated and wild varieties. However, at a later stage of high leaf area index, the dry matter production of cultivated sugar beets far exceeds that of the wild varieties.

Dry matter production as a function of radiation and leaf area index. Black (1963) examined the relationship of solar radiation, leaf area index, and crop growth rate for subterranean clover. His experiments were repeated three times in the summer, in early winter, and in the spring. In each experiment, radiation was varied by two levels of shading, and the leaf area index was varied by eight densities of sowing. The combined results are presented in the generalized curves in Figures 28 and 29. At any given radiation intensity, the growth rate rises with the leaf area index to a maximum and falls off thereafter. With the increase of the leaf area index, the response of growth to increasing radiation steepens. At a radiation intensity of 600 langleys/day, the optimum leaf area index is between 6 and 7. Maximum growth can be maintained at a leaf area index in excess of 7 only at radiation intensities greater than 600 langleys per day. The data also show the dependence of the compensation point upon the leaf area index. At a low leaf area index, the compensation point is exceedingly low; at a leaf area index of 9, the compensation point is reached at a radiation intensity of about 240 langleys per day.

Stern and Donald (1963) also showed that the optimum leaf area index is a function of radiation intensity (Figure 30). When the radiation is 50 langleys per day, the optimum leaf area index for clover is about 3.5. It rises to 5.5 at a radiation intensity of 200 langleys per day. The isopleth for a zero crop growth rate indicates the value of the leaf area index and the radiation at the compensation point. Below the zero isopleth, the plant is losing weight.

The theoretical computation of an optimum leaf area index would be rather simple if all leaves had the same capacity for photosynthesis. Since the photosynthetic rates of many plant leaves decline with age, shading

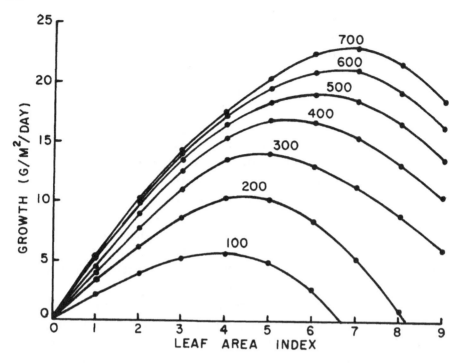

FIGURE 28. Interrelationship of growth, leaf area index, and solar radiation (after Black).

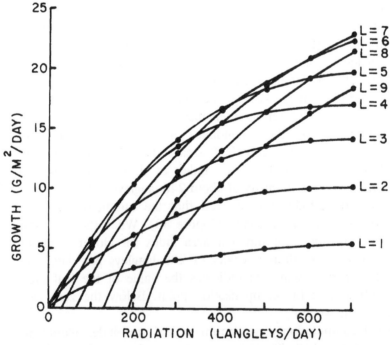

FIGURE 29. Interrelationship of growth, leaf area index, and solar radiation (after Black).

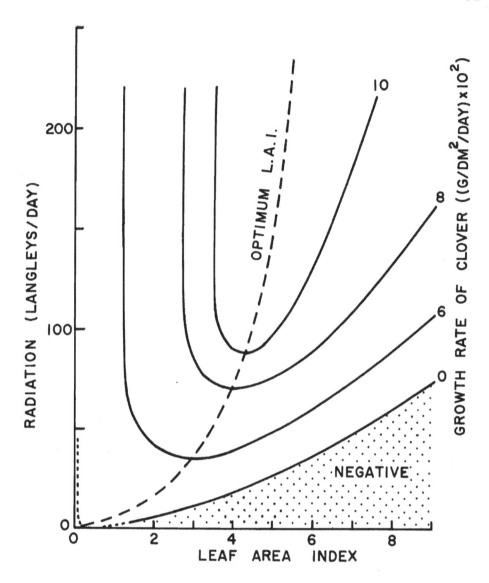

FIGURE 30. Relationship between radiation, leaf area index, and growth rate of clover (after Stern and Donald).

leads to earlier senescence when new leaves form from the base of the plant than at the top of the crop.

Variation of leaf area index throughout the crop cycle. The dry matter accretion of a crop is very much dependent upon the development of its leaf area. Gaastra (1958) compared the growth of a sugar beet crop with the change of its leaf area index (Figure 31). In May and June, when the leaf area index was less than 1, the production of organic matter was very low. The growth rate increased sharply in July

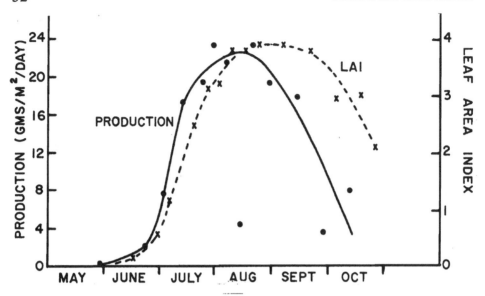

FIGURE 31. Leaf area index and organic matter production (gms/m²/day) in a field crop of sugar beets (after Gaastra).

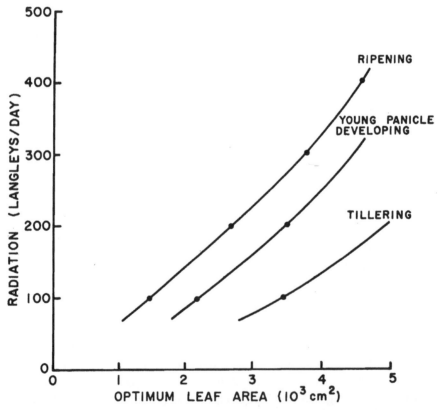

FIGURE 32. Relationship between radiation and optimum leaf area (after Takeda).

corresponding with the rapid rise in the leaf area index. At the end of the season, the production decreased more quickly than the leaf area, indicating that the leaf area might have exceeded the optimum value at that time.

As the ratio between the total leaves and the nonphotosynthetic organs changes with crop development, the optimum leaf area index should change accordingly. The curve in Figure 32 illustrates the relationship between radiation and optimum leaf areas for a rice crop at three stages of crop development (Takeda, 1961). The optimum leaf area is largest at the tillering stage and smallest during the ripening stage.

The seasonal variation of the leaf area index has an important bearing on the planting date. Ideally, the maximum leaf area index should be developed when climatic conditions are most favorable for photosynthesis. Watson (1947) studied the seasonal change in net photosynthesis and the leaf area index for four crops planted at different times of the year (Figure 33). As he pointed out, although the leaves of sugar beets and potatoes had greater net photosynthetic rates than the cereals at comparable times in the seasonal cycle, this difference was offset by the leaf area of the cereals being greatest in May and June, approximately when net photosynthesis was maximal; whereas the leaf areas of potatoes and sugar beets were minimal at the time of seasonal peak in net photosynthesis. Similarly, the slight superiority of potatoes over sugar beets in leaf area index in June and July, when net photosynthesis was high, was sufficient to counterbalance the much greater deficiency of potatoes in leaf area index in September and October when net photosynthesis was low. This study makes it obvious that, in the absence of other complicating factors, both sugar beets and potatoes should be planted earlier to take full advantage of the high summer radiation.

Watson's study also shows considerable differences among several crops in their pattern of leaf area developments. In wheat and barley, the maximum value of about 3 occurred at the time of rapid shoot elongation; by the time of ear emergence, it had fallen to half the peak value, and at harvest it was zero. Potatoes and sugar beets showed maximum values of leaf area index much later in the crop cycle and still had considerable leaf at harvest. It is also notable that the winter wheat had a leaf area index below 1 during 75 per cent of its growing period.

Leaf area index as a guide to cultural practices. Because the optimum leaf area index varies not only seasonally but also diurnally, it is impossible to maintain crop density at that level at all times. The system that comes closest to this ideal will be the most efficient. Under normal field conditions, light wastage is particularly high during the early stages of crop growth. It is desirable to promote extensive leaf area by cultivation methods such as immediate fertilizing, raising the irrigation water tempera-

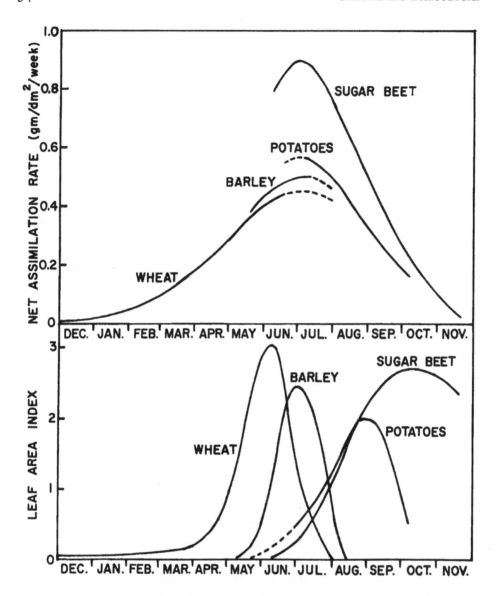

FIGURE 33. Seasonal change in net assimilation rate and leaf area index of different crops (after Watson).

ture, and the like. Light wastage can be minimized by properly adjusting row spacing and plant population and by selecting the most advantageous time of planting.

Another method is transplantation, which traditionally has been employed with rice, tobacco, and tomato crops. Certainly, this method could be adopted for many other crops, especially if the seeds are small and delicate and where the growing season is relatively short. Anderson, Dubetz, and Russell (1958) reported yield increases of 36 to 52 per cent

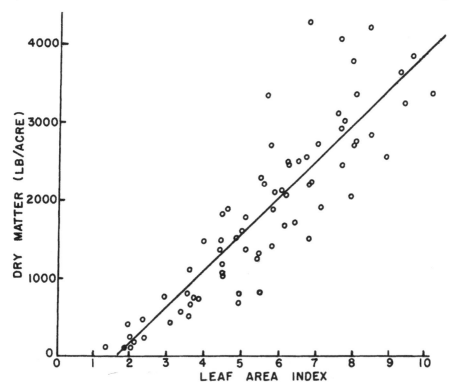

FIGURE 34. Relationship between leaf area index and herbage yield (after Brougham).

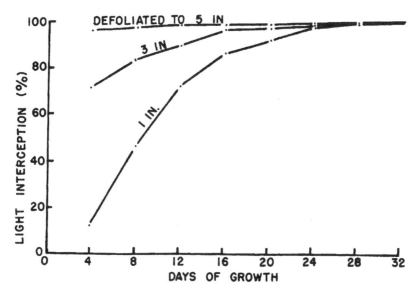

FIGURE 35. Percentage of light intercepted by herbage above the one-inch level (after Brougham).

with transplanted sugar beets in Canada. The advantages of growing cotton by transplantation have been discussed by Christidis (1962). Even with the high labor cost in Hawaii, transplanting may well be a profitable practice for sugar cane culture; it could reduce the crop age by three to four months.

The application of the optimum leaf area concept is best utilized in pasture management. Brougham (1956) showed that in areas of high radiation, an increase of one square foot of leaf area yielded a 468-pound dry matter up to the highest leaf area index of 10 (Figure 34). Obviously, grazing should not be so heavy as to reduce the crop height greatly. Figure 35 shows how the light wastage can be minimized in order to maintain optimum growth potential. When the initial defoliation height was one inch, little light was available at first for dry matter production. Light interception did not reach the 95 per cent level until after 24 days. When the pasture was cut down initially to three inches, 95 per cent light interception was obtained sixteen days after cutting. When the initial cutting was at five inches, sufficient herbage was left to intercept nearly all the incident light immediately following cutting.

Since the optimum leaf area index will vary considerably between summer and winter, and between temperate and tropical regions, procedures for pasture management should change accordingly. Growth rates will decline if the radiation is reduced when the leaf area is high, or if the leaf area is reduced when the radiation is high. The former situation is found when the pasture with a high leaf area index is carried into a low radiation period. Excessive grazing with a rising radiation level illustrates the latter situation. Thus, in winter, the time interval between grazing should be lengthened, and the amount of herbage remaining after grazing should be appreciably lower than in the summer.

CHAPTER 6

RADIATION UTILIZATION BY FIELD CROPS

EFFICIENCY OF RADIATION UTILIZATION BY FIELD CROPS

Chapter 3 noted the inefficient utilization of radiation by leaves for four reasons:

(1) Part of the incoming radiation is reflected, and only the visible light is photosynthetically active.

(2) The quantum requirement for photosynthesis is high.

(3) Leaf light saturation is reached far short of full sunlight.

(4) Respiration reduces net photosynthesis.

In the field, the efficiency of solar radiation utilization by crops should be increased because of the multiple leaf layers, but it is actually further reduced for two reasons: (1) because the soil surface is not completely covered by the crop, wasting a large portion of the radiation, and (2) because there are varying deficiencies in water and mineral nutrients, pest damage and unfavorable temperatures. Another factor contributing to the apparent low efficiency is the exclusion of information on root material, which may account for as much as 30 per cent of the total dry matter in harvest data (Schuurman and Makkink, 1955).

The efficiency of radiation utilization can be computed by comparing the caloric value of the organic matter produced per unit of cultivated area with the incident radiation on the same area during the same period. Some studies on the efficiency of radiation utilization are summarized in Table 4. The data are not strictly comparable. For example, the efficiency values reported by Wassink were originally expressed in terms of visible light, but they were converted into total radiation by assuming that the former accounts for 41 per cent of the latter. These data indicate that an ordinary crop converts less than 1 per cent of solar radiation. Spoehr (1956) estimated that under optimum conditions, corn can convert about 1.5 per cent of the incident radiation into organic matter including shelled corn, cobs, leaves, stalks, and roots during a four-month growing period.

TABLE 4
DRY MATTER PRODUCTION AND THE EFFICIENCY OF
RADIATION UTILIZATION FOR SEVERAL AGRICULTURAL CROPS

Crop	Vegetation period	Yield (tons/hectare)	Efficiency (per cent)	Reference
Potatoes	Apr.-Aug.	9.60	.50	Wassink (1948)
Winter wheat	Nov.-Aug.	10.45	.52	Wassink (1948)
Sugar beets	May-Oct.	16.00	.90	Wassink (1948)
Fodder beets	May-Oct.	16.00	.90	Wassink (1948)
Swedes	May-Oct.	11.00	.62	Wassink (1948)
Carrots	May-Oct.	6.86	.39	Wassink (1948)
Chicory	May-Oct.	9.00	.54	Wassink (1948)
Turnips	May-Nov.	3.60	.51	Wassink (1948)
Corn	June-Sept.	15.52	1.05	Transeau (1926)
Sugar cane	22 months	129.48	1.43	Burr, Hartt, Brodie, Tanimoto, Kortschak, Takahashi, Ashton, and Coleman (1957)

COMPUTATION OF POTENTIAL PHOTOSYNTHESIS

De Wit (1959) presented a method for calculating growth based on incident light energy. He used Gaastra's photosynthesis curve for sugar beet leaf (Figure 10) as the model for a number of agricultural crops. He also assumed that the photosynthetic rate is not affected by the range of temperatures normally encountered, and that the reflection and absorption of light by crop leaves are independent of the angle of incidence. He used solar radiation values, derived from date and latitude, to calculate the growth potential in terms of grams CH_2O per unit area. However, the potential photosynthesis, thus calculated, can be realized only when other conditions are nonlimiting.

According to de Wit, the potential photosynthesis would be 5 g CH_2O/m^2/day in December, and 29 g/m^2/day in June for the average radiation received in the Netherlands. The latter value is comparable with the maximum rate of 20 g/m^2/day for sugar beets in the Netherlands (Alberda, 1962), and the highest rate of 26 g/m^2/day for tropical grasses in Puerto Rico (Vincente-Chandler, Silva, and Figarella, 1959). Most crop plants do not exceed 20 g/m^2/day in dry matter production.

De Wit's method has been applied by Stanhill (1962c) to calculate alfalfa growth at Gilat, Israel. The calculated values agree well with the harvest data when allowance was made for respiration loss, light wastage, and root growth.

At best, the method proposed by de Wit is crude. To improve it, I suggest:

(1) Different photosynthetic curves should be used for different crops or even different varieties.
(2) The temperature dependence of respiration should be incorporated.
(3) Measurements of light distribution with the canopy during successive stages of crop development should be used whenever possible.

The determination of potential photosynthesis can be used by farmers not only to interpret the fluctuation of yield from year to year but to guide in the selection of a proper planting date. The calculation of potential photosynthesis also provides a quantitative basis for estimating the agricultural potential for different parts of the world. Bonner (1962) is of the opinion that the upper limit of crop yield is already being approached today in parts of Japan, of western Europe, and of the United States. On the other hand, Noffsinger (1962) painted a much brighter picture. He contended that the existing arable land can support many times the present world population with ease. The controversy over the agricultural potential of the world cannot be resolved until the model for computing potential photosynthesis is perfected. This is by no means an easy task.

One limitation to the method of calculating potential photosynthesis is that it can estimate only the total dry matter but not the yield of a specific part of the crop. The seed of a cereal is more highly esteemed than straw. The rice husk is worthless and even a nuisance at harvest. An increase in yield, resulting from the use of better varieties, may be limited to a shift in the distribution of dry matter to more valuable organs without an increment in the total net photosynthesis. This has been the case in the sequence of plant breeding for several wheat varieties.

COMPARISON BETWEEN THE TROPICS AND TEMPERATE REGIONS

It is a common misconception that tropical regions have a very high agricultural potential and that the low productivity there is largely the result of technological incompetence. For example, Lee (1957) remarked: "[In the tropics] plant growth and multiplication are, in general, greatly favored by the high temperatures and humidities." Such a statement betrays an unfamiliarity with the basic principle of photosynthesis.

In addition to the disadvantage of high nighttime temperatures, which accelerate respiration, tropical regions, especially the wet tropics, usually have lower radiation than temperate zones during their normal growing season. In July, the average daily radiation is 440 langleys at Madras,

India (13°N), as against 680 langleys at Fresno, California (36°N), and 450 langleys at Fairbanks, Alaska (64°N). A rough calculation by Best (1962) indicates that during the seven months of the summer growing season, the average radiation in temperate regions is approximately 1.5 times that of the tropics. This ratio also applies when a sunny growth season in the tropics is compared with the sunny summer conditions in the Po Valley or Suecca, Spain. He further explained:

> From this consideration, it can be said that in respect of the light factor the potential production of annual crop plants is often about one and a half times as high in the temperate zone as in the tropics. The actual yields vary much more, however, for example, rice is one of the few crops which are grown in both equatorial and temperate zones. The average yields of this crop vary in the tropics round the equator from 1.1 to 1.8 tons/hectare; whereas in Spain and Italy they are 4 to 5 times as high. Naturally there are, apart from light, many different factors that are responsible for the phenomenon, but results from experimental stations in the equatorial tropics show yield maximum in the order of 5 tons/hectare against 12.5 tons/hectare in Bologna, Italy.

Although the productivity of annual crops is usually higher in temperate regions, the tropics have the advantage of year-round crop production. Tropical lands are best suited for crops with a long vegetative period able to intercept light to a maximum extent. The average annual yields for such crops as sugar cane, cocoa, and oil palm are indeed impressive, and they rank among the highest.

RADIATION UTILIZATION DURING SUCCESSIVE STAGES OF CROP DEVELOPMENT

The efficiency of radiation utilization varies throughout the crop life cycle. Gaastra (1963) has compared the seasonal variation of the dry matter production of a sugar beet crop with the potential photosynthesis calculated according to the method proposed by de Wit (Figure 36). The crop accumulated the greater part of its dry matter in the middle of the season when the plant was actively growing and the cover was nearly complete. About 86 per cent of the final yield was produced in the middle period covering 44 per cent of the total growing season. During the middle season, the actual yield nearly matched the potential predicted by de Wit. The efficiency of energy conversion was 6.1 per cent for visible light, and about 2.5 per cent for the total incident radiation. Under optimum conditions, the maximum efficiency of radiation utilization as high as 4 per cent has been recorded during the middle season for a corn crop (Lemon, 1963).

The extremely low efficiency in the early stage reflects the large amount of light wastage in an incompletely covered stand. In the late stage,

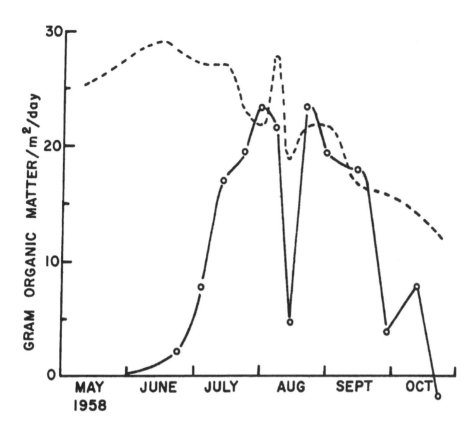

FIGURE 36. Seasonal variation of daily dry matter production by a sugar beet crop (solid curve) and of potential photosynthesis (dotted curve), calculated according to the method proposed by de Wit (after Gaastra).

the rate of dry matter production is reduced because photosynthesis has already reached a plateau after full development of canopy; whereas the respiration rate continues to rise with the increase of nonphotosynthetic organs in the plant. Thus, from planting to harvest, the dry matter accumulation follows a sigmoid curve similar to that in Figure 37 for a barley crop (Gregory, 1926).

In a study of the leaf area and net assimilation rate of flax during the summers of 1958 and 1959, Larsen (1960) found that the total unfolded leaf area reached a peak slightly more than a month after planting or approximately when flowering began (Figure 38). However, the effective assimilation area falls sharply after flowering. Withering eventually exceeds the growth increment in leaf area as the crop reaches senescence. At ripening, although the total leaf area remained high, the net assimilation rate reaches zero.

Many crops are harvested when the net photosynthesis drops below a certain point. The proper age of harvest varies not only with plant species but also with the climatic environment. One remarkable example of the

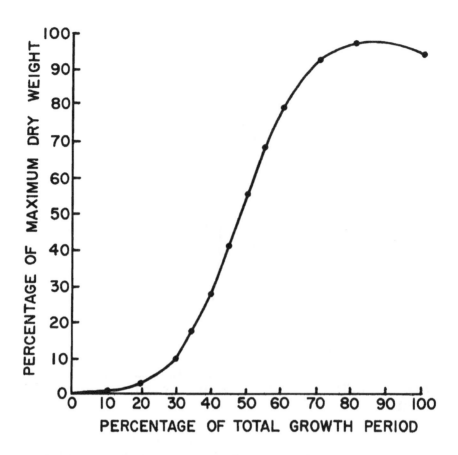

FIGURE 37. Curve of dry weight increase for barley (after Gregory).

relation between climatic conditions and harvest age is the sugar cane culture in Hawaii where it is usually a 22- to 24-month crop in lowland areas. However, in one highland area of Kau at the 3,000-ft. elevation, crop maturity is delayed until it is three years old. There, net photosynthesis proceeds at a reduced rate but for a longer period because of low radiation and temperature. Barnes (1953) states that the average age of a sugar cane crop in East Africa increases seven months in rising from sea level to 4,500 ft.

EMPIRICAL RELATIONSHIP BETWEEN
RADIATION AND CROP YIELD

The relationship between radiation and dry matter production can best be studied by a method similar to that proposed by de Wit. However, the data needed for such a study are rarely available, and a simple correlation between radiation and crop yield is often adequate for the solution of many practical problems.

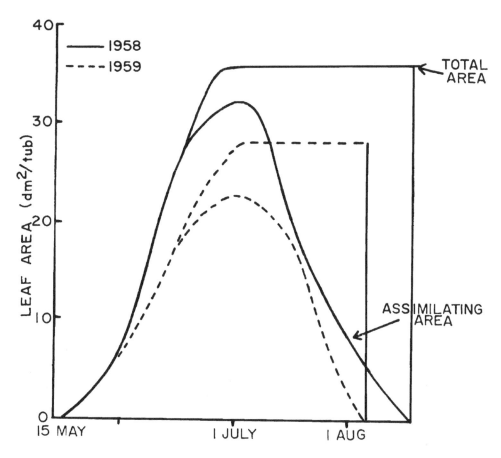

FIGURE 38. Total unfolded and assimilating leaf area of flax as function of growing time for two summers (after Larsen).

In the absence of water deficits, mineral deficiencies, and other limiting factors, a good relationship can be expected between radiation and net photosynthesis even on an hourly basis. A good example is the work by Baker (1965) in Figure 39. Moss, Musgrave, and Lemon (1961) also reported that 90 per cent of the hourly fluctuations in net photosynthesis in a corn stand could be explained by light fluctuations alone. Murata and Iyama (1960) found that the correlation between radiation and photosynthesis rose with an increasing leaf area index of the plants. Therefore, a close relationship can be expected between the total radiation and the yield of a dense crop with a consistently small light wastage. On the other hand, such a correlation cannot be expected for a crop that has a highly variable light wastage, and whose ground coverage is far from complete. This might be the case reported by Salter (1960) for a cauliflower crop.

For a dense crop like Hawaiian sugar cane, which intercepts a good deal of the sunlight throughout the greater part of its two-year growing season, radiation plays a dominant role in determining the final yield in

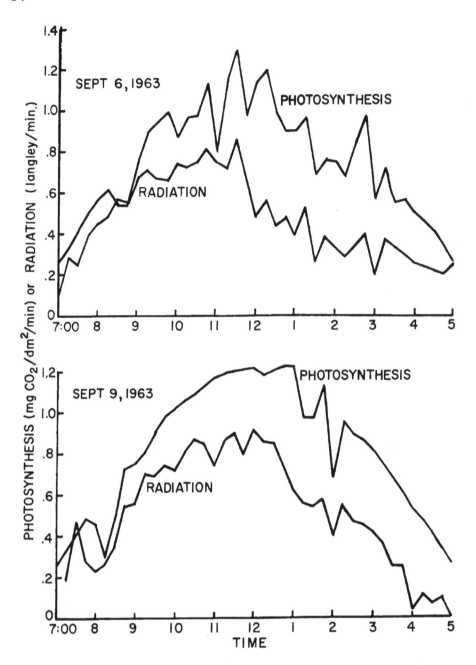

FIGURE 39. Daily variation of photosynthesis and radiation (after Baker).

areas of adequate water supply. Figure 40 shows the relationship between radiation and tons of sugar per acre at the Pepeekeo plantation, where sunlight may limit growth even in years of high radiation. Understanding the role played by radiation can serve as a useful guide for making many management decisions. For instance, other things being equal, fertilizer applications should be increased in years of high radiation to meet the demand of a foreseeably heavy crop.

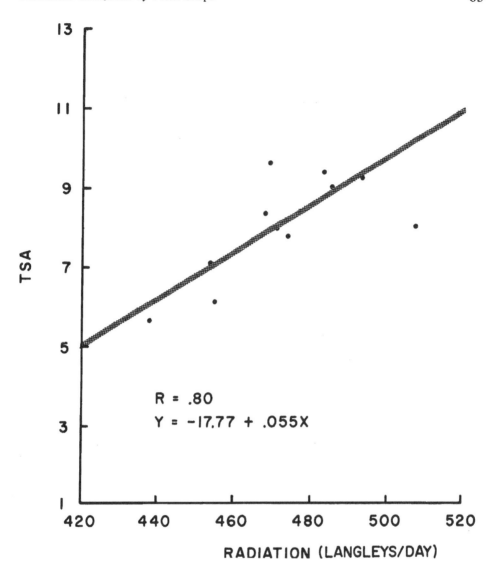

FIGURE 40. Relationship between radiation and tons of sugar per acre on an Hawaiian sugar plantation (data obtained from Hawaiian Sugar Planters' Association).

SHADE EXPERIMENTS AND ARTIFICIAL LIGHT

Under normal cultural practices, many plants can take advantage of full sunlight. Benedict (1951), working in Wyoming, found that shading reduces the growth of three species of range grasses. His investigation was undertaken with the knowledge that the summer light intensities in Wyoming, among the highest in the world, could reach 14,000 foot-candles. Similar results have been reported for sugar cane in Hawaii where cutting sunshine in half reduces yield by about the same proportion. In a shade experiment on corn, Earley, Miller, Reichert, Hageman, and Seif (1966) found a significant decrease in measured components (grain, stover, protein, total oil, etc.) as light was decreased. The reduction of protein with decreasing light was nearly linear. Light intensity is an important factor for fruit bud formation in the sultana. (May and Antcliff, 1963). Shade reduces berry size as well as yield.

Some plants grow better under shade. In a shade experiment in Yonkers, New York, Arthur and Stewart (1931) reported that sunflowers produced maximum yield under 78 per cent sunlight, dahlia and buckwheat under 58 per cent sunlight, and tobacco under 35 per cent sunlight. Certain types of drug plants such as ginseng and goldenseal also grow well under partial shade. However, unless the photosynthetic rate of the leaf shows a drop at high light intensity, the higher yield obtained under the shade may be caused by other indirect factors. In the tropics, many plants grow better under shade because it reduces moisture loss from the soil, suppresses weeds, preserves soil structure, maintains high humidity, and protects the plant from pest damage and nutrition deficiency. For instance, leaves on plants grown under shade show higher percentage contents of nitrogen, phosphorus, and potassium than similar plants grown unshaded. Thus, with a poor nutrient supply, plants grown in full sunlight may show marked signs of nutrient deficiency, particularly nitrogen.

Cocoa, for example, is grown traditionally in the shade of the forest. The usual explanation is that it is an understory species in its home in the Amazon rain forests. However, experiments by Cunningham and Lamb (1959) have demonstrated conclusively that, when the application of water and nutrients are adequate, cocoa will grow far more vigorously and yield up to four times as much in the open as in the shade. The growth of shade trees is also a common practice in the many tea estates devoted to the variety of *Camellia sinensis assamica*. But experiments in Africa (McCulloch, Pereira, Kerfoot, and Goodchild, 1965, 1966; Laycock and Wood 1963a) have shown that shade trees caused a substantial reduction in yield (Figure 41). Tea quality was also adversely affected.

The photosynthetic rates of coffee fall at high light intensity, and, as a result, it is one of the very few plants that grows well under shade. Investigations at Kabanyola, Uganda, indicate that both *Robusta* and *Arabica* varieties grow best under moderate shading. Full midday sunlight in excess of 1.5 langleys/minute hastens the breakdown of chlorophyll in the coffee leaf (Huxley, 1965).

Stoughton (1955) pointed out that in general usage, the term growth has two different connotations, namely, elongation growth and dry matter production. While the latter is promoted by high light intensity, the former is often suppressed by it. Mason (1925) found that date palm elongation stopped soon after sunrise and was not resumed near sunset. Elongation growth was resumed within a few minutes if the plant was

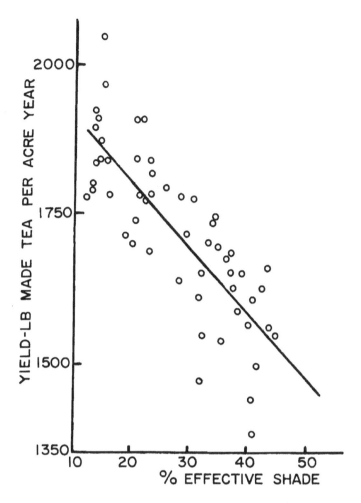

FIGURE 41. Yield of tea as function of percentage of effective shade at Limuru, Kenya (after McCulloch, Pereira, Kerfoot, and Goodchild).

covered during the day. In general, the less light available to the plant, the more spindly and elongated it will become, so that plants normally prostrate become erect when shaded. Any attempt to use elongation as an indicator of plant growth and, hence, as a guide to irrigation interval control or other cultural practices should be made with utmost caution.

Plants grown in partial shade develop morphological structures common to shade plants. The leaves become larger and thinner. The plants are more succulent but lack the vigor of those grown in full sunlight, and the roots are usually less well developed.

Shade also affects fruit quality. For example, strawberries exposed to sun have a higher content of ascorbic acid than those in the shade. Hansen and Waldo (1944) studied the problem by shading the entire plant, or the solitary berries, with muslin bags and concluded that under the former conditions, a much lower amount of ascorbic acid was synthesized, because leaves were its primary manufacturers.

Artificial light has been used to supplement sunlight either to increase the photosynthetic rate or to modify the photoperiod. The latter effect will be discussed in the next chapter. Although the longer wavelengths of the visible spectrum (e.g., the red and orange region) are most efficient in photosynthesis, the shorter wavelengths at the blue end of the spectrum are important in ensuring a normal growth in many species. Therefore, to obtain the best overall reaction, white light is preferred. The current available light sources provide light throughout the visible spectrum but differ widely in the proportions emitted in each of the wavelength regions. The relative merits of various light sources have been summarized by Stoughton and Vince (1957)):

> In terms of luminous efficiency, availability, and initial cost, the range of possible sources is very limited. The common incandescent tungsten-filament lamp has a spectral emission which is very rich in the wanted region but unfortunately this is accompanied by a much greater proportion of the longer, or thermal, wavelengths. Its use, therefore, would entail some method of screening the plant from this heat radiation, and this is not practicable on a commercial scale.
>
> The neon discharge tube also possesses a very suitable emission spectrum but suffers from the disadvantage of high initial cost, largely due to the complicated starting gear which is required for the types of lamp at present available.
>
> The fluorescent tube is a more promising light source; by choice of a suitable phosphor, the emission spectrum can be varied over a wide range; moreover, it has a high luminous efficiency and it is practically free from thermal radiation. Its disadvantage is its low intrinsic brightness, which means that a number of tubes must be used in efficient reflectors to provide adequate light intensity. Again, unless some mechanism is provided for removing the lights when not in use, a considerable loss of natural daylight results from shading. Nevertheless, it has been and is widely used.
>
> Until recently, however, the most popular light source for supplementing daylight has been the high-pressure mercury vapor lamp (HPMV) of 400 watts input. This lamp is readily available, has a fair luminous efficiency, is reasonable in cost, and is a compact source. Specially designed simple reflectors are made for the particular purpose. Its chief disadvantages are that a high proportion of emission is in the green region of the spectrum, which is least efficient in photosynthesis and that, since there is a fairly high thermal emission, care must be taken that overheating of the plants does not occur.
>
> In recent years, new types of lamp, exemplified by the mercury fluorescent lamp, have been developed. These combine the high intensity and compactness of the HPMV lamp with the lower thermal emission and the better spectral composition of the fluorescent tube. It is possible that these may come to replace the other types for this particular purpose.

For economic reasons, the use of artificial light has been restricted to a few high-priced crops, such as cucumber, lettuce, and chrysanthemum. Often, artificial lighting is used only during the seedling stage to accelerate growth beneficial for subsequent plant development.

CHAPTER 7

PHOTOPERIODISM

HISTORICAL BACKGROUND

The length of a day is known as a photoperiod and the responses of the plant development to a photoperiod is called photoperiodism. The first study on photoperiodism was published by Garner and Allard (1920). They noticed that a late-flowering variety of tobacco, Maryland Mammoth, which had been developed in the southern United States, did not flower during the growing season further north than Washington, D.C. The plants were always cut off by frost in the autumn before flowering occurred. Plants grown in the greenhouse during the winter, however, flowered and fruited freely. Garner and Allard also found that the plants in the greenhouse could be prevented from flowering in the winter by extending the day with artificial light.

Subsequently, many other plants were tested by Garner and Allard. In addition to the effect on the formation of flower, fruit, and seeds, the photoperiod was found to influence vegetative growth, the formation of bulbs and tubers, the character and extent of branching, leaf shape, abscission and leaf fall, pigment formation, pubescence, root development, dormancy, and death. The light intensity required for photoperiodic response is so low that even the twilight before sunrise and after sunrise is effective.

CLASSIFICATION

In their early work, Garner and Allard divided plants into three groups on the basis of their response to the photoperiod, namely, long-day plants, short-day plants, and day-neutral plants. Long-day plants flower only under daylengths longer than fourteen hours. In short-day plants, flowering is induced by short photoperiods of less than ten hours. The day-neutral plants can form their flower buds under any period of illumination.

Later, Allard (1938) added a fourth group, which he designated as intermediate. The intermediate plants flower at a daylength of twelve to fourteen hours but are inhibited in reproduction by daylengths either above or below this duration. The photoperiodic responses of a number of plants are listed in Table 5 (Spector, 1956).

These major photoperiodic groups can be subdivided in a number of ways. For example, some long-day plants have a quantitative response to daylength. The longer the daylength or the greater the number of long days to which the plants are exposed, the sooner the flowers form. Such plants will eventually flower, even in short days. Plants of this kind are known as "facultative long-day plants," in contradistinction to the "strict long-day plants," which have a well-defined minimum daylength below which flowering does not occur. A similar division may be made within the short-day group.

Plants within the same group may also differ in their response to daylength, subsequent to flower initiation. For example, the strawberry is a short-day plant for floral initiation, but it is a long-day plant for fruit formation. Other short-day plants such as the soybean, prefer a short photoperiod throughout.

Later experiments, however, turned up a surprising fact. If a plant requires a certain length of the day to flower, darkening the plant for part of the day did not interfere with its flowering. On the other hand, illumination at night affected the plant's flowering. Thus, the critical factor in photoperiodism is not the length of the day but the length of night; strictly speaking, plants should be classified as long-night and short-night rather than long-day and short-day.

Photoperiodic induction. In their studies of the influence of various photoperiods on the reproduction of soybeans, Garner and Allard (1923) observed that an exposure of ten short-days was all that was required to bring about flower formation, which was continued when the plants were exposed to long-days thereafter. Another example, the Mexican sunflower (Tithonia species), requires exposure to long-nights of approximately fourteen hours for 14 to 21 days to enable it to initiate flower buds. Once initiated, flowering continues even in long-days in all subsequent shoots. (Stoughton and Vince, 1954). Thus, Tithonia plants exposed to short-days in the seeding stage will subsequently flower throughout the summer. Untreated plants, though, will not flower until autumn when the night is long. This effect, known as photoperiodic induction or photoperiodic aftereffect, has also been observed for long-day plants.

Relation to temperature. The influence of daylength on plant development is often modified, and sometimes even inhibited, by other environmental factors, particularly temperature. Many plants do not respond to critical photoperiod unless their thermal requirements are met (Coleman and Belcher, 1952). Most biennial plants will fail to flower until they have

TABLE 5
Photoperiodic Responses of Selected Plants

Species	Photoperiodic class and light period	Species	Photoperiodic class and light period
Fruit and Vegetable Crops		**Legumes and other field crops**	
Artichoke (Helianthus tuberosus)	s,N	Alfalfa (Medicago sativa)	l
Bean, lima (Phaseolus Iunatus)	N,S	Beet, sugar (Beta vulgaris)	L
Bean, string (P. vulgaris)	N,S	Clover (Trifolium spp)	l
Beet, garden (Beta vulgaris)	l	Clover, red (T. pratense)	L(>12 hr)
Cabbage (Brassica pekinensis)	l	Cotton (Gossypium hirsutum)	N,s
Chicory (Cichorium intybus)	L	Lespedeza (Lespedeza stipulacea)	S(<13.5 hr)
Carrot (Daucus carota)	N	Soybean, Biloxi and Mandarin	
Celery (Apium graveolens)	N	(Giycine soja)	S,s
Cucumber (Cucums sativus)	N	Soybean, Mandell (G. soja)	s
Dill (Anethum graveolens)	L(>11 hr)	Sweetclover (Melilotus alba)	L
Lettuce (Lactuca sativa)	l	Tobacco (Nicotiana tabacum)	N
Onion (Allium cepa)	l,s,N	Tobacco, Havana	l
Pea (Pisum sativum)	N,l	(N. tabacum)	
Pepper (Capsicum annuum)	N,s	Tobacco, Md. Mammoth	S(<14 hr)
Potato (Solanum tuberosum)	l,s,N	(N. tabacum)	
Radish (Raphanus sativus)	L	Vetch, spring (Vicia sativa)	l
Spinach (Spinacia oleracea)	L(>13 hr)		
Strawberry (Fragaria chiloensis)	S(<10 hr)	**Ornamental Plants**	
Strawberry, everbearing	l,N	Althea (Hibiscus syriacus)	L(>12 hr)
(F. chiloensis)		Aster (Callistephus chinensis)	l
Sweet potato (Ipomoea batatas)	S	Azalea, coral bell	N
Tomato	N,l,s	(Rhododendron sp)	
(Lycopersicon esculentum)		Balsam (Impatiens balsamina)	N
Turnip (Brassica rapa)	l	Begonia (Begonia semperflorens)	N
		Bryophyllum	S(<12 hr)
Grasses		(Bryophyllum pinnatum)	
Barley, spring	l	Cactus (Zygocactus truncatus)	s
(Hordeum vulgare)		Chrysanthemum	L
Barley, winter (H. vulgare)	L(>12 hr)	(Chrysanthemum frutescens)	
Beardgrass (Andropogon gerardii)	S(<18 hr)	Chrysanthemum (C. indicum)	S(<15 hr)
Bentgrass (Agrostis palustris)	L(>16 hr)	Cineraria (Senecio cruentus)	s
Bluegrass, annual (Poa annua)	N	Cornflower (Centaurea Cyanus)	l
Bluegrass, Kentucky	l	Cosmos (Cosmos bipinnatus)	s
(P. pratensis)		Cosmos, Klondyke	S(<14 hr)
Bromegrass (Bromus inermis)	L(>12.5 hr)	(C. sulphureus)	
Broomsedge	s(12-14.5 hr)	Cosmos, orange flare	N
(Andropogon virginicus)		(C. sulphureus)	
Canary grass	L(>12.5 hr)	Foxglove (Digitalis purpurea)	l
(Phaloaris arundinacea)		Fuchsia (Fuchsia hybrida)	N
Cloudgrass (Agrostis nebulosa)	L(>13 hr)	Gardenia	N
Corn (Zea mays)	N,S	(Gardenia jasminoides fort.)	
Fescue (Festuca elatior)	L	Geranium (Pelargonium	N
Foxtail (Alopecurus pratensis)	L(>9 hr)	hortorum)	
Oat (Avena sativa)	L(>9 hr)	Holly, English (Ilex aquifolium)	N
Orchardgrass	L(>12 hr)	Hydrangea	N
(Dactylis glomerata)		(Hydrangea macrophylia)	
Rice, summer (Oryza sativa)	N	Kalanchoe	S(<12 hr)
Rice, winter (O. sativa)	S(<12 hr)	(Kalanchoe blossfeldiana)	
Rye, spring (Secale cereale)	l	Larkspur (Delphinium cultorum)	L
Rye, winter (S. cereale)	l	Morning glory	S
Ryegrass, Italian	L(>11 hr)	(Ipomoea hederacea)	
(Lolium italicum)		Morning glory (I. purpurea)	S
Ryegrass, early perennial	L(>9 hr)	Orchid (Cattleya trianae)	S
(L. perenne)		Pansy (Viola tricolor)	N
Ryegrass, late perennial	L(>13 hr)	Petunia (Petunia hybrida)	l
(L. perenne)		Phlox (Phlox paniculata)	L
Sorghum (Sorghum vulgare)	l	Poinsettia	S(<12.5 hr)
Sudan (Holcus sudanensis)	s	(Euphorbia pulcherrima)	
Sugar cane	s	Salvia (Salvia splendens)	s
(Saccharum officinarum)		Sedum (Sedum spectabile)	L(>13 hr)
Sugar cane, var. 28NG 292	IM(12-14 hr)	Snapdragon (antirrhinum majus)	l
Timothy, Hay (Phleum pratensis)	L(>12 hr)	Stock, German	l
Timothy, Pasture (P. nodosum)	L(14.5 hr)	(Matthiola incana)	
Wheat, spring	l	Tephrosia (Tephrosia candida)	IM(10-13.2 hr)
(Triticum aestivum)		Violet (Viola papilionacea)	S(<11 hr)
Wheat, winter (T. aestivum)	L(>12 hr)		
Wheatgrass (Agropyron smithii)	L(>10 hr)		

L=long day required; l=long day favorable; S=short day required; s=short day favorable; N=day-neutral; IM=intermediate. Where there is more than one symbol of classification, varietal differences occur, the most common class being entered first. Classification is followed in parentheses by light period for flowering (>12 hr should be interpreted as 12 hours or more; <12 hr, as 12 hours or less).

passed through a period of low temperature. Temperature may also modify the action of daylength in other ways. For example, the temperature must be above 55° to 60° F. when the chrysanthemum buds are forming in short-days, or they will fail to continue their development to flowering.

Photoperiodism as a factor in plant distribution. Photoperiodism is an important factor in the natural distribution of plants. In general, plants that have originated in low latitudes require short-days for flowering, while those of high latitudes are long-day plants. When the latter are moved to low latitudes, they will not produce blossoms. When low-latitude plants are grown in the long photoperiods of high latitudes they will continue to grow vegetatively until killed by frost. Some wild varieties of sugar cane flower only in the tropics. Spinach, on the other hand, never flowers in the tropics, because it requires fourteen hours of daylight for at least two weeks. Maize is a short-day plant that has difficulty in adapting to the long photoperiods.

When some plants are transferred to other latitudes, their life cycle may be unduly speeded. Biennials commonly produce seed during the first year in Alaska. In tropical and subtropical regions with their short days, Dutch potato varieties develop too rapidly and are unable to maintain the yield of the temperate zone (van Dobben, 1962).

Within plant species, some "photoperiodic ecotypes" have adapted themselves to the photoclimate of their environment during evolution. Katschon (1949) found that pines from the Alps seem to have different photoperiodic reactions depending on the elevation of the seed sources. Crowder, Ramirez, and Chaverra (1961) noted marked differences in photoperiod responses between the temperate species and the tropical species of pasture. When grown together in Colombia, the tropical species had a greater number of floral stems and produced higher quantities of seed than did those from high latitudes. In fact, many pasture varieties originating above the 50° latitude did not develop floral stems.

Response of tropical plants. In their early studies, Garner and Allard surmised that the seasonal differences in daylength in the tropics was too small to be an important factor in controlling plant behavior. Their view has been refuted by subsequent experimental evidence that tropical plants might be more sensitive than temperate plants to small differences in photoperiod. Njoku (1959), working at Ibadan, Nigeria (7° N), where the daylength varies from 11 hours and 40 minutes in December, to 12 hours and 33 minutes in June, experimented with a number of tropical plants. In about half of these, flowering was dependent on photoperiod, and for some of them a difference of only fifteen minutes could determine whether a plant flowers or not. Dore (1959) found that a Malayan variety of rice was sensitive to small differences of photoperiod.

Tropical plants do not necessarily prefer a photoperiod of about twelve hours. For example, *Rauwolfia vomitoria Afzel* and *Theobroma cacao L.*

grow more vigorously on sixteen-hour days than on twelve-hour days, even though sixteen-hour days are never encountered in the tropics (Borthwick, 1957).

PRACTICAL APPLICATIONS

Photoperiodism has many practical applications. The selection of a plant or a variety for a given locality requires knowledge of its interaction with the photoclimate. The United States Department of Agriculture now routinely determines photoperiod requirements before new plants are introduced.

In horticulture, artificial irradiation has been used to control flowering seasons and to increase the production of greenhouse crops. In middle and high latitudes, certain long-day plants will not flower in time for Christmas when the demand is greatest. If they are exposed to artificial illumination for a proper length of time, the flowering date can be specified. In the Netherlands, artificial light has been used to hasten the blooming of tulip and to retard the sprouting of seed potatoes (van Sluis, 1952).

In plant breeding work, flower initiation has greatly reduced the time span from germination to maturity. New varieties can be developed more rapidly. The artificial flower induction also makes it possible to cross plants that flower at different seasons in natural conditions.

Photoperiodism can be used to advantage in selecting the date of sowing. For some field crops, the length of the juvenile stage largely determines the size that a plant will reach at flowering, which in turn decides the ultimate size at harvest. If the crop is planted during a photoperiod resulting in a short juvenile stage, the yield will be reduced. Rasumov (1930) pointed out that late sowing in spring, during longer days, speeds up development, reducing yield of long-day plants but having a reverse effect on short-day crop.

Crop yield is not only reduced by planting in a season that will cause plants to flower early, but also by planting in a season that will cause very late flowering. Thus, certain varieties of rice in Ceylon, with a normal vegetative period of about five to six months, may extend their life to more than a year when planted in the wrong season, causing almost complete loss of yield (Chandraratha, 1949). In Nigeria, cowpeas will flower early and produce many seeds only when planted in day lengths of twelve hours or less (Njoku, 1959).

CHAPTER 8

AIR AND LEAF TEMPERATURE

CARDINAL TEMPERATURES

Regardless of how favorable light and moisture conditions may be, plant growth ceases when the temperature drops below a certain minimum value or exceeds a certain maximum value. Between these limits, there is an optimum temperature at which growth proceeds with greatest rapidity. These three points are known as cardinal temperatures.

Parker (1946) has pointed out that the physiological complexity of the plant precludes the precise determination of the cardinal temperatures, for different processes have different temperature requirements. However, the approximate values of cardinal temperatures for most plant species are known. With typical cool-season crops, such as oats, rye, wheat, and barley, the points are all comparatively low: minimum 0° to 5° C. (32° to 41° F.), optimum 25° to 31° C. (77° to 87.8° F.), and maximum 31° to 37° C. (87.8° to 98.6° F.). For hot-season crops, such as melons and sorghums, the temperatures are much higher: minimum 15° to 18° C. (59° to 64.4° F.), optimum 31° to 37° C. (87.8° to 98.6° F.), and maximum 44° to 50° C. (111.2° to 122° F.).

The cardinal temperatures also may vary with the stage of development. Certain plants require a period of low temperature during germination and early seedling stages for optimum growth. Many biennial plants must receive a cold treatment at the end of the first year's growth in order that the formation of flower buds and subsequent flowering during the second year may be induced. Apparently some substance destroyed by high temperatures accumulates during the cold period to trigger the reproductive cycle. Cold treatment (a temperature of about 32° F.) of the germinated seeds before sowing, or of seedlings before transplanting, is known as vernalization. First discovered by a Russian scientist, Lysenko (1925), vernalization has many practical applications for cold-climate plants. For instance, plant breeders have used this procedure to transform winter rye into the spring type because cold treatment makes it possible for winter cereals to flower even when they are planted in late spring.

THE VAN'T HOFF LAW

Some investigators believe that between the minimum and the optimum temperatures, dry matter production follows the Van't Hoff Law. That is, for every 10° C. rise in temperature, the rate of dry matter production approximately doubles. A good example of the linear relationship between logarithms of the yield and the mean temperatures during the growing period is that for irrigated pastures, cited by Prescott (1948) in Figure 42. The correlation coefficient is as high as 0.986. Other investigators, however, have stated that the Van't Hoff Law is only valid in the temperature range of 20° to 30° C. (Lehenbauer, 1914). Still others find no evidence to support the law.

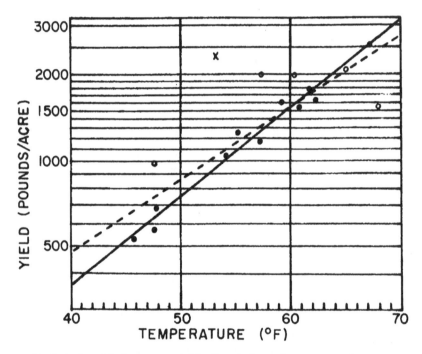

FIGURE 42. Relationship between yield of an irrigated pasture and temperature at Woods Point, Australia. Black circles exclude data for zenith growth period, which are indicated by open circles. Continuous line gives the regression for black circles; broken line is for all observations excluding September-October yield indicated by cross (after Prescott).

This controversy can be easily resolved if one understands the basic principle of net photosynthesis. Temperature exerts an opposite effect on photosynthesis and respiration, and even under comparable climatic conditions, the relationship between temperature and yield must vary with the stand density. Therefore, the Van't Hoff Law cannot have general validity.

DEGREE-DAY

The concept of degree-day dates back over 200 years. It holds that the growth of a plant is dependent upon the total amount of heat to which it was subjected during the lifetime, expressed as a degree-day. A degree-day is a measurement of the departure of the mean daily temperature above the minimum threshold temperature for a plant. If the minimum temperature is 50° F., and the mean daily temperature is 60° F., then we would have ten degree-days. The minimum threshold temperature varies with plants (e.g., 40° F. for peas, 50° F. for corn, and 55° F. for citrus fruits).

The degree-day concept assumes that the relationship between growth and temperature is linear instead of logarithmic as predicted by the Van't Hoff Law Again, the criticism of the latter applies to the former. In addition, the degree-day concept suffers from the following weaknesses:

(1) It makes no allowance for threshold temperature changes with the advancing stage of crop development.

(2) Brown (1960) contended that the method gave too much weight to high temperatures over 80° F.: for such high temperatures may even be detrimental.

(3) One cannot differentiate between the combination of a warm spring and a cool summer from a cold spring and a hot summer. Wang (1962) demonstrated that in Wisconsin, the former combination signified a good yield for peas; whereas the latter signified a poor yield.

(4) No consideration is given to the diurnal temperature range, which is often more significant than the mean daily value.

In spite of its lack of theoretical soundness, the simplicity of the degree-day method has been widely used to guide agricultural operations and the planning of land usage. The success of the method depends on a close relationship between radiation and temperature. At best, it is an empirical method that should be used with care. Eventually, it will have to be replaced by a method of potential photosynthesis calculation, similar to, but more refined than that proposed by de Wit.

Brooks (1951) reported that the degree-day method can be used for predicting the harvesting date of Blenheim or Royal apricots in Contra Costa County, California. During an experiment of seventeen years, the method was accurate to within two days half of the time, and within four days every season. Predictions were made six weeks after the full bloom of the apricots. Since the fruits take approximately eleven to sixteen weeks to develop from full bloom to maturity, the forecast is available five to ten weeks before harvesting, giving the farmer enough time to plan his schedule. During the period of study, the shortest time lapse between full bloom and harvest time was 89 days, and the longest was 112 days, a difference of 23 days. The same approach has also been applied to apples, pears, cherries, peaches (Anstey, 1966), French prunes, and bananas (Hord and Spell, 1962), all with good results.

Similarly, the planting date can be selected by the degree-day method. The accumulation of temperature over 40° F. is the primary factor affecting the maturity of oat varieties (Wiggans, 1936). The early maturing varieties require fewer degree-days to reach maturity than late maturing varieties. For each variety, the value is almost a constant. Thus, oats planted in early spring, when the temperature is low, will not ripen in the same length of time in the fall as oats planted when the temperature is high. At Ames, Iowa, a three- to four-day delay in seeding represents approximately a one-day delay in maturity.

In the southeastern United States, the dry matter production of ryegrass is closely related to the cumulated mean daily temperature (Weihing, 1963) (Figure 43). This relationship could aid in pasture management decisions, such as whether to increase or decrease the amount of livestock, or to provide supplementary food.

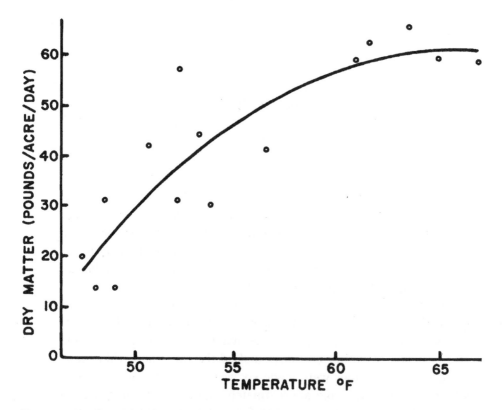

FIGURE 43. Growth of ryegrass as a function of temperature. The regression equation is $Y = -495.72 + 17.0285\,X - 0.1306\,X^2$ (after Weihing).

Air and Leaf Temperature 79

THERMOPERIODICITY

One of the objections to the use of the degree-day is that it does not take into consideration the diurnal temperature regime known as thermoperiodicity. Chapter 3 discussed the general favorableness of a large diurnal temperature range to net photosynthesis. Experiments conducted in the phototron have demonstrated that the development of some plants does not depend on the degree-day, but on the proper diurnal temperature variation (Went, 1950). A constant temperature during day and night—referred to as phototemperature and nyctotemperature, respectively—compared with a high phototemperature and low nyctotemperature, results in considerably less growth in the tomato.

Night temperature exerts a greater effect on growth and development of the tomato plant than the day temperature. In Figure 44, the growth rate of tomato plants is shown as a function of the night temperature (Went, 1956). At high night temperatures, the vegetative growth is accelerated, but the flower size and fruit development are inhibited. For the entire plant, the optimum night temperature is 17° C., with a possible deviation of 1° to 2° C. due to varietal differences. However, the process of sugar translocation from the leaves through sieve tubes of the stem to the growing zones has a different temperature requirement: it is accelerated by decreasing temperatures. According to Wellensiek (1957), very ambitious tomato growers in the Netherlands rose in the middle of the night to fire their

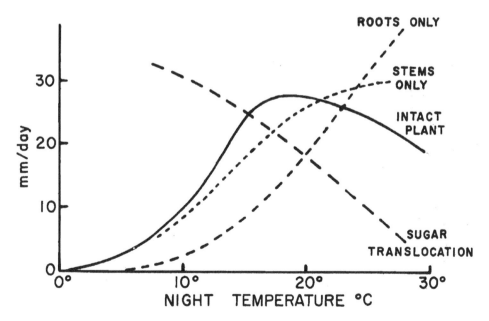

FIGURE 44. Rates of different processes plotted as a function of the night temperature for tomato plant (after Went).

greenhouse furnaces. Nevertheless, they had poorer plants than their lazy neighbors who let the night temperatures fall. The importance of low temperatures explains why tomato production is inconsequential in the tropics.

Night temperature is also the dominant factor in the growth of potatoes, chili peppers, tobacco, and other plants. The most successful potato-growing areas are in the middle and high latitudes because the optimum temperature for tuber formation is about 12° C. On the other hand, the day temperatures are critical in peas and strawberries for growth and fruit set. Regardless of whether the daytime or nighttime temperature is more important, Went has suggested that the response of plants to thermoperiodicity is a general phenomenon.

However, it has been pointed out recently that the experiments by Went, all carried out in a 24-hour cycle, may have only limited significance. In order to completely understand the effect of temperature on tomato growth, we will have to repeat his experiments under different lengths of the circadian rhythm. For many plants, the translocation rate increases instead of decreases with temperature, as shown in Figure 44. Fortanier (1957) has also demonstrated that peanuts do not respond to diurnal thermoperiodicity.

The quality of fruits and seeds is often affected by thermoperiodicity. The sucrose concentration of both sugar beets and sugar cane increases with the decrease of night temperature. In an experiment in California, Ulrich (1951) reported an increase in the sucrose content of beets from 7 per cent at 86° F. to 12 per cent at 36° F. In Hawaii, the sugar cane plantations with a large diurnal temperature range and low night temperature are known to have better juice quality (Das, 1931). In strawberries, the daytime temperature is critical. Went (1957a) explained:

> When strawberry plants are grown in warm or moderate temperatures, the fruits are red, sweet and slightly acid, but they have no strawberry flavor! To develop flavor, they must ripen at daytime temperatures of about 50 degrees. By various experiments, we learned that the plants have to be exposed to the right light and temperature conditions for at least a week to acquire the full strawberry aroma.
>
> These results explain why generally the first strawberries of the season taste best. They ripen while the early morning temperature is about 50 degrees. Later in the spring and during the summer the strawberry crop is practically without aroma, because the ripening fruit does not receive the proper temperature at any time during the day. At high latitudes, or far north, low morning temperatures occur even during summer, and strawberries from Alaska, northern Sweden, or the high Rockies taste marvelous at any time.

A knowledge of the ideal combination of daytime and nighttime

Air and Leaf Temperature 81

temperatures for the growth of a plant is essential for its successful introduction and selection of the proper growing season. Went (1956) has illustrated how this knowledge can be put into practical use. In Figure 45, he plotted the optimum growing conditions for eight garden plants, as well as the monthly daytime and nighttime temperatures for Pasadena, California. Note that the climatic ellipse of Pasadena transects the optimal growing conditions of the different garden plants at different time of the year. Whenever the climate coincides with the temperature requirements of a plant, it will grow best at that time. The African Violet, Stock, and Petunia would not grow well in Pasadena during any season of the year.

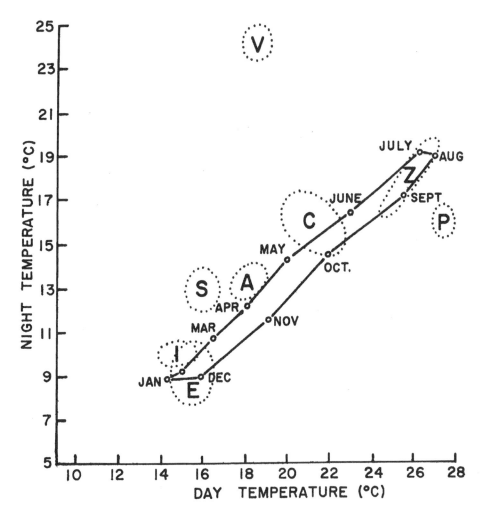

FIGURE 45. Optimal growing range of different plants in relation to day and night temperature. V = Africa Violet, Z = Zinnia, P = Petunia, C = China Aster, A = Ageratum, S = Stock, I = Iceland Poppy, E = English Daisy. Drawn line shows progression of average day and night temperatures in Pasadena throughout year (after Went).

Thermal regime in the tropics. A humid tropical climate is characterized by its small temperature range both diurnally and annually. Many cool-weather plants would die within a relatively short time in the tropics, simply because they cannot adapt to the thermal regime. Plants grown in the equable high temperatures differ in morphology from their counterparts in the temperature climate. The experimental results obtained by van Dobben (1962) are most instructive:

> When the temperature is raised from 16° to 25° C., vegetative growth in temperate-zone plants (such as small grains, peas and flax) is distinctly accelerated in the seedling, but only slightly in the following stage, whereas, development is speeded up considerably. As a result, the plants remain smaller at 25° as compared to 16° C.
>
> For crops of subtropical origin (maize and beans) reverse relations prevail. When the temperature rises from 16° to 25° C., growth is accelerated relatively more than development, so that the plants finally become larger. Hence, maize and beans have the faculty of compensating and even overcompensating the acceleration of development at higher temperatures by a relatively still greater increase in daily growth rate. This faculty is probably typical of species originating from warm climates. Without it, annuals in tropical regions would never reach large proportions.

The high temperatures in the tropics usually result in an increase in the shoot-root ratio (Khalil, 1956) and are unfavorable for the ripening of many annual crops. Even with the same total dry matter production, the grain yield of a rice crop is usually higher in a temperate climate because of the low autumn temperature. With low temperature, the transportation of carbohydrates to the grain is slow but lasts for a longer period. Senescence of leaves occurs late and the ripening period of the grain is prolonged, resulting in a higher grain weight than under high temperature conditions.

TEMPERATURE RECORDS

Air temperature is usually measured in the standard shelter at a height of 4.5 feet above the ground. In applying temperature records to agricultural problems, the two questions most frequently asked are: How long a recording period is necessary? How representative is the shelter temperature of the condition in the microclimatic layer near the ground?

Landsberg and Jacobs (1951) suggested that a five- to fifteen-year span of temperature records is needed to obtain a stable frequency distribution in tropical regions, and 10 to 25 years in extratropical regions. The length of the recording period should be increased from island locations to mountain areas in the interior of the continent. In a later study covering annual, monthly, and daily values, Enger (1959) concluded that the optimum length of record to estimate temperatures one and more years in the future is about twenty years.

Air and Leaf Temperature

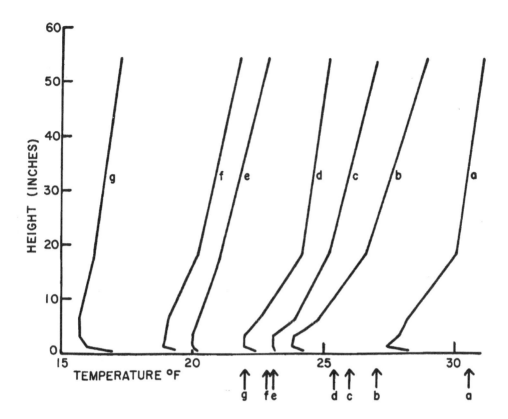

FIGURE 46. Profiles of mean temperatures during night of March 19-20, 1955: (a) 1830 GMT, (b) 1900, (c) 1932, (d) 1950, (e) 2115, (f) 2200, (g) Minimum. Arrows indicate surface temperatures (after Lake).

The temperatures measured at the shelter height of 4.5 feet are representative of the readings of a general area over horizontal distances, sometimes as far as several miles away, but they may differ materially from those near the ground at the same locality. During the day, the temperature decreases with height. At night, the temperature profile is often reversed due to radiational cooling, a phenomenon known as temperature inversion. On nights of great atmospheric stability, the minimum air temperature is not necessarily reached at the surface, but above it. Ramdas (1935) has observed repeatedly in India that the minimum temperature occurred at one foot or more above the surface. The temperature profile over bare soil observed by Lake (1956) in England during the nights of March 19 and 20, 1955, suggests that, as the duration of stable conditions increased, the layer of maximum cooling occurred further away from the surface (Figure 46).

LEAF TEMPERATURE

Until the recent development of the infrared radiometer, leaf temperature had been measured by means of a thermocouple inserted into the leaf. The latter method is slow and disturbs the leaf. On the other hand, the infrared radiometer, such as the Stoll-Hardy model (1955), permits quick reading and is accurate to within 0.3° C.

The radiation load on a leaf is disposed of in three ways: transpiration, heat loss to the air, and heat storage in the leaf. For a leaf that is not actively transpiring, the cooling of leaves heated by sunshine is in large part a result of conduction from the leaf to the air.

Loomis (1965) cited an example:

> Assume that the leaf is absorbing 0.7 gcal/cm^2/ min and is heated 10° C. above air. Typical transpiration rates for such a leaf would be 1 or less commonly 2 g/dm^2/min, which would dissipate 0.1 or 0.2 gcal of the 0.7 g absorbed and 0.6 or 0.5 gcal/cm^2/min would be lost by emission. Only the improbably high transpiration rate of 7 g/dm^2/min would be capable, acting alone, of holding the leaf temperature constant at air temperature.

Thus, the temperature of a leaf exposed to sunlight is often well above that of the air, especially for a thick leaf that is not transpiring in still air. In Ceylon, under conditions of high insolation and high humidity, leaves commonly have temperatures 15° C above air temperature (Smith, 1909). In Palestine, leaf temperatures as high as 52° C. or some 10° C. above air temperature, have been recorded (Konis, 1949).

For plants that are exposed to the sun but do not suffer from water stress, the difference between air and leaf temperatures is usually small. Linacre (1964) has collected and compared 40 observations by previous workers (Figure 47). He noted a tendency for equality at a temperature of about 33° C. (91.5° F.). Below that temperature, leaves tended to be warmer than the air; above that temperature, the reverse was true. His findings were pursued further by Priestley (1966) from general considerations of the energy budget. He argued that the maximum temperature observed over any wet surface, including plants, should be about 92° F. Leaves whose surface temperature exceeds 92° F. suffer from partial water deficit.

The temperature of a sunlit leaf responds quickly to the change in wind speed and cloud cover. Gates (1962) related his experience:

> A typical example is the case of sunlit Populus acuminata leaves when the temperature of the leaf was 10° C. above the air temperature with no breeze and, with a breeze of about 3 mph, the temperature dropped to about 4° C. above air temperature. The air temperature in this instance was 33.5° C. A similar behavior is observed when the sun is obscured by a dense cloud. As an example, during an observation of Quercus macrocarpa, the temperature of the sunlit leaves in still air was 49° C. and air temperature was 28° C. A dense cumulus cloud came across the sun and the leaf temperature dropped to 30.5° C. in about 30 seconds. Comparisons were also made between the

Air and Leaf Temperature 85

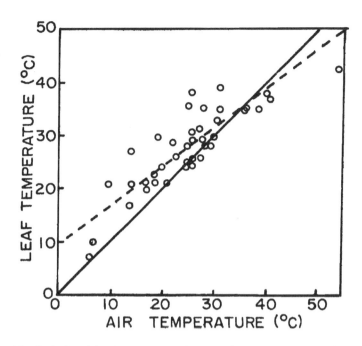

FIGURE 47. Relationship between leaf and air temperature when leaves are exposed to bright sunshine (after Linacre).

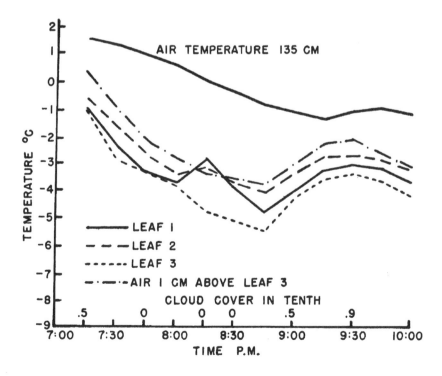

FIGURE 48. Leaf and air temperatures the night of October 5-6, 1952 (after Shaw).

temperature of the upper and lower surfaces of leaves. In many instances for most species, the temperature was within ± 0.2° C. for the upper and lower surfaces. However, for Quercus macrocarpa, many measurements under strong solar radiation showed the temperature of the upper sunlit surface to be as much as 2° C. higher than the temperature of the lower surface. This difference can be accounted for by the stomatal transpiration taking place primarily from the lower surface of the leaf. For Elmus americana the upper sunlit surface was occasionally 1° C warmer than the lower.

Leaves shaded from direct sunlight are usually about 1° C. warmer than the air (Ansari and Loomis, 1959).

On a clear, calm night, the leaf temperature can be considerably lower than air temperature. Shaw (1954) compared leaf temperatures of tomato plants with air temperatures in the standard instrument shelter in Iowa on the night of October 5 and 6, 1952 (Figure 48). Air temperatures in the shelter were 3° to 4.5° C. warmer than leaf temperatures when the sky was clear. A leaf was about 1° C. colder than air one centimeter above it. With increasing cloud cover, the difference between air and leaf temperature was reduced. In Hawaii, Noffsinger (1961) reported a case of the leaf temperatures of both pineapple and papaya plants exceeding the air temperature during a period of dense cloud cover at night.

A knowledge of the difference between air and leaf temperature should help farmers in adopting protective measures. In middle and high latitudes, frost often occurs before the air temperature in the instrument shelter drops to freezing point. Likewise, in summer, the heat injury to plants might be much more serious than air temperature records suggest. For instance, farmers in Taiwan have made it a common practice to shade the pineapple fruit, in order to prevent heat damage.

CHAPTER 9

SOIL TEMPERATURE

SIGNIFICANCE OF SOIL TEMPERATURE

In many instances, soil temperature is of greater ecological significance to plant life than air temperature. Beech, oak, and ash trees can withstand an air temperature of $-25°$ C., but their finer roots succumb to cold from $-13°$ C. to $-16°$ C. (von Mohl, 1848). Soil temperature, more responsive to the local effects of insolation, topography, and the like, may differ greatly from air temperature. Many localities in polar areas and on high mountains would certainly be devoid of vegetation were it not for the considerably higher temperature of the soil than that of the air, especially during the period of the sun's radiance. Soil temperature brings out, more closely than air temperature, the contrasting growing seasons between different mountain aspects and slopes.

Soil temperatures, particularly the extremes, influence the germination of seed, the functional activity of roots, the rate and duration of plant growth, and the occurrence and severity of plant diseases. The living tissues of representative conifer seedlings of western America are killed quickly when a surface temperature of about $54°$ C. is reached (Baker, 1929). Extremely high soil temperatures also have a harmful effect on roots and may cause destructive lesions on the stems of plants. On the other hand, low temperatures impede the plant mineral nutrient intake. At $1°$ C., the soil moisture ceases to become available to plants (Kihlman, 1890). Persistently cold soil results in dwarfed growth.

The ecological significance of soil temperature is obviously of vital concern to those engaged in agriculture. An unfavorable soil temperature during the growing season may retard or even ruin the crops. A vegetable grower would place soil that warms up rapidly in the spring at a premium. Much effort has been directed by plant cultivators to modify soil temperature. On many a farm, the weal and woe of the tiller are closely knitted to the soil temperature variations.

Methods of measurement and records. Instruments for measuring soil temperature include the mercury thermometer, resistance thermometer, and thermocouple thermometer. The latter is considered the best. In addition to low initial and maintenance costs, the thermocouple possesses a high degree of accuracy and is thinner than other thermometers.

The World Meteorological Committee (1956) recommended that the standard depths for earth temperature measurements should be 10, 20, 50, and 100 centimeters. Where proper equipment is lacking, the U. S. Weather Bureau encourages measurements at the ten-centimeter depth (Landsberg and Blanc, 1958). However, measurements at two depths are preferred.

Temperature measurements below the surface of the soil are not nearly as inaccurate as those above the surface because swift changes are counteracted by the great heat capacity of the soil. Therefore, it is considered sufficient for most agricultural purposes to measure only the daily maximum and minimum temperatures, especially at great depths.

Chang (1958) collected the soil temperature data from 780 stations throughout the world and then constructed maps for January, April, July, and October at three depths of 10, 30, and 120 centimeters (Chang, 1957a). He also has mapped the global distribution of the annual range in soil temperature at the same depths (Chang, 1957b).

THERMAL PROPERTIES OF SOILS

The specific heat of a substance is the amount of heat required to raise the temperature of one gram of the substance by 1° C. The specific heat of all mineral soils differs only slightly, being on the average about 0.18-0.20 cal/gm deg. Humus has a much higher specific heat of about 0.45 cal/gm deg.

The heat capacity of a substance is the amount of heat required to raise the temperature of one cubic centimeter by 1° C., and hence is sometimes called volumetric specific heat. The heat capacity of a soil varies greatly according to its moisture content. In a dry state, humus soils have a lower heat capacity than mineral soils, because of the low density of the former. In the field, however, organic and fine-textured soils, owing to their high water-holding capacity, ordinarily have a higher heat capacity than coarse soils. Most soils have a heat capacity ranging from 0.3 to 0.6 cal/cm^3 ° C.

The rate of heat flow in the soil is determined by its temperature gradient and thermal conductivity. The thermal conductivity of a substance is defined as the quantity of heat flowing per unit of time through a unit area of a plate of unit thickness when a unit temperature difference is maintained between the two opposing faces. The thermal conductivity of the soil can be accurately determined by any heat flow probes as designed by de Vries and Peck (1957), Buettner (1955), Lachenbruch (1957), or

Blackwell (1954). Other things being equal, the greater the thermal conductivity of the soil, the smaller is the surface temperature variation and the more effective its role as a heat reservoir.

The thermal conductivity of a soil is determined primarily by its porosity, moisture, and organic matter content. For a given moisture content, thermal conductivity decreases from coarse- to fine-textured soils as the porosity increases. However, in the natural fields, fine-textured soils ordinarily have a higher water content, which greatly increases the thermal conductivity of the soil. Organic matter does not transfer heat as readily as the mineral soil. According to a measurement by Karsten (1912), the thermal conductivity of a dry, fine, sandy soil was 0.00046 cal/cm sec. deg. in comparison with 0.00027 for a fine humus soil.

While the rate at which heat is transferred in a body is dependent upon the thermal conductivity of the substance, the temperature rise that this heat will produce will vary with its heat capacity. Therefore, the quotient of thermal conductivity and heat capacity is an index of the facility with which a substance will undergo temperature change, and is known as the thermal diffusivity. In other words, it is the change in degrees C. that occurs in one second, when the temperature gradient changes one degree per cubic centimeter. The thermal diffusivity of a soil may be determined either by a direct measurement of samples or by a computation from the amplitude or phase lag at different depths in the soil according to equations discussed in the next section.

The thermal diffusivity of a soil rises with increasing moisture content, reaches a maximum, and then decreases. A small amount of water in the soil between the grain breaks the air insulation. As more water is added the heat capacity of the soil is rapidly raised, lowering the temperature rise produced by a given quantity of heat. Patten (1909) found in one experiment that the maximum diffusivity is reached at a moisture content of about 15 per cent for sand, 12 per cent for sandy loam, and 32 per cent for muck.

Organic matter lowers the thermal diffusivity, while packing the soil increases it. The thermal diffusivities of soils lie between 10^{-2} and 10^{-3} square centimeters per second. Keen (1932) suggested that the average value of the thermal diffusivity of soils in moist temperate regions can be taken as 0.004 square centimeters per second.

PHYSICAL LAWS GOVERNING
THE CHANGE OF SOIL TEMPERATURE

Heat at the ground surface is propagated downward in the form of waves with the amplitude decreasing rapidly with depth. The temperature range for any point below the surface is given by the equation:

$$R_z = R_o e^{-z\sqrt{\frac{\pi}{KhP}}}$$

where R_o and R_z are the temperature ranges at surface and depth z, respectively, P is the oscillation period, and Kh is the thermal diffusivity of the soil.

According to this equation, the amplitude of each wave decreases with depth exponentially. The amplitude decreases more slowly with depth if the diffusivity is greater or the period longer. Thus, the annual temperature range is reduced less rapidly with depth than the diurnal range. The ratio between the respective depths is $\sqrt{365.25}$ or 19.1.

The time lag of the maximum and minimum of the heat cycle in the soil is expressed by:

$$t_2 - t_1 = \frac{z_2 - z_1}{2}\sqrt{\frac{P}{Kh\pi}}$$

where t_2 and t_1 are the times when the maximum or minimum temperatures are observed at depths z_2 and z_1 respectively.

The velocity of the propagation of the temperature wave is directly proportional to the square root of the thermal diffusivity of the soil, and it is inversely proportional to the square root of the period.

The time lag of temperature waves is practically linearly proportional to the depth in a homogeneous soil. Decker (1955) plotted the lag of daily maximum temperature as reported by a number of investigators (Figure 49). The graph shows that the maximum temperature is delayed for about 3 hours at a depth of 10 centimeters, 12 hours at 30 centimeters and 33 hours at 60 centimeters. However, the lag in daily maximum is usually greater than that in the minimum.

SOIL TEXTURE

Because of lower heat capacity, lower thermal conductivity, and less evaporative chilling, sandy soils warm up more rapidly in the spring than do clay soils. Light soils not only thaw first, but they also become warm enough for plant growth to commence distinctly earlier than do heavier soils. King (1899) showed that the difference in the evaporation rate of clay soil with a water-holding capacity of 40 per cent and a sandy soil with a capacity of 5 per cent—when both are well drained—is often sufficient to produce a minimum lower by 3.9° C. in the clay at a depth of one foot. This difference would be still greater in the surface layer. These differences

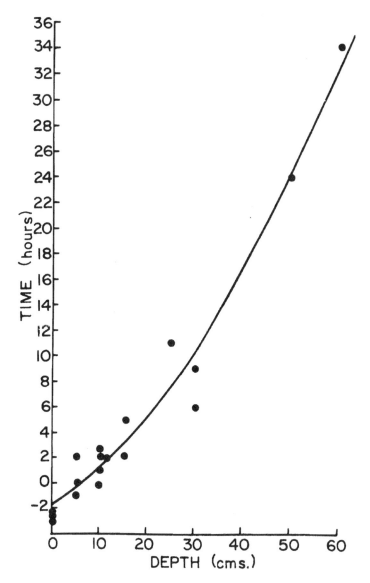

FIGURE 49. Time lag of diurnal maximum temperature between air temperature at five feet above the surface and temperature at selected depths below the surface (after Decker).

in soil temperature in the Netherlands account for the harvest of lettuce in a cold greenhouse in spring on a sandy soil about fourteen days earlier than on clay soil (Abd El Rahman, Kuiper, and Bierhuizen, 1959). Conversely in the autumn when the light soils are being cooled by radiation and by the percolation of cold rain, heavy soils remain relatively warm. Coarse soils are more responsive to weather changes.

Because of the poor thermal conductivity of sandy soil, the energy received by it is concentrated mainly in a thin layer; and because of its small heat capacity, there is a large corresponding rise in temperature. Consequently, a sandy soil heats up to an extraordinary degree in its upper layers on a summer afternoon. Fugures 50 and 51 show the daily courses of temperature in different types of soil as observed by Yakuwa (1946). Note that the radiational cooling at night is greater in light soils than in heavy ones. In the top layer, sand has the greatest temperature range, followed by loam and clay. The decrease of range with depth is more rapid

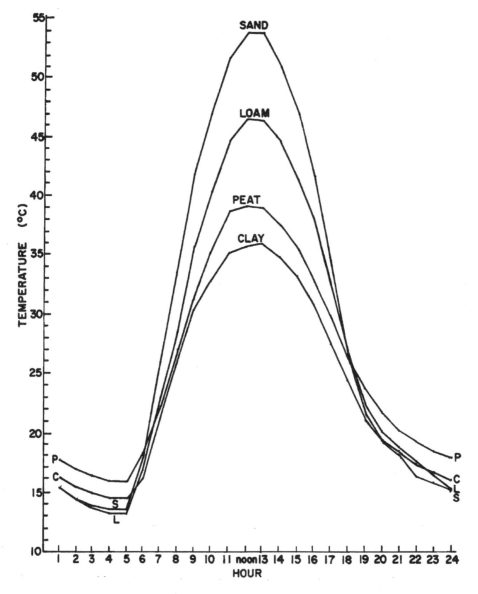

FIGURE 50. Daily course of surface temperature on clear summer days at Sapporo, Japan (after Yakuwa).

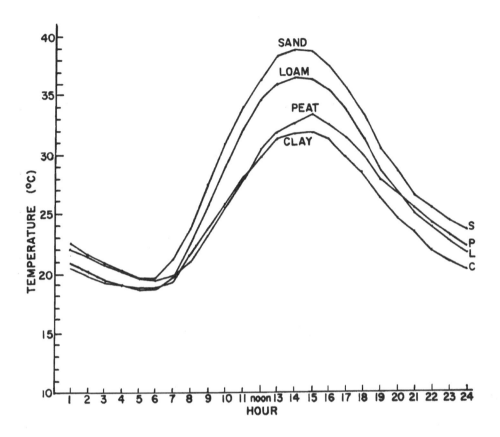

FIGURE 51. Daily course of soil temperature at five cm. depth on clear summer days at Sapporo, Japan (after Yakuwa).

in light soils than heavy ones when they are dry, but slower when they are wet. The effect of percolating water in equalizing the temperature throughout the profile is more pronounced in light soils.

The organic matter reduces the heat capacity and thermal conductivity of the soil, increases its water-holding capacity, and has a dark color, which increases its absorptivity. In a humid climate, peat and muck, because of their large water content, are much colder than the mineral soils in spring and summer, but they may be slightly warmer in autumn and winter. If the organic soils are dry, their top layers are ordinarily warmer than the mineral soils by day, and colder by night. The high surface temperature of the dry peat is caused by its high absorption, its virtual lack of evaporation, its low heat capacity, and its low conductivity. At night, the low thermal conductivity of the organic soil prevents a rapid flow of heat from the subsoil to replenish the heat lost by radiation. The low heat capacity permits expansive cooling from little radiation, and the surface of relatively dry peat becomes excessively cold, even though the temperature in the lower layers is much higher than that of the mineral soils. The large diurnal surface range of a dry peat soil diminishes rapidly with depth. The diurnal range of a wet peat is usually small and is restricted to a thin layer.

ASPECT AND SLOPE

Slope exposure is of little significance in low latitudes, but it is important outside the tropics. In the middle and high latitudes of the northern hemisphere, the southern slopes receive more insolation per unit area than the northern exposure. To evaluate the insolation received by the slopes, one must separate the direct solar radiation and the diffuse sky radiation. Direct insolation is a function of both aspect and slope, while diffuse radiation, being essentially uniform in all azimuths, is affected only by the latter. A 10° north slope receives just as much diffuse sky radiation as a 10° south slope. The greater the ratio of diffuse sky radiation to total radiation, the less is the difference in energy received by various slope exposure. On an overcast day when there is no direct solar radiation, the effect of aspect is minimized.

In general, the ratio of diffuse to total radiation is high in polar regions, owing to the high cloudiness and the low altitude of the sun; similarly, the ratio is higher in winter than in summer. Therefore, exposure is a more important factor in middle latitudes than in polar regions and more so in summer than in winter.

In the absence of clouds and other complications, the southwest slope is usually warmer than the southeast slope. Not only does the direct beam of sunshine on the southeast slope occur shortly after the prolonged cooling at night, but the evaporation of the dew in the morning also requires energy.

The greatest difference in temperatures between the southern and northern slopes occurs during the spring and summer. In spring, the south slope warms up rapidly while the north slope remains cool and water-logged. At the height of summer, Pool (1914) reported a difference in surface temperature of 29.7° C. (62.8° C. and 33.1° C.) between a southern and a northern aspect of a small sandy hill in northwestern Nebraska. This difference exceeded the contrast in air temperature. Aikman's measurements (1941) on a hill of a 20° slope at Floris, Iowa, indicated that in July, 1939, the greatest difference in soil temperature at the two-inch depth between the four aspects was 3.8° C. as against 1.7°C. for air temperature. The northern slope may receive so little radiation that the ground surface is often colder than the air.

The difference in the minimum temperature between the northern and southern aspects is smaller than the difference in maximum temperature. Shreve (1924) found that on Santa Catalina Mountain at Tucson, Arizona, both the maximum and minimum temperatures at three inches depth were higher on the south than on the north aspect, the maximum by 13° to 20° F. (7.2° to 11.1° C.) and the minimum by 4° to 11° F. (2.2° to 6.1° C.). The diurnal temperature range is consequently larger on the south slope. Furthermore, the maximum temperature on the north aspect often comes later than that on the south slope. On Santa Catalina Mountain, the delay is about half an hour.

In winter, the contrast in ground temperature between the north and south slopes is small. With the advance of the season, the greater warmth of the south slope is gradually asserted. Crops and vegetation start earlier on southern exposures than on northern ones. Paradoxically, for some stone fruits, the delay in blossoming and the consequent reduction of the danger of frost is an advantage on the relatively cold north slope.

The degree of slope determines the amount of insolation received per unit area. By changing the degree of slope, the effect of latitude is simulated on a small scale. The temperature differences between exposure are usually accentuated by the slope. Outside the tropics, in the northern hemisphere, a gentle, south-facing slope is warmer than level land. Alter (1913) stated that land in southern Idaho that slopes 5° to the south is in the same solar climate as level land some 300 miles south in Utah. On the other hand, ground sloping 1° to the north lies in the same solar climate as level land 70 miles further north. The warmest slope is the one most nearly perpendicular to the sun's rays during the growing season. Its steepness, therefore, increases with latitude.

THE EFFECT OF TILTH

By loosening the top soil and creating a mulch, tillage reduces the heat flow between the surface and the subsoil. Since the soil mulch has a greater exposed surface than the undisturbed soil and no capillary connection with moist layers below, the cultivated soil dries up quickly by evaporation, but the moisture in the subsoil underneath the dry mulch is conserved.

The diurnal temperature wave of the cultivated soil has a much larger amplitude than that of the uncultivated. At East Lansing, Michigan, the air one inch above the cultivated soil is often 5° to 10° C. hotter than that over an untilled plot on a summer afternoon (Bouyoucos, 1913). At night, the loosened ground is colder and more liable to frost than the uncultivated soil. Below the soil mulch, the condition is reversed; the cultivated soil has a smaller temperature fluctuation.

SOIL TEMPERATURE AND CROP YIELD

In many instances, soil temperature is more important than air temperature to plant growth. In the tropics, high soil temperature causes degeneration of the tubers in potatoes. The optimum soil temperature for the potato is 17° C.; tubers will not grow at all at soil temperatures above 29° C.

In central Iowa, corn yield has been found to be closely related to soil temperatures at planting time (Riley, 1957). At that time, if the soil temperature is below normal and the outlook is for continued moist, cool weather, the corn yield prospect is low. Allmaras, Burrows, and Larson (1964) found that the corn yield increases almost linearly as the four-inch soil temperature rises from 60° F. to 81.3° F. (Figure 52). However, the crop yield decreases at soil temperatures above 81.3° F. The temperature of the surface soil can be raised by ridge planting followed by shallow cultivation to aerate the ground, thus minimizing the effects of one of the worst threats to corn—a cold late spring.

Similarly, Wang (1962a) reported that in Wisconsin, soil temperature was most significant in the early stages of corn, but air temperature assumed greater significance in the reproductive stages. In analyzing the

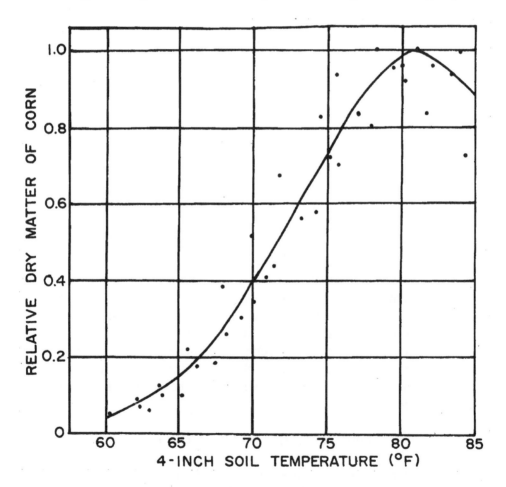

FIGURE 52. Relative yield of dry matter in young corn plants as affected by average four-inch soil temperature (after Allmaras, Burrows, and Larson).

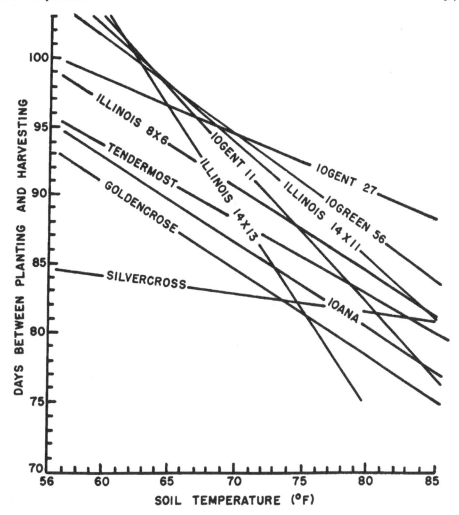

FIGURE 53. Days between sweet corn planting and harvesting as affected by variety and soil temperature (after Wang).

experimental data of ten sweet corn varieties at Ames, Iowa, from 1938 to 1950, he has derived relationships between the average soil temperature for the first 11 days after planting, and the date of maturity. In Figure 53, some varieties have steep slopes where soil temperature exerts a great effect on the length of the period to maturity. Where soil temperature has less effect on the days to maturity, the slopes are more level. Wang claimed that from these relationships he was able to predict the date of maturity for the canning industry with a reasonably high degree of accuracy.

METHODS OF MODIFYING SOIL TEMPERATURE

The two most important approaches for modifying soil temperature are: regulating the incoming or the outgoing energy or altering the thermal properties of the ground in order to produce a different manner of energy consumption in the ground.

There are four general ways of regulating the effect of incoming energy: (1) placing an insulating layer on or near the ground surface such as paper, straw mulch, screens, glass, polystyrene, trees, etc., (2) heating, (3) changing the absorptivity of the ground, or (4) varying the air temperature by the use of a wind machine or the erection of a shelter belt.

The net outgoing terrestrial radiation can be reduced by utilizing various kinds of insulating material and by generating smoke or fog in the air.

The thermal properties of the ground can be modified: (1) by increasing the absorptivity of the ground, (2) by changing the thermal conductivity by cultivation, rolling, and irrigation, (3) by altering the heat capacity by adding or draining water, or (4) by varying the rate of evaporation by the removal of weeds, regulating soil moisture, and placing mulch, screens, or sand on the ground surface.

Some of the methods of modifying the soil temperature will be discussed in the next chapter, as they are primarily designed for frost protection. Here we will discuss only the effect of mulch and carbon black.

Any material spread over and allowed to remain on the ground surface is referred to as mulch. Paper, straw, and polyethylene have all been used in mulching. The first large commercial use of paper as a soil covering was made in 1914 on a sugar plantation in Hawaii. Later, it was extensively adopted by the pineapple industry there. Although its primary function is to conserve moisture and to check weed growth, the effect of paper mulch on soil temperature has been investigated by Shaw (1926), Smith (1931), and Hagan (1933). Since the thermal conductivity of a mulch is usually much lower than the soil, there is less heat gain or loss under a mulch. In summer, the area covered with mulch paper is generally cooler by day and warmer by night than the unmulched plot. In the winter, mulch protects the soil from excessive cooling. The effect of paper mulch on soil temperature varies with the color, the presence or absence of perforation, and the prevailing weather conditions.

Hanks, Bowers, and Bark (1961) compared the effect of different mulches on soil temperature in the summer. They found that soil temperature was highest under plastic mulch, followed by bare soil, black painted gravel, aluminum painted gravel, and straw mulch, in that order. The "greenhouse effect" caused the high temperature under the plastic cover. The high soil temperature of the bare soil in comparison to the gravel and straw mulch plots was a result of the effectiveness of the latter as an insulator.

For increasing soil temperature in the cool spring months, mulches that transmit radiation are the best. Waggoner, Miller, and de Roo (1960) have reported the benefit of such mulches for the growth of strawberries:

This was shown in the early blossoming and the early and abundant fruit and runners produced by strawberries mulched with natural or green translucent film. The parts of the plant that actually grew in the warmed environment, the roots, grew more than those beneath other films or bare soil.

In a three-year investigation on the effect of plastic mulch on corn yields in Illinois, Pendleton, Peters, and Peek (1966) found that white ground cover, by virtue of its high reflectivity, could increase the light absorption by the plants, thereby producing a higher yield of some 6 per cent than that covered by nonreflective black mulch.

The addition of carbon black to the surface soil promotes absorptivity, thereby increasing the temperature. Everson and Weaver (1949) measured, for more than a year, the ground temperature of different portions of the same plot, part of which had been treated with 4,000 pounds of carbon black per acre. Both daily maximum and minimum temperatures were increased. The effect lasted for several years. A report from Kazakhstan, U.S.S.R., claims an advance in cotton ripening by a month or more by applying 100 pounds of coal dust per acre (Brooks, 1955). Carbon black has also been used to accelerate the melting of snow.

CHAPTER 10

FROST PROTECTION

THE DAMAGING EFFECTS OF FREEZING TEMPERATURE

The protoplasm in living plants can function properly only within a restricted temperature range. No growth of agricultural importance takes place when the temperature approximates the freezing point. Many tropical and subtropical plants may be killed at temperatures as high as 5° C. Molisch (1897) has called low temperature damage, in the absence of freezing, "chilling injury" as opposed to frost injury. Those plants that are subject to chilling injury are usually killed by the first touch of frost. On the other hand, many cold-climate plants may be frozen solid at low temperatures without injury. Between these two extremes all gradations occur. After reviewing the pertinent literature, Levitt (1956) states that, in general, the larger the cell size, the more probable that the plant will be damaged by frost.

Even for the same plant, the frost killing temperature may vary widely with the manner of the temperature change, the season, the physiological state of the plant, and the like. Killing may occur at higher temperatures if the freezing is rapid, rather than gradual. Greater injury to the plant may occur after long-continued freezing than that after short freezing periods at the same temperature. Injury may also occur after two or more freezings at the same temperature that failed to injure the plant in one freezing. Some plants that survive the cold in winter may be killed by a very slight freezing during spring. In spite of these complications, Ventskevich (1961) has compiled in Table 6 the critical temperatures harmful to 43 crops in different developmental phases.

FROST WEATHER

Frosts may be divided into two types: radiation frost and wind or advection frost. Radiation frost occurs on clear nights with little or no wind when the outgoing radiation is excessive. Under such conditions, the air

TABLE 6
Resistance of Crops to Frost in Different Developmental Phases

	Temperature (°C) harmful to plant in the phases of		
	germination	flowering	fruiting
Highest resistance to frost			
Spring wheat	-9,-10	-1,-2	-2,-4
Oats	-8,-9	-1,-2	-2,-4
Barley	-7,-8	-1,-2	-2,-4
Peas	-7,-8	-2,-3	-3,-4
Lentils	-7,-8	-2,-3	-2,-4
Vetchling	-7,-8	-2,-3	-2,-4
Coriander	-8,-10	-2,-3	-3,-4
Poppies	-7,-10	-2,-3	-2,-3
Kok-saghyz	-8,-10	-3,-4	-3,-4
Resistance to frost			
Lupine	-6,-8	-3,-4	-3,-4
Spring vetch	-6,-7	-3,-4	-2,-4
Beans	-5,-6	-2,-3	-3,-4
Sunflower	-5,-6	-2,-3	-2,-3
Safflower	-4,-6	-2,-3	-3,-4
White mustard	-4,-6	-2,-3	-3,-4
Flax	-5,-7	-2,-3	-2,-4
Hemp	-5,-7	-2,-3	-2,-4
Sugar-beets	-6,-7	-2,-3	-
Fodder beets	-6,-7	-	-
Carrot	-6,-7	-	-
Turnip	-6,-7	-	-
Medium resistance to frost			
Cabbage	-5,-7	-2,-3	-6,-9
Soy beans	-3,-4	-2,-3	-2,-3
Italian millet	-3,-4	-1,-2	-2,-3
European yellow lupine	-4,-5	-2,-3	-
Low resistance to frost			
Corn	-2,-3	-1,-2	-2,-3
Millet	-2,-3	-1,-2	-2,-3
Sudan grass	-2,-3	-1,-2	-2,-3
Sorghum	-2,-3	-1,-2	-2,-3
Potatoes	-2,-3	-1,-2	-1,-2
Rustic tobacco	-2,-3	-	-2,-3
No resistance to frost			
Buckwheat	-1,-2	-1,-2	-0.5,-2
Castor plant	-1,-1.5	-0.5,-1	-2
Cotton	-1,-2	-1,-2	-2,-3
Melons	-0.5,-1	-0.5,-1	-1
Rice	-0.5,-1	-0.5,-1	-0.5,-1
Sesame	-0.5,-1	-0.5,-1	-
Hemp mallow	-0.5,-1	-	-
Peanuts	-0.5,-1	-	-
Cucumbers	-0.5,-1	-	-
Tomatoes	0,-1	0,-1	0,-1
Tobacco	0,-1	0,-1	0,-1

temperature generally increases with height in the microlayer near the ground. In frost protection research, when the temperature 40 to 50 feet above the ground is 13° F. or more warmer than it is a few feet above the soil surface, the temperature inversion is considered to be strong (Goodall,

Angus, Leonard, and Brooks, 1957). When the temperature difference is less than 5° F., the inversion is considered weak. The Brunt formula has been utilized sometimes to estimate the radiational cooling and, hence, the frost risk in crops. Apart from the inaccuracy of the formula, Monteith (1958) has cautioned that the method would be dangerous if no account were taken of the temperature differences between the leaves and ambient air and of the temperature gradients in the crops. The use of the net radiometer would be an improvement over the Brunt formula.

Wind frost occurs at any time of the day or night, whatever the state of the sky, when a wind speed usually in excess of four miles per hour brings the air from the cold regions. In some instances, a wind frost may be intensified by a radiation frost, and the two types may occur simultaneously. Most frost protection schemes can only raise the temperature by a few degrees C., while some are effective only against radiation frost. In the following sections, several important frost protection methods will be discussed.

HEATER

Heaters were utilized in the first successful attempts to prevent frost injury in California toward the end of the last century. They have since been acknowledged as the best frost protection measure.

Heating is most effective on a night with a strong temperature inversion. With an ordinary inversion ceiling of between ten to fifteen meters above the ground, the depth of air to be heated is rather shallow. In the absence of a temperature inversion, the principal value of direct heating is to radiate heat to the plants and the ground surface and to produce a pall of humid smoke, which constitutes a moderating screen against net radiation loss from the ground. Angus (1955) investigated the effect of the temperature inversion on temperature responses to 188 heaters per acre in a pineapple plantation in southern Queensland, Australia (Figure 54). The area over which a temperature rise of a given amount was produced was found to increase linearly with the strength of the temperature inversion.

In general, a large number of small heaters is more effective than a few larger heaters. The large heaters are ineffective because the upward movement of the warmed air rising from them is unchecked. A current is often set up so that the cold air is drawn in from above and the valuable warm ceiling is broken up. For protection against radiation frost, heaters should be placed in such a manner that each tree in the orchard can "see" a heater. However, for the protection against advective frost, heaters should be placed in heavier concentrations along the upwind border. In hilly country, heaters should be concentrated mainly in the valleys; for the heat produced will move upward along the slope.

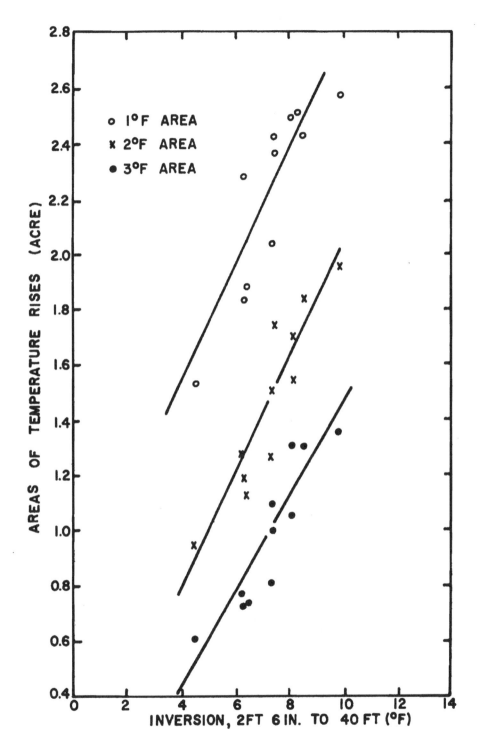

FIGURE 54. Effect of temperature inversion on temperature response to 188 heaters per acre (after Angus).

Oil heaters are widely used in the United States. In Germany, coal, briquettes, and wood are the principal fuels. Oil is by far the most efficient because other fuels cannot often be ignited quickly enough to avert the frost damage. Moreover, once lit, the other fuels cannot be put out if the temperature rises, but must be left to burn wastefully to the end. Because of the high cost of heating, the few growers who tried the system in England soon gave up the practice. Even in California, heating is not as common as it was some years ago. It is used only for a few high-priced crops such as citrus fruits.

WIND MACHINE

Frost protection by a wind machine came into use in the United States in the 1920s. Although it does not afford reliable protection, its uses have increased steadily, largely because the operating cost is only about 20 per cent of that for heaters. According to a study by Brooks, Kelley, Rhoades, and Schultz (1952), California alone had more than 2,800 wind machines.

Wind machines break up the nocturnal temperature inversion by mechanically mixing the air. In effect, the eddy conductivity of the air is increased many times to bring back to the ground heat that was carried upward during the day. The wind machine is not expected to afford protection in cases of a freeze with cold daytime conditions, cold soil, and no relatively warm air overhead on a clear, cold night. The effectiveness of the wind machine increases with the strength of temperature inversion. Goodall, Angus, Leonard, and Brooks (1957) presented the results of twelve test runs in an almond orchard in Chico, California, in Figure 55. For a given amount of temperature increase, the response area increases linearly with temperature inversion.

Even under strong temperature inversion conditions, the gain in ground surface temperature by this method is rather small, usually less than 3° to 5° C. Therefore, it is a common practice to install both wind machines and heaters in the field. On strong inversion nights, the wind machine may be operated alone; on weak inversion nights, it may be supplemented by heaters. In cases of wind frost, only the heaters are effective.

FLOODING AND SPRINKLING

Flooding and sprinkling prevent excessive cooling of the ground by increasing the thermal conductivity and heat capacity of the ground and by releasing latent heat when the water freezes. The temperature of the plant will not fall below the freezing point as long as a change of state from water to ice is taking place. An objection to flooding is that, while preventing a fall of temperature during the night, water also retards an increase in warmth during the day, so that while frost may be warded off the first or second night, danger from a third or fourth night may be

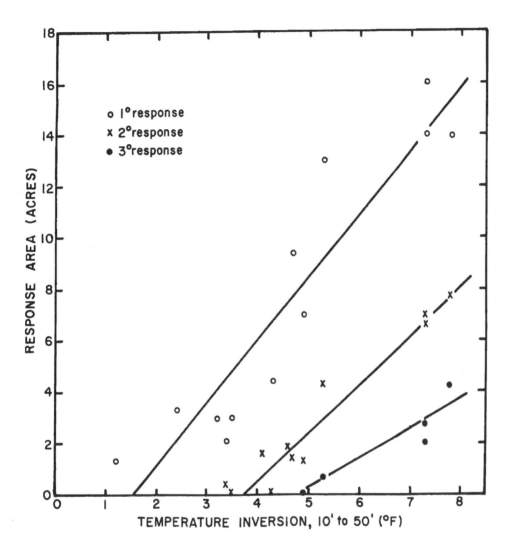

FIGURE 55. Results of twelve test runs of wind machine in Chico almond orchard. Graph plots acreage within isoline for 1° F., 2° F., and 3° F. responses against inversion strength in deg. F. Straight lines are calculated lines of best fit and provide reasonable evaluation of wind machine performance under these conditions (after Goodall, Angus, Leonard, and Brooks).

slightly enhanced. Thus, unless a large amount of water is available for repeated applications, the method has serious limitations. Flooding has been adopted in the cranberry bogs of Wisconsin (Cox, 1910) and Cape Cod, and in the vineyards of Australia (Foley, 1945). In some instances, flooding may be harmful to the plant. In Cape Cod, cranberry growers prefer to ward off light frost with wind machines and flooding only at times of a heavy freeze (Gunness, 1941).

Sprinkling possesses an advantage over direct flooding in that the water particles suspended in the air check net outgoing radiation. On the other hand, the plant temperature declines immediately when sprinkling stops, and the ice formation on the sprayed plants often causes damage to the blossoms or even breaks the tree branches. Tests by Schultz and Parks (1957) in Santa Cruz, California, showed that blueberry stalks can carry an ice coating only .25 inch in thickness without damage.

The success of sprinkling depends, to a large extent, on the amount and frequency of water application. Ideally, the speed of rotation should be such that all the water has just turned to ice at the next pass of the sprinkler. Thus, in general, the speed of rotation should increase with the severity of frost. Wheaton and Kidder (1965) tested the effect of application frequencies of 20, 60, and 120 seconds for an application rate of 0.11 inch per hour under wind frost conditions in Michigan (Figure 56). They explained:

> For a wind of one mph, this application rate gave protection to almost 30° F. using a 120 second repeat frequency and down to about 25.5° F. for a 60 second frequency. A lowering of the safe temperature in excess of 4 degrees was accomplished by changing the repeat frequency from 120 to 60 seconds. In a wind of 3 mph the same application rate protected down to about 30° F. with a 60 second frequency, and to about 27° F. when a 20 second frequency was used . . . In conclusion, under windborne freeze conditions, the safe temperature level was lowered from 1.5 to 4 F. deg. by increasing the repeat frequency of application, other factors remaining constant.

Davis (1955) also reported that small sprinklers, turning at the rate of one revolution per 12 to 20 seconds, gave satisfactory results while larger sprinklers, turning one revolution per 90 seconds or more did not adequately protect a tomato crop at 24° F. Shorter frequencies of repeated applications result in less temperature variation of the leaf surface.

Many observers have noted that, under light radiation frost conditions, very limited damage occurs in unsprinkled areas; whereas serious damage occurs in plants sprinkled with an insufficient amount of water. Businger (1965) gives two reasons for the enhanced damage inflicted by deficient sprinkling:

> (1) When the air is dry, the sprinkled leaf will approach the wet bulb temperature, which may be significantly lower than the dry bulb temperature. Quantitatively, this effect of evaporation . . . may be very small or zero for a dry leaf and be significant for a wet leaf. (2) The small amount of ice that forms on the leaf will prevent the undercooling of the cell solution and also may dilute the cell solution, thereby raising the freezing temperature.

For a hydrophobic leaf or fruit surface, it may be necessary to supply the water with a wetting agent in the beginning of the sprinkling in order

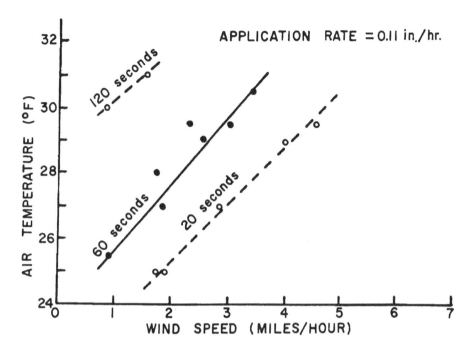

FIGURE 56. Effectiveness of sprinkling in frost protection as a function of application frequencies (after Wheaton and Kidder).

to create a more uniform film of water. This is particularly helpful in the prevention of frost damage for orange and other citrus fruit with oily and unwettable surfaces.

BRUSHING

Brushing is a frost protection scheme extensively used for tomatoes, squash, and other vegetable crops in the Coachella Valley, California. The field looks "brushy" as shields of brown kraft paper are attached to arrowed stems on the north side of the east-west rows leaning over the plants. No plants are located on the shaded side, which is used for irrigation ditches. During the day, the shields deflect radiation to the plant and soil. They also act as a windbreak against the cold north wind. At night, the shields reduce radiation loss to the sky. Brushing is more effective in protecting against radiation frost than wind frost. It is also more effective for small plants that have not outgrown the height of the paper.

Hart and Zink (1957) studied the use of reflective materials such as aluminum foil for shields. They found that, although aluminum foil increased the heat input to the soil by day, it was less effective than kraft paper in conserving heat at night. The germination rate of tomato seedlings was slightly speeded under the aluminum foil brushing; yet the plants did not produce a subsequently higher yield to justify the use of the more expensive aluminum foil.

SANDING

Cranberry growers in Cape Cod generally add a thin layer of sand to the ground every few years. A sandy surface warms up easily and cools only slowly by radiation. Sand also minimizes evaporation, because of its low water content. Sanding is capable of raising by several degrees the temperature of loam and clay, and even more for organic soils, thus diminishing the frost hazard.

WINDBREAKS

By excluding or diminishing the inflow of cold air and by shielding the field from "seeing" all the skies at night, windbreaks protect against frost. Brooks (1960) reported a case after a severe frost in Santa Paula, California, where the only marketable fruits in an orchard were found within a few tree rows of tall windbreaks. The effects of windbreaks or shelterbelts on microclimate will be further discussed in Chapter 22.

CHAPTER 11

WIND PROFILE NEAR THE GROUND

The wind profile data provide a measurement of momentum flux and are essential for evaluating the transfer of water vapor and CO_2 by the so-called aerodynamic approach. The wind profile data also aid in estimating the wind speed at one height from the measurement of another. For instance, in overhead irrigation, the distribution of water in the field is very much dependent upon the wind speed at the height of the nozzle. Therefore, in designing the sprinkler for a given area, the wind data taken at two meters, or some other fixed height, should be adjusted.

THE LOGARITHMIC EQUATION

At or near the surface of the ground, the horizontal wind velocity is zero and increases with the height above the surface. The wind gradient above the ground depends on the wind speed as well as the surface conditions. Halstead and Covey (1957) have presented generalized wind profiles in a corn field with varying crop heights up to 80 centimeters (Figure 57). At a height of about 640 centimeters the wind speed is not affected by the surface conditions.

The wind profile over short crops may be expressed by the logarithmic equation:

$$u = \frac{1}{k}\sqrt{\frac{\tau}{\rho}} \ln \frac{z}{z_0}$$

where u is the wind velocity at the height z; k is von Karman's constant having the value of 0.4; ρ is air density; τ is the shearing stress; and z_0 is the roughness parameter.

According to this equation, the wind speed near the ground increases exponentially with height over a very smooth surface. Over a moderately

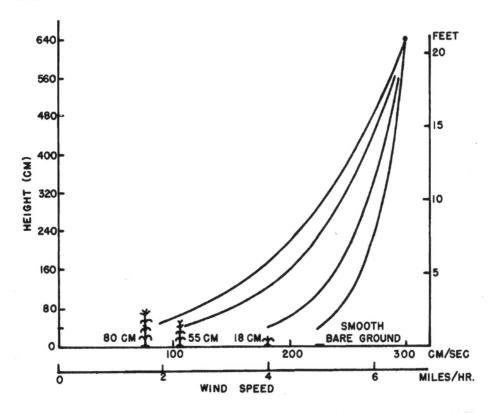

FIGURE 57. Effect of growing corn on wind profiles at Shirley, New Jersey, 1952 (after Halstead and Covey).

rough surface (e.g. short grass), the logarithmic wind profile holds true only above a hypothetical height z_o, known as the roughness of the surface. The equation may be solved by measuring wind speeds at three heights with very high precision.

Two problems arise with the logarithmic equation: (1) its validity depends on the constancy of the shearing stress with height up to approximately 30 meters above the ground, and (2) the equation holds true only under neutral conditions.

The constancy of the shearing stress with height was first proved by Ertel (1933) and was confirmed by Calder (1939), but Scrase (1930) presented evidence that was in disagreement with this concept. Recently, Tanner (1963a) stated that if the temperature gradient above the surface is zero, and if sufficient "fetch" exists, then the shearing stress is constant with the height above the crop. His explanation of the effect of "fetch" is of particular importance in selecting a proper site for measurement.

As air moves from one type of surface to another, the wind structure changes from that resulting from properties of the first surface to one which depends on the properties of the second. This change to a structure that is in equilibrium with the new surface does not occur immediately throughout the entire profile. Near the lead edge of the surface, that part of the wind profile which represents the new surface may be only a few centimeters high. The depth of this representative boundary layer grows with the downwind distance traversed over the new surface. Profile measurements representing the new surface must be made within the wind layers with properties developed from that surface, which means there must be sufficient fetch to permit growth of a well-developed profile to the height of measurement.

The logarithmic equation is employed under neutral conditions when no heat is added to, or subtracted from the surface. Over the land surface, neutral conditions are rarely observed. In the daytime, the lapse rate is usually superadiabatic; while at night, inversion is common. The logarithmic equation is least accurate under unstable conditions when the surface is warmer than the air above, but the deviations are generally small over moist vegetation.

DEACON'S EQUATION AND THE RICHARDSON NUMBER

In an attempt to improve the logarithmic equation by taking into consideration the stability of the atmosphere, Deacon (1949) proposed the following wind profile relationship:

$$\frac{du}{dz} = \alpha z^{-\beta}$$

where α and β are constants. Deacon showed that the parameter β is related to the Richardson number.

The Richardson number can be evaluated by

$$R_i = \frac{\frac{g}{t}\left(\frac{dt}{dz} + \Gamma\right)}{\left(\frac{du}{dz}\right)^2}$$

where R_i is the Richardson number; g is the acceleration of gravity; and Γ is the dry adiabatic lapse rate.

The Richardson number is dimensionless and is a measure of the importance of buoyance forces (thermal convection) in producing turbulence, as compared with frictional forces. The Richardson number equals zero under neutral conditions, is negative under superadiabatic conditions, and is positive during temperature inversion. According to Lumley and Panofsky (1964), atmospheric turbulence cannot be maintained at Richardson numbers greater than 0.25. If the Richardson number were independent of height, the Deacon's equation could accurately describe the wind profile. However, the absolute magnitude of the Richardson number normally increases with height, and forced convection near the ground is usually replaced by free convection at greater heights. The transition from forced to free convection usually takes place at Richardson numbers between -0.02 and -0.05. The variation of wind profile with stability remains a troublesome problem.

WIND PROFILE OVER TALL CROPS

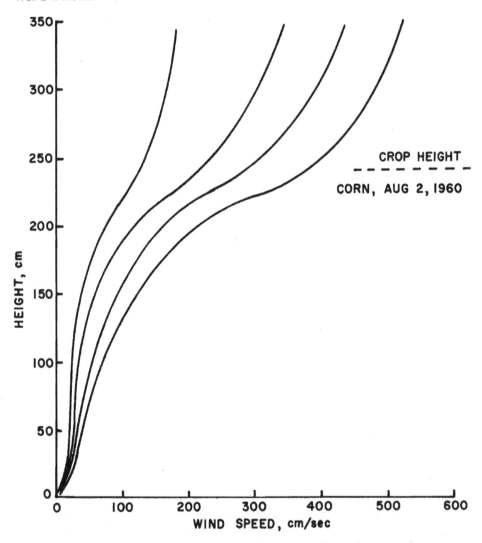

FIGURE 58. Representative wind profiles above and through corn crop canopy (after Lemon).

Wind Profile near the Ground

The wind structure over a tall crop is different from that over a short crop. In Figure 58, Lemon (1963) presented wind profiles in a corn field. The wind profile changes abruptly at a height of about 220 centimeters, slightly below the canopy. Above that height, the logarithmic relationship seems to hold; below it, the wind speed is greatly reduced. Therefore, the wind profile equation for tall crops should be modified to the following form:

$$u = \frac{1}{k}\sqrt{\frac{\tau}{\rho}} \ln\left(\frac{z-d}{z_0}\right)$$

where d is known as the zero plane displacement. The zero plane displacement is roughly the order of the depth of the layer of air trapped among the plants. In other words, it is a datum level, above which the normal turbulent exchange takes place freely. The zero plane displacement may also be regarded as the sink for momentum.

As mentioned in the previous section, profile measurements over a short crop should be taken near the surface. However, over a tall crop, especially a row crop, the nonhomogeneity of the surface is so pronounced that measurements taken near the canopy level may not be representative of the whole field. Therefore, readings at more than three heights over tall crops are preferred, with the highest recording level at considerable distance above the canopy.

ROUGHNESS AND ZERO PLANE DISPLACEMENT

Both z_0 and d are geometric constants of the surface. Their meaning can be further clarified by the schematic diagram in Figure 59. Under neutral conditions, the relationship between the wind velocity and the logarithm of height is linear, and the intercept is the roughness. However, over a tall crop, the relationship is curvilinear, and the departure from the straight line is known as the zero plane displacement. Under both lapse and inversion conditions the relationships between wind speed and the logarithm of height are also curvilinear, as shown in Figure 60.

For short vegetation, or relatively smooth surfaces, the roughness parameter is relatively constant over a range of wind speeds. Typical values of roughness are as follows: ice, 0 centimeter; sand, 0 to 0.1 centimeter; open water, 0.02 to 0.6 centimeter; snow surface, 0.1 to 0.6 centimeter; short grass, 0.6 to 4.0 centimeters; and long grass, 4.0 to 10.0 centimeters. In general, roughness increases with the height of the vegetation. Kung and Lettau (see Tanner and Pelton, 1960) have found that the log-log plot of the vegetation height versus roughness is approximately linear (Figure 61). In the absence of actual measurements, this graph should provide a good estimate of the roughness parameter.

For tall crops, which bend and weave with wind, both z_0 and d are

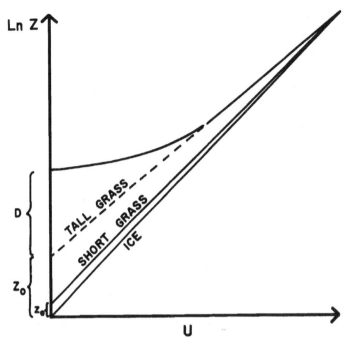

FIGURE 59. Schematic diagram showing relationship between wind speed and logarithm of height. z_o is the roughness parameter; d is the zero plane displacement.

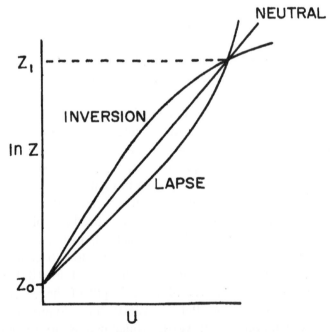

FIGURE 60. Form of wind profile over surface of roughness length z_o and with same wind speed at height z_1 in lapse, neutral, and inversion conditions, plotted against logarithmic height scale.

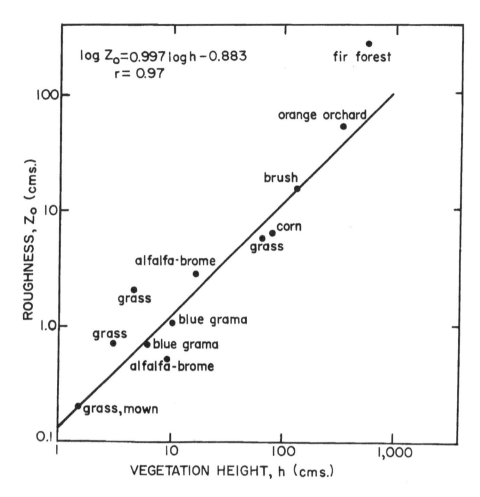

FIGURE 61. Graph for estimating roughness length from vegetation height (after Kung and Lettau).

subjected to considerable variations. Inoue (1963) observed that for a rice crop of 90 centimeters in height, the value of d varies from 35 to 90 centimeters and z_o from 7 to 18 centimeters for wind speeds up to 10 centimeters per second at a height of 150 centimeters (Figure 62). This complicated relationship is the result of the bending of the stalks and the thickening of the canopy, which causes a variation in the drag coefficient. Variations of the d value have also been reported by Penman and Long (1960) for rice and wheat, by Lemon (1960) for corn, and by Rider (1954a) for oats. Rider plotted the variations of crop height with time and plotted values of the zero plane displacement divided according to wind speed groups (Figure 63). In his experiments, the value of d increased with wind speed. He also noticed a tendency for d to vary with the wind direction. When the wind was directed along the drill lines, the value was

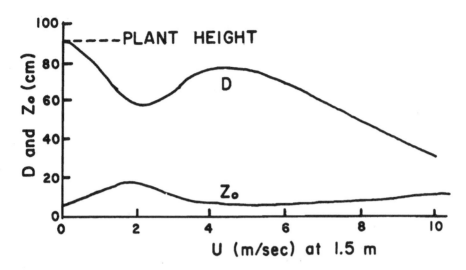

FIGURE 62. Change of z_o, roughness, and d, zero plane displacement, with wind velocity at height of 1.5 m. over paddy fields (after Inoue).

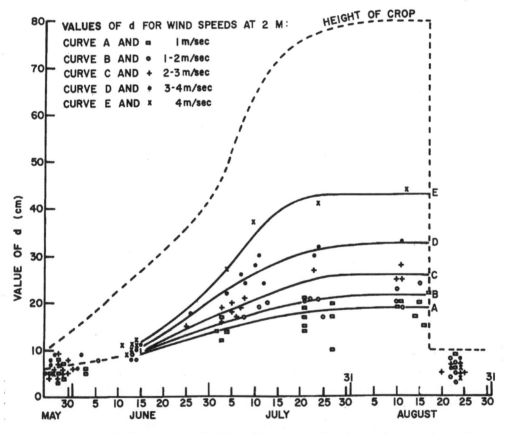

FIGURE 63. Variation of crop height with time and values of zero plane displacement divided according to wind speed groups (after Rider).

often smaller than when the wind direction was at right angles to these lines.

The value of the zero plane displacement is a function of the height as well as the density, and the mechanical properties of the plant. Plants that stand erect in the wind should have a higher value of d. The value for oats reported by Rider averaged less than 50 per cent of the plant height; that of rice, reported by Inoue, averaged about 60 per cent. In Hawaii, for a mature sugar cane crop approximately four meters tall, the zero plane displacement has been found to be 3.5 meters, or about 88 per cent of the plant height (Chang, 1961).

SIGNIFICANCE OF ROUGHNESS

The roughness of a surface has several implications in the micrometeorological study of plant environment. First, Lettau (1952) has shown that, other things being equal, an increase in roughness will cause lowering of maximum temperature during the daytime and a rise in the minimum temperature at night.

Secondly, the rougher the surface, the greater the mixing and swirling. According to the theory of turbulent transport, the rate of mixing, expressed as a coefficient of diffusion, or the Austausch coefficient, does not depend upon the wind speed, but upon the rate of change with height of the wind speed. Over a rough surface, heat and water vapor are readily transferred, even though the wind speed may be fairly low. Sellers (1965) pointed out that the transfer coefficient can increase by about 50 per cent with an increase of the roughness length from 0.2 centimeter (mowed grass, 3 centimeters high) to 0.7 centimeter (mowed grass, 7.5 centimeters high). Therefore, other things being equal, the evapotranspiration of a rough surface will exceed that of a smooth surface, especially in areas of strong advection (Tanner and Pelton, 1960).

Thirdly, it would be difficult to determine the transfer of water vapor, CO_2, and other properties by the aerodynamic method for a crop whose roughness and zero plane displacement vary greatly. Any equation that requires the use of a single value of z_o and d cannot be applied with confidence to tall crops.

CHAPTER 12

WATER IN RELATION TO PLANT GROWTH

GENERAL EFFECTS

Warming (1909) distinguished three groups of plants on the basis of their water relationships: hydrophytes, mesophytes, and xerophytes. Hydrophytes normally grow in water or swamps—mangrove, bullrush, and paddy rice being good examples. Most field crops belong to the mesophyte group. Paltridge and Mair (1936) further divided mesophytes into two subgroups. Those that wilt permanently after losing 25 per cent of their water content are called true mesophytes; those that wilt after losing from 25 to 50 per cent of their water are xerophytic mesophytes. Xerophytes are capable of enduring even more severe drought. They wilt permanently only after losing from 50 to 75 per cent of their total water content.

This classification provides a primary approximation of crop adaptation and distribution in various moisture regimes. However, it is far too crude to serve as a useful guide for irrigation and other cultural practices. For those purposes, it is essential to have an understanding of the various physiological plant responses to water.

In a recent review, Kramer (1963) pointed out that water is: (1) the major constituent of physiologically active plant tissue; (2) a reagent in photosynthesis and in hydrolytic processes, such as starch digestion; (3) the solvent in which salts, sugar, and other solutes move from cell to cell and organ to organ, and (4) an essential element for the maintenance of plant turgidity, necessary for cell enlargement and growth. In addition, water is needed for transpiration, which, though serving no direct, useful function in plant growth and development, has several beneficial effects.

Almost every process occurring in plants is affected by water. However, no simple and uniform relationship exists between water stress and the various aspects of plant function. The relationship varies with plant characteristics, stages of development, and soil and climatic conditions.

Water deficiencies not only reduce the yield, but also change the pattern of growth. In general, effective rooting depth decreases as the soil moisture level increases (Weaver and Himmel, 1930). Roots developed under limited moisture conditions are finer and have more and longer branches than roots developed under favorable soil moisture conditions (Kmoch, Raming, Fox, and Koehler, 1957). Thus, frequent irrigation often leads to shallow and horizontally spread root systems, which are a disadvantage when dry conditions occur. The ratio of root to shoot usually is increased by water stress. Leaf area is often reduced but leaf thickness is increased. The quality of the economic yield, flower formation, and seed production are all influenced by moisture conditions.

EFFECT OF WATER ON PHOTOSYNTHESIS

Although it is commonly estimated that less than 1 per cent of the water that passes into a plant is utilized in photosynthesis (Meyer and Anderson, 1952), water deficits in the plant have a profound effect on the rate of photosynthesis. Plant water stress directly reduces the photosynthesis process because dehydrated protoplasm has a lowered photosynthesis capacity. Indirectly, once the leaf loses its turgidity, the stomata guard cells close, thus preventing any further intake of CO_2 for photosynthesis. In general, the respiration rate of a plant tends to increase as moisture decreases (Schneider and Childers, 1941, and Chrelashvili, 1941). This change is, however, much smaller than the accompanying changes in photosynthesis.

Plant species differ markedly in their ability to withstand a water shortage before their photosynthesis rate is seriously reduced. They also differ in their ability to recover when the water shortage ceases. Most experiments deal with the effect of soil moisture on photosynthesis, but very little information is available on the relationships between photosynthesis and the water deficit in leaves, largely because of the lack of a satisfactory method for measuring plant water stress.

In general, the rate of photosynthesis declines noticeably after a reduction of approximately 30 per cent in the water content of the leaves. When 60 per cent of the leaf moisture is lost, photosynthesis usually ceases. Thoday (1910) reported that the turgid leaves of the sunflower (*Helianthus annuus L.*) carried on photosynthesis at a rate ten times that of wilted leaves. Brilliant (1924) found that the photosynthesis of English ivy (*Hedera helix L.*) and balsam (*Impatiens parviflora Bedd.*) almost stopped when leaf moisture was reduced by 41 to 63 per cent. A water loss of 34 per cent in the leaves of Jerusalem sage (*Phlomis pungens Willd.*) caused a reduction of 13 per cent in the rate of photosynthesis. (Iljin, 1923).

Experimental results by Upchurch, Peterson, and Hagan (1955) on ladino clover, by Allmendinger, Kenworthy, and Overholser (1943) on apples, by Loustalot (1945) on pecans, and Aston (1956) on sugar cane, indicate that photosynthesis is little affected until most of the soil moisture is depleted, or the permanent wilting point is almost approached. On the other hand, the data obtained by Schneider and Childers (1941) on apples, by Bordeau (1954) on oak, and by Moss, Musgrave, and Lemon (1961) on corn, showed that photosynthesis began to drop when the soil moisture content was relatively high. Kramer (1963) cautioned that experiments to determine the effect of soil moisture on photosynthesis are not easy to design and that some of them might be faulty. One reason that certain investigators observed no reduction in photosynthesis—until the permanent wilting point was approached—was that the tops of the plants were enclosed in containers, which greatly reduced transpiration.

The study by Ashton on the effect of soil moisture on photosynthesis of sugar cane was well documented. His data in Figure 64 showed that photosynthesis did not drop until the soil moisture content was depleted to less than 40 per cent, not much above the permanent wilting point. Photosynthesis was slightly reduced at high soil moisture content. Gaastra (1962) explained that photosynthesis is increased by slight soil moisture stress, possibly as a result of the "hydropassive" stomata opening. Ashton's study has immediate implications in determining irrigation intervals. Ideally, irrigation should be applied just prior to the sharp decline in photosynthesis. In the case of sugar cane, frequent irrigations of short intervals would not increase the yield. Waterlogged conditions should also be avoided. For any crop whose photosynthesis declines with a slight drop in soil moisture—at a soil moisture content of 60 per cent or higher, let us say—irrigation of the crop should be frequent, or the crop should be grown in water to obtain the maximum yield.

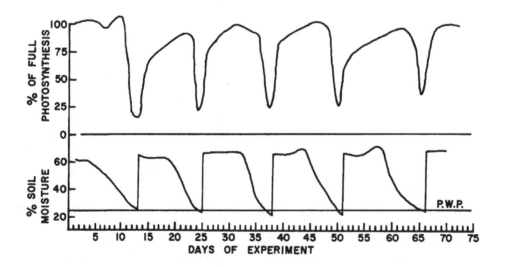

FIGURE 64. Rates of photosynthesis and percentages soil moisture during five drying cycles (after Ashton).

As the soil moisture tension reaches the permanent wilting point, the photosynthesis rate of sugar cane falls to about 25 per cent of the potential (Burr, Hartt, Brodie, Tanimoto, Kortschak, Takahashi, Ashton, and Coleman, 1957). The recovery of full photosynthesis after irrigation is slow, requiring about two days according to the experiment by Ashton, as shown in Figure 65. The rate of recovery varies with the plant species, soil and atmospheric conditions, and the method of water application. Schneider and Childers (1941) also reported that upon wilting the photosynthesis of the apple tree dropped to 13 per cent of the optimum, and that after irrigation, several days were needed to return to normal. On the other hand, Verduin and Loomis (1944) observed full recovery within a few hours after rewatering temporarily wilted maize leaves whose photosynthesis had been reduced to 5 per cent of the optimum. Overhead irrigation, by applying water to the plant, has an advantage over the surface irrigation as far as the rate of recovery of photosynthesis is concerned.

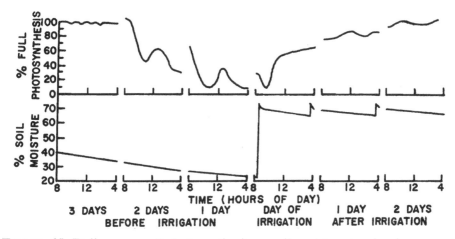

FIGURE 65. Daily course of photosynthesis as soil moisture tension increases to permanent wilting percentage (after Ashton).

A somewhat different experiment on the effect of water stress on photosynthesis was carried out by Larsen (1960) for flax. He observed that a ten-day period of water shortage not only caused a temporary halt in the growth of leaves, but rendered half of the leaves ineffective in producing dry matter (Figure 66). The plants took about a week to recover after rewatering. If the drought had continued longer, they would presumably have been unable to resume growth. Premature ripening would have occurred and the growth period ended about the middle of July.

TRANSPIRATION

Transpiration takes place in leaves through the stomata. Some plants (e.g., maize and many grasses) have nearly equal numbers of stomata in

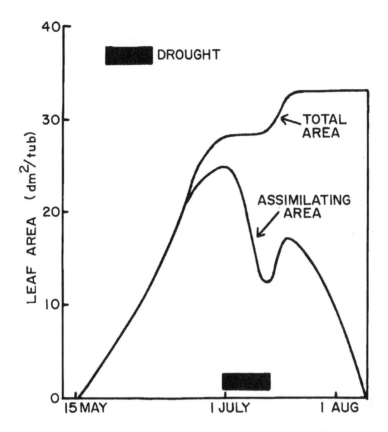

FIGURE 66. Effect of drought on total unfolded and assimilating leaf area of flax (after Larsen).

the upper and lower epidermis of their leaves. Others, such as broad leaves of trees, have practically no stomata in their upper epidermis. The number of stomata may be as high as 20,000 in one square centimeter of leaf surface. The opening and closing of the stomata and, hence, the rate of transpiration are controlled by guard cells. During the day, the guard cells at each side of the pore in a well-watered leaf separate as the swollen regions at each end enlarge. This movement exposes the moist leaf interior to the drier atmosphere outside. Normally, the stomata are closed at night. The stomata can be forced to close partially by spray application of several chemicals (Zelitch, 1961, 1964; Zelitch and Waggoner, 1962, and Angus and Bielorai, 1965). However, chemical reduction of transpiration often causes plant damage.

Transpiration has often been regarded as a necessary evil with no useful function in the plant growth or development. However, Winneberger (1958) observed that plant growth was reduced or stopped when the plants were grown under high (almost 100 per cent) relative humidities, and transpiration was completely stopped. He suggests that transpiration may be the energy source for all translocation, except that which results from diffusion.

Transpiration prevents excessively high temperatures, which would otherwise have an adverse effect on other processes. Gates (1965b) pointed out that a transpiration rate of only .0005 of a gram of water per square centimeter per minute gives rise to an energy loss of approximately 0.3 of a calorie. This is enough to lower the temperature of a transpiring leaf by as much as 15° C. Kinbacker (1963) demonstrated in an experiment that winter oats exposed to high temperatures of 44° to 45° C. and 100 per cent relative humidities were damaged more severely than plants exposed to those same temperatures but with a relative humidity of only 50 per cent. Apparently, the cooling effects of high transpiration rates under low humidities were of considerable benefit to plant growth and survival.

If transpiration exceeds water absorption, then the water balance of the plant becomes negative. Under such conditions, plants have several means of avoiding an increasing water deficit. They may intensify the water absorption through extension of the root system, or increase their suction forces to take up the more firmly held soil water. Plants may also reduce transpiration through the shedding of leaves, or through an increase in the diffusion resistance of leaves, such as the closing of the stomata or the incipient drying of the cell membranes. When the latter conditions occur, plant growth would be restricted.

In general, transpiration decreases with the increase of soil moisture tension. However, the relationship between the two is a controversial subject and will be discussed in Chapter 13. During periods of high radiation and low humidity, even those plants growing in soils nearing field capacity may be subject to severe water stress. The phenomenon known as "midday depression" or "midday water deficits" is a good example. Because of the adverse effect of midday water deficits, Stocker, Leyer, and Vieweg (1954) found in their experiments that a heavy sprinkling at midday increases growth considerably.

TRANSPIRATION AND DRY MATTER PRODUCTION

For a single leaf, the net assimilation, or net photosynthesis, increases with light intensity to the saturation point and then levels off. The transpiration rate will, however, increase linearly with radiation to a much higher intensity. Thus, the ratio between transpiration and photosynthesis will vary according to the radiation intensity in a manner postulated by de Wit (1958), as shown in Figure 67. This same relationship was later quantitatively presented by Bierhuizen (1959) as shown in Figure 68. The high ratio occurring at extremely low radiation intensity is because transpiration has some value, whereas photosynthesis first has to compensate the respiration. This high ratio is of little significance because of the low rates of both processes. The lowest ratio is reached at a radiation intensity of 0.1 to 0.2 langley per minute. Such low radiation intensities are

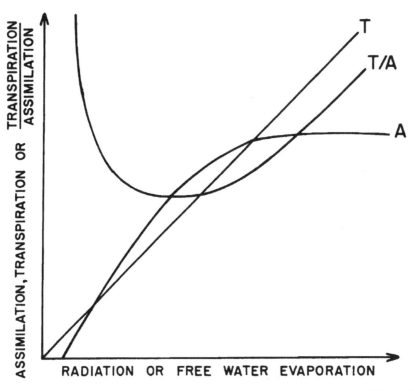

FIGURE 67. Relationship between net assimilation (A), transpiration (T) and the transpiration to assimilation ratio (T/A) for leaves of plants as a function of the radiation or free water evaporation (after de Wit).

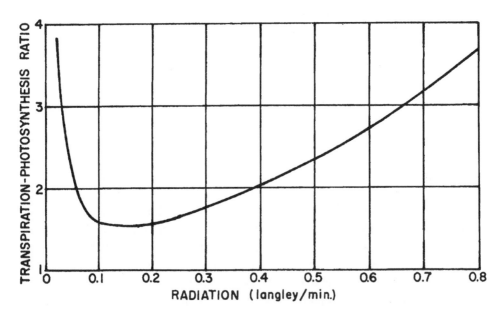

FIGURE 68. Relationship between radiation and the transpiration-photosynthesis ratio (after Bierhuizen).

observed only in the early morning and late afternoon. As the radiation intensty increases, beyond 0.2 langley per minute, the ratio of transpiration to photosynthesis for a single leaf increases nearly linearly. Thus, for a single leaf, the efficiency of water use in the production of dry matter will be lower in areas of high radiation, such as in the arid tropics.

In the field, the relationship between radiation and the ratio of transpiration to photosynthesis will be quite different than that for individual leaves. If the leaf area index is sufficiently high, the photosynthesis of the plant will increase with the radiation to a high intensity. In that case, the rate of transpiration will closely parallel that of photosynthesis, and the ratio between the two will not vary greatly with radiation. The classical work by Brigg and Shantz (1913), as shown in Figures 69 and 70, clearly demonstrated a close relationship between transpiration and dry matter production. In their experiments, the linear relationship holds for different varieties.

De Wit (1958) and Arkley (1963) reviewed the studies on the relationship between transpiration and dry matter production for ten crops. In every instance, the relationship is linear. This linear relationship has immediate applications in irrigation and water management studies. However, four points need to be clarified. First, a considerable controversy exists as to whether the relationship is of a cause-and-effect nature. Penman (1956b) stated that, "The maintenance of maximum transpiration rate is a necessary condition for maximum growth." However, Milthorpe (1961) considered it advisable to modify the statement to remove the implication that transpiration and growth are causally related. He rephrased the sentence in this way: "Transpiration is maintained at the potential rate until such time as growth is reduced by shortage of water in the leaves." Even though a cause-and-effect relationship may not exist between transpiration and dry matter production, their close correlation can be explained by the fact that net radiation, which determines to a large extent the transpiration rate, and solar radiation, which determines photosynthesis, are linearly related (Monteith, 1965b).

Secondly, the close relationship is obtained only when the plants are actively growing. During the periods of senescence or ripening, the relationship may be somewhat different. Lemon, Glaser, and Satterwhite (1957) reported that the rapid drop in the transpiration rate of cotton in late August coincided with the maturation of the bolls. After that, the transpiration rate was much reduced while growth continued until the December frost. On the other hand, there are also instances when transpiration continued even after maturing while the dry matter production stopped with the shedding of leaves.

Thirdly, under unfavorable temperatures or other environmental conditions, transpiration may desiccate protoplasm to such an extent as to damage the plant growth. Curry and Church (1952) reported a case in the

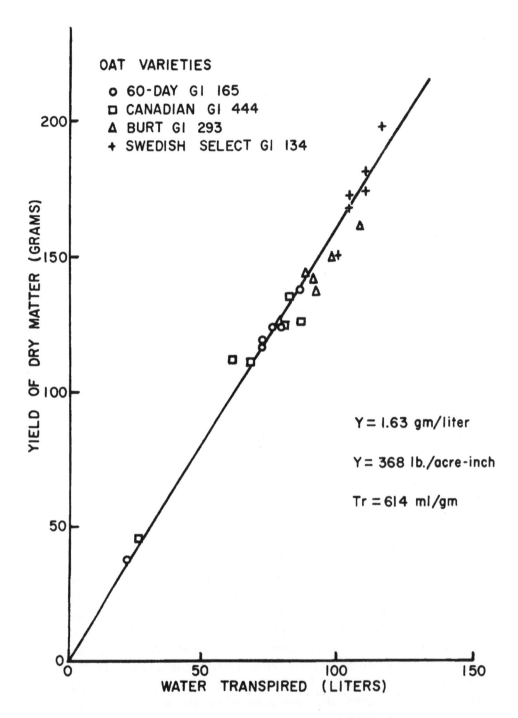

FIGURE 69. Relationship between yield of dry matter and amount of water transpired. Data obtained by Briggs and Shantz (after Arkley).

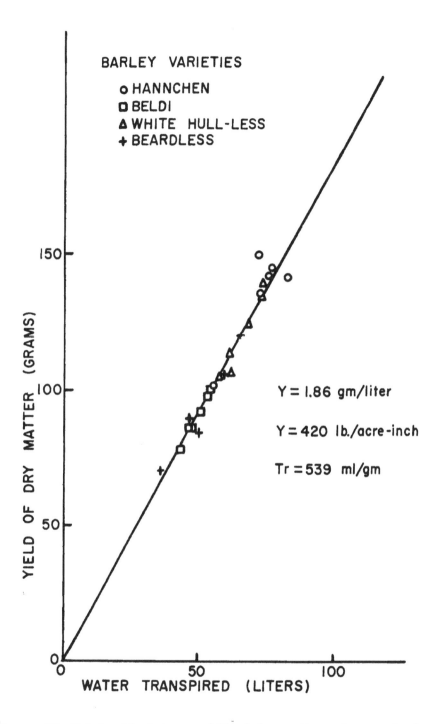

FIGURE 70. Relationship between yield of dry matter and amount of water transpired. Data obtained by Briggs and Shantz (after Arkley).

Adirondacks where the excessive transpiration in early spring far exceeded what the plants could absorb from the frozen soils. The leaves of all the mountainside conifers dried and became discolored.

Fourth, the close relationship between transpiration and dry matter production does not imply that the old concept of the transpiration ratio is acceptable. The concept of the transpiration ratio predicts that the ratio of the weight of water transpired to the weight of dry matter produced is a constant. Penman and Schofield (1951) suggested that the constant ratio might be valid in a limited case, whereas de Wit has conclusively demonstrated that the regression equation between transpiration and dry matter production varies with a number of factors. In areas of high radiation, he proposed a relationship:

$$M = bW/E_o$$

where M is total dry matter production, W is total transpiration, E_o is free-water evaporation, and b is a constant. In the Great Plains of the United States, the value of b for sorghum, wheat, and alfalfa is 20.7, 11.5, and 5.5 grams dry matter per kilogram water per day, respectively.

In areas of low radiation, his formula changes to:

$$M = cW$$

where c is a constant. In the Netherlands, the value of c for beets, peas, and oats is 6.1, 3.4, and 2.6 grams per kilogram water per day, respectively.

The values of b and c are more dependent on the climatic conditions than on the nutrient level of the soil and the availability of water, provided that the nutrient level is not too low and the availability of water not too high. These values are also independent of the degrees of mutual shading, provided that the leaf mass is not "too dense." Where these conditions are not fulfilled, the b and c values are larger.

The equations derived by de Wit are based exclusively on experiments in containers, and they refer only to transpiration. Their application to field conditions will be discussed in Chapter 20.

CHAPTER 13

EVAPOTRANSPIRATION

DEFINITION

Evapotranspiration is the combined evaporation from all surfaces and the transpiration of plants. Except for the omission of a negligible amount of water used in the metabolic activities, evapotranspiration is the same as the "consumptive use" of plants.

Because the rate of evapotranspiration from a partially wet surface is greatly affected by the nature of the ground, it is advisable to first consider the case when the water supply is unlimited. This leads to the concept of potential evapotranspiration, which Penman (1956b) defines as "the amount of water transpired in unit time by a short green crop, completely shading the ground, of uniform height and never short of water." This definition, though basically correct, needs qualification and amplification on three counts.

First, the "short crop" is not specified. Penman argued that when the cover is complete, the potential evapotranspiration is determined primarily by weather and is not affected by plant species. This has been substantiated for most plant species. For example, van Bavel, Fritschen, and Reeves (1963) showed that transpiration by well-watered sudan grass is regulated by meteorological factors rather than physiological factors. Ogata, Richards, and Gardner (1960) found this to be true for irrigated alfalfa. Fritschen (1966) suggested that the small differences in water use by various plants may be explained, at least in part, by their physical properties such as albedo and roughness. Broad leaves, being aerodynamically rougher than grasses, are capable of extracting more energy from the air and, hence, have a higher rate of transpiration.

Since Maximov (1929) rejected the idea that drought-hardy plants had low transpiration rates when supplied with adequate water, high transpiration rates by xerophytes have been recorded by Turrell (1944) and Abd El Rahman and Batanouny (1965). With a high moisture level, the transpiration rates of plants with different degrees of drought resistance are more or less the same (Salim and Todd, 1965). This removes a common misconception that desert plants, having a better survival mechanism under drought conditions, should have a much lower rate of potential evapotranspiration. It is also significant to note that, when water supply is unrestricted, the water need of a cotton plant whose root knots are infected by nematodes is comparable to that of an uninfected plant (O'Bannon and Reynolds, 1965). As soil moisture stress increases, the evapotranspiration rate of the infected plant is significantly less than that of the uninfected plant. However, a very few plants, such as pineapples (Ekern, 1965a) in the tropics and lichen (Nebiker, 1957) in the arctic, do behave quite differently from conventional vegetation because of their physiological peculiarity. Pineapples have an extremely low evapotranspiration rate because their stomata are closed throughout the day.

Secondly, the criterion for a complete ground cover is not clear. Theoretically, a vegetation cover may be described as completely shading the ground only if it intercepts all the radiation energy. In reality, even a tall, dense crop with a high leaf area index can hardly absorb 95 per cent of the incident radiation. Fortunately, after the canopy is reasonably well developed, large differences in vegetative growth can cause only relatively small differences in evapotranspiration rate when soil moisture is adequate. This has been found to be true for wheat crop by Fischer and Kohn (1966). According to a study by Stern (1965), the evapotranspiration rate of well-watered safflower remains nearly constant after the crop reaches a leaf area index of 4 and intercepts more than 80 per cent of the radiation (Figure 71).

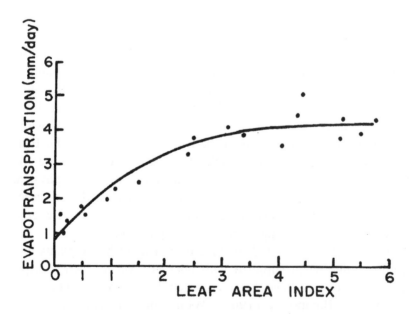

FIGURE 71. Relationship between leaf area index and evapotranspiration. Curve represents equation $E = 4.20 - 3.12\ e^{-0.698F}$ (after Stern).

Thirdly, Penman's definition does not specify the size of the field or the condition of the surrounding areas. In other words, no provision is made for the effect of advected energy. In a humid climate, where evaporation usually takes place in what Penman (1965b) calls the "midocean" environment, advected energy is not a serious problem. In an arid or semiarid climate, however, the existence of large amounts of advected energy renders the concept of potential evapotranspiration, as defined above, inexact and unrealistic. If potential evapotranspiration requires an extended evaporating surface upwind or the absence of any advected energy, then the climate is no longer arid.

Recently, Pruitt (1960b) designated the term "potential maximum evapotranspiration" to describe the situation when advected energy is present. This would certainly remove any previous confusion. Thus, one should not expect an empirical formula for potential evapotranspiration derived in a humid climate to be adequate for estimating the potential maximum evapotranspiration in an arid climate.

METEOROLOGICAL FACTORS DETERMINING
POTENTIAL EVAPOTRANSPIRATION

The rate of potential evapotranspiration depends on evaporative power of the air as determined by temperature, wind, humidity, and radiation. Mukammal and Bruce (1960) have found that the relative importance of radiation, humidity, and wind, in determining the pan evaporation, are in the ratio 80:6:14 respectively. They neglected the air temperature presumably because it was affected, to a large extent, by radiation. Although this analysis was based on pan evaporation data, the results would be approximately the same for potential evapotranspiration. In either case, radiation is the dominant factor.

Evaporation is a diffusive process, partly turbulent and partly molecular. The turbulent process is the dominant mechanism, except in the thin layer near the evaporating surface. According to the theory of turbulence, the upward flow of water vapor is equal to the product of the vertical gradient of vapor pressure and the rate of mixing. The latter does not depend upon the wind speed at any height, but upon the rate of change of the wind speed with height. Thus, any method of estimating potential evapotranspiration that employs only wind speed at one height must rely upon an extremely crude measurement of turbulence. Penman realized that the wind speed gradient near the surface is a function of its roughness. This is why he stated in his definition that the vegetation must be of uniform height.

It is commonly believed that evaporation is proportional to the vapor pressure deficit. This is true only when the temperature of the air equals that of the evaporating surface, a condition rarely observed in nature. In the absence of this equality of air and surface temperatures, evaporation is proportional to the vapor pressure gradient between the evaporating surface and the air.

Mather (1959) constructed a diagram showing the relationship between the air temperature and the relative humidity and vapor pressure of the air (Figure 72). He further explained how the relative humidity and vapor pressure deficit are often misused:

> Consider, for instance, an example of the rate of evaporation from two moist surfaces having exactly the same temperature into two air samples both having a relative humidity of 60% and with temperatures of 60° F. and 80° F., respectively. While it might at first appear that the rate of evaporation would be greater into the air at 80° F., such is not the case..... The air at 80° F. and 60% relative humidity has a vapor pressure of just less than 16 mm [of mercury]. This is the same vapor pressure which would exist at a water surface at a temperature of about 64.5° F. If the surface water temperature were below 64.5° F., the vapor pressure at the surface would be below the vapor pressure of the air and condensation onto the surface would be occurring instead of evaporation.
>
> The air at 60° F. and 60% relative humidity has a vapor pressure of about 8 mm [of mercury], which is equal to the vapor pressure at a water surface at a temperature of about 45.5° F. Thus, if the moisture surface in

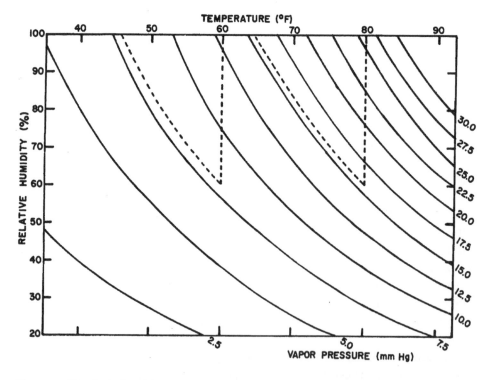

FIGURE 72. Relationship between temperature and moisture content of air (after Mather).

our example had a temperature between 45.5° F. and 64.5° F. there would be evaporation into the air sample at 60° F. and condensation from the air sample at 80° F. Only if the moist surface is above 64.5° F. will evaporation into both samples occur. Then the rate would be greater into the air at 60° F. than the air at 80° F. because the evaporation is proportional to the vapor pressure gradient.

EVAPORATION AND TRANSPIRATION DIFFERENCES BETWEEN

Transpiration differs from free water evaporation. The transpiration process may be physically described in terms of a resistance to a diffusive and turbulent vapor flux in the external air—a similar diffusive resistance that results from the internal leaf geometry, including the stomata, and parallel to the latter, a resistance to vapor diffusion through the cuticle (Raschke, 1956). In contrast, the last two resistances do not exist in the evaporation from an open water surface.

When the stomata are open, transpiration is determined primarily by the available energy. With the development of a water deficit and the partial closing of the stomata, the transpiration rate is reduced. Brouwer (1956) demonstrated that, for a given stomatal aperture, the transpiration rate is proportional to the radiation energy. Transpiration may continue through the cuticle, even after complete stomatal closure at night, at a very low rate. Tanner (1957) reported a case in which after nightfall closed the stomata, the transpiration rate was only 5 to 10 per cent of the daytime value. At night, the evaporation rate is relatively high, using in part the energy stored during the day. However, for a period of a day or longer, the transpiration curve often closely parallels the evaporation curve.

Penman and Schofield (1951) gave three reasons why the potential evapotranspiration of a short crop is less than open water evaporation: (1) the higher albedo of the vegetation, (2) the closure of the stomata at night, and (3) the diffusion impedance of the stomata. Subsequently, Neumann (1953) argued, from the standpoint of the turbulent theory, that the potential evapotranspiration of a short crop is approximately 75 per cent of the open-water evaporation.

ACTUAL EVAPOTRANSPIRATION

When soil moisture is plentiful, the evapotranspiration rate is maintained at the potential rate, determined largely by the prevailing weather conditions. As the soil dries out, the actual evapotranspiration will, at some stage, fall below the potential rate. Considerable controversy exists as to the effect of the soil moisture tension on the depletion rate.

Veihmeyer and Hendrickson (1955) presented a thesis that evapotranspiration proceeded at the potential rate up to the wilting point and fell sharply thereafter. They argued that the equal availability of water between field capacity and the permanent wilting point can be explained by the extremely small amount of energy required to remove a gram of water at the permanent wilting point and transport it to the transpiring leaf surface. Subsequently, Veihmeyer, Pruitt, and McMillan (1960) showed that the evapotranspiration of perennial ryegrass in the large weighing lysimeter at Davis, California, did not fall until the moisture tension in the upper half of the 35-inch soil profile reached thirteen to fifteen bars. Similar results were obtained by Lowry (1956) in the forest soils of California, and by

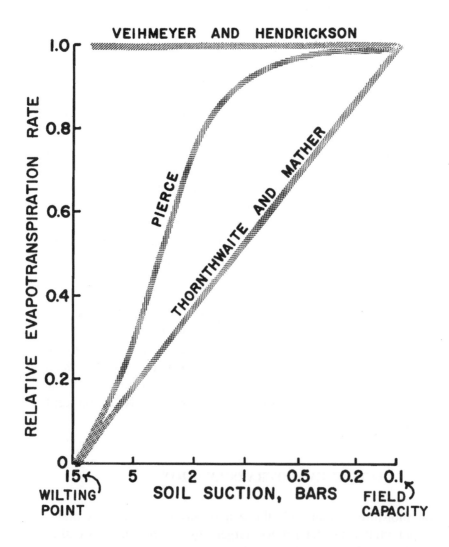

FIGURE 73. Various relationships between relative evapotranspiration rate and soil moisture tension.

Evapotranspiration 135

Glover and Forsgate (1964) for closely matted Kikuyu grass in a floating lysimeter at an altitude of over 6,000 feet in Kenya. The latter observed:

> The most striking feature of the experiment is that the grass did not wilt until all the soil in the lysimeter was depleted of available water. Although most of the available moisture was removed from the top foot of soil by the middle of August, from at least the top two and a half feet by the end of September, it was only when the soil at the bottom of the tank was dried out that the plants wholly lost turgidity. The collapse was sudden and complete.

However, Thornthwaite and Mather (1955b) presented an entirely different conclusion. They suggested a linear decline of evapotranspiration with increasing tension, based on the vapor pressure and temperature profile measurements at O'Neill, Nebraska.

Still others, probably the majority of research workers, propose a compromise between these two extremes. Their data indicate that the actual evapotranspiration proceeds at the potential rate for some time, and then decreases rapidly in an exponential manner. Pierce's curve (1958) in Figure 73 is a good example. However, there are considerable discrepancies in the data as to where the actual evapotranspiration begins to drop.

The conflicting state of evidence is not a result of experimental errors, but is attributable to differences in experimental conditions. Only recently have research workers begun to realize that the relationship between the

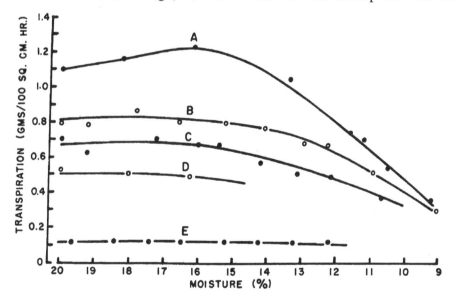

FIGURE 74. The transpiration of kidney beans in gms/100 cm^2/hr versus the moisture percentage of the soil at various light intensities at a temperature of 20° C. and a relative humidity of 40%. A: light intensity 4.5 x 10^4 ergs/cm^2 sec; B: 2.4; C: 1.4; D: 0.66; E: results from Veihmeyer and Hendrickson (after Bierhuizen).

evapotranspiration rate and the soil moisture tension depends upon a number of factors, such as soil texture, moisture tension characteristics, hydraulic conductivity of the soil, rooting depth, crop density, and atmospheric conditions.

Probably the most important factor is the evaporative power of the atmosphere. Bierhuizen (1958) measured the transpiration rate of kidney beans versus the soil moisture percentage at various light intensities, at 20° C. and 40 per cent relative humidity (Figure 74). He observed:

> At the highest light intensity . . . the transpiration rate was 0.3 gms/100 sq. cm hr near the wilting point, then it increases gradually with the increase of the available moisture content until it attains a maximum of 1.2 gms/100 sq. cm hr at a moisture content of 16%. At the moisture content above 16% the transpiration rate exhibits a slight fall. The decrease in transpiration at low moisture tension may be due to lack of oxygen, which strongly affects water absorption.
>
> . . . a decrease of the actual evapotranspiration . . . with increasing moisture tensions only occurs at high evaporation conditions. At low evaporation conditions no such decrease was observed. The lack of the relation between transpiration rate and availability of soil moisture in Veihmeyer and Hendrickson's results may be due to the fact that they carried their experiments under a limiting potential evapotranspiration.

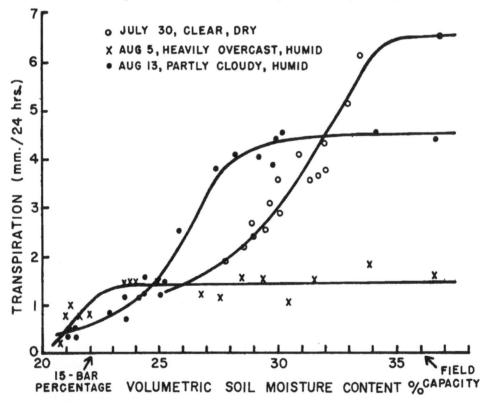

FIGURE 75. Actual transpiration rate as function of soil moisture content (after Denmead and Shaw).

Denmead and Shaw (1962), working with corn in Iowa, reported that on a clear day, when the potential transpiration rate was as high as six to seven millimeters per day, the decline in the transpiration rate occurred at a very low moisture tension, close to the field capacity (Figure 75). On a heavily overcast day, when the potential transpiration rate was only 1.4 millimeters per day, the transpiration rate did not decline until the soil moisture reached twelve bars, not much above the permanent wilting point.

Other studies by Scholte-Ubing (1961b), Holmes and Robertson (1963), and Makkink and van Heemst (1956) also confirmed that under low evaporation conditions, the evapotranspiration rate drops below the potential at a higher soil moisture tension than under high evaporation rate conditions. The study by Scholte-Ubing, as shown in Figure 76, is of particular interest because he measured net radiation as well.

Under the same radiation and temperature conditions, the evaporative powers of the air will increase with the decrease of humidity. Therefore, the transpiration rate will begin to fall at a lower moisture tension as the relative humidity decreases. Closs (1958) demonstrated this by

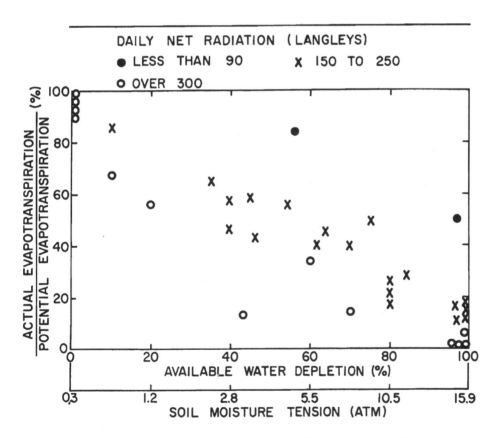

FIGURE 76. Relative daily actual evapotranspiration rate from short grass as function of moisture depletion from the root zone and soil moisture tension averaged over the total root depth (after Scholte-Ubing).

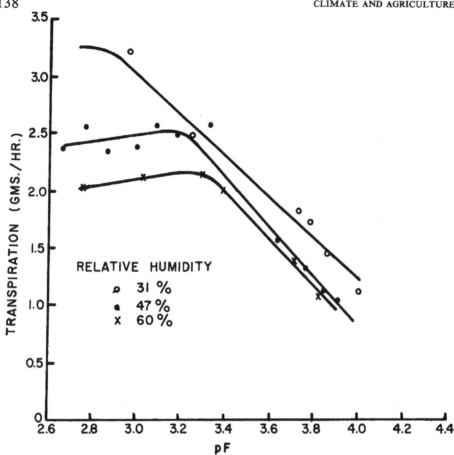

FIGURE 77. Effect of soil moisture tension on transpiration of mustard plants (after Closs).

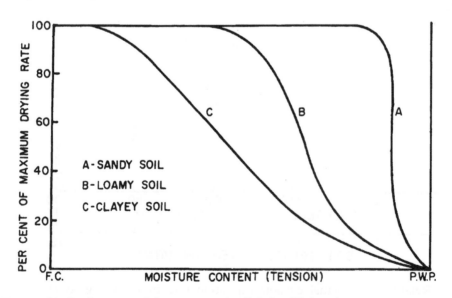

FIGURE 78. Drying rate of three types of soil (after Holmes).

measuring the transpiration rates of mustard plants at an air temperature of 27° C., and relative humidities of 60, 47, and 31 per cent (Figure 77).

The moisture depletion curve may vary with the soil texture and rooting depth because they affect the rapidity of water movement. Holmes (1961) has presented idealized curves for three soils (Figure 78). Curve A is typical of the sandy soils that are permeated with roots. The low moisture-holding capacity and the low colloid content permit a rapid removal of much of the soil moisture. The depletion rate remains at a high level almost to the wilting point. On the other hand, Curve C represents the depletion rate of a heavy clay soil, where the moisture cannot be removed rapidly. The transport mechanism usually breaks down early in the drying cycle.

Other things being equal, the shape of the depletion curve changes during the course of a growing season on a particular soil as the roots of the plants ramify. This is shown schematically in Figure 79 by Holmes and Robertson (1959). An increase in the density of plant roots has the effect of shortening the path that the water must travel through the soil. Thus, with a dense root system, evapotranspiration can be maintained at the maximum rate for a longer period. The increase of ground cover has the same effect as the ramification of roots. In general, actual evapotranspiration equals the potential over an increasingly larger part of the available range of moisture, as the leaf area increases.

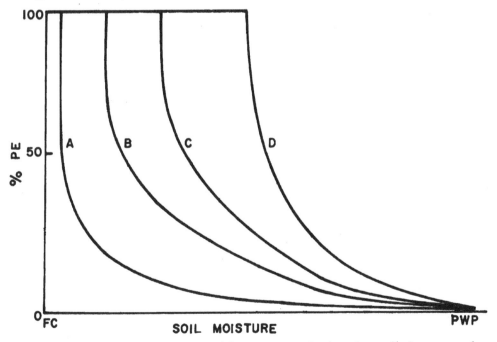

FIGURE 79. Adjustment of potential evapotranspiration for soil dryness and rooting depth of crops. Curves A to D correspond to increases in rooting depth of crop (after Holmes and Robertson).

The relationship between the soil moisture tension and actual evapotranspiration is of utmost importance in an irrigation operation. Other factors being constant, the irrigation interval can be lengthened if the actual evapotranspiration drops at a higher soil moisture tension.

EVAPORATION FROM BARE SOIL

On the basis of laboratory experiments, Penman (1941) stated that, under summer conditions in England, evaporation from a bare soil took place at the same rate as evaporation from a free water surface until a certain critical deficit was reached, approximately at 0.35 inch. Later observations by Fritschen and van Bavel (1962) in Arizona indicate that, if the ground is kept wet all the time, evaporation from bare ground may even exceed free water evaporation. In reality, soil cannot be kept wet all the time by applying irrigation water continuously. With high radiation conditions, a surface mulch of dry soil may develop within a few hours after the soil has been watered. The dry soil mulch acts as a capillary break and, hence, greatly reduces the evaporation rate. Under cloudy conditions with low initial rates of evaporation, with continuous liquid flow, the rate of loss may approximate that of a free water surface for a much longer period of time and may possibly lead to much greater total loss.

Soil structure and particle size also influence the rate of evaporation. The measurements by Burov (1952) indicate that evaporation is at a maximum from soils whose particles range from 0.5 to 3 millimeters. In finer soils, the capillary movement of water is greater. In coarse soils, air has freer access to the evaporating surface. The presence of salt in the soil water depresses the initial rate of evaporation, but this decreased rate is maintained for a longer period (Penman, 1941).

ADVECTION

Advection has been defined by Philip (1957) as the exchange of energy, moisture, or momentum as a result of horizontal heterogeneity. If the area upwind of an irrigated field is hot and dry, then sensible heat will be transferred to the irrigated field and its evapotranspiration rate will be increased. On the other hand, if the advected air is colder than the vegetation, then the evapotranspiration rate will be relatively low. Advection is a serious problem in arid and semiarid climates.

Advected energy resolves itself into the "clothesline effect" and the "oasis effect." When warm air blows through a small plot with little or no guard area, a very severe horizontal heat transfer occurs, which Tanner (1957) calls the clothesline effect. Inside a large field, the vertical energy transfer from the air above to the crop is called the oasis effect by Lemon, Glaser, and Satterwhite (1957). The clothesline effect represents either the experimental bias because of the small size of the field or the border conditions unrepresentative of the large field as a whole. The clothesline effect cannot be tolerated in agronomic or climatological investigations. On the other hand, the oasis effect must be reckoned with as a climatic characteristic since it affects the evapotranspiration rates many miles into an irrigated field. In some instances, the oasis effect reflects the large-scale upper air subsidence.

The clothesline and the oasis effects blend into each other in such a way that their transition takes place gradually in a broad zone. The clothesline effect diminishes downwind from a dry-wet border in an exponential manner. Theoretically, the clothesline effect may be considered negligible where the transformation of wind, temperature, and humidity profiles from those of the upwind field to the vegetated surface is complete. The distance required to develop the representative profiles of the new surface varies with the temperature and humidity differences between the two fields, and the contrasting roughness characteristics. Raney (1959) estimated that in humid regions, the effect extends to a distance equivalent to approximately 40 times the crop height. In arid regions, the distance is much greater. Halstead and Covey (1957) cited an extreme example. They assume that a completely moist area is surrounded by a completely dry area with a net radiation of 0.1020 langley per second, and a wind gradient of $u_{2z} - u_z = 100$ centimeters per second, and a roughness of $z_0 = 5$ centimeters and that the surrounding area has a temperature and vapor pressure, at 160 centimeters height, of 30° C. and 8 millibars. The temperature at 160 centimeters and the rate of evaporation would then vary with the size of the plot according to their calculations as follows:

Size	T (160 centimeters)	Evaporation (centimeters per hour)
6 foot tanks	30.0° C.	0.45
50 foot plot	28.6° C.	0.26
300 foot field	23.8° C.	0.19
1 mile field	20.0° C.	0.13

In this example, the clothesline effect extends more than 300 feet into the field. Thornthwaite and Mather (1955a) have since offered a rule of thumb. They consider an upwind guard ring of 50 meters as necessary to minimize the clothesline effect in a humid climate. In desert areas, even a distance of 400 meters would not be too large.

Plants on the upwind edge of a field, subject to a strong clothesline effect, require more water for optimum growth than those in the interior of a field. Millar (1964) measured the evaporation rate and relative turgidity of ladino clover grown in lysimeters at four distances downwind from the leading edge of a small irrigated field adjoining a drier area in Australia. Both evaporation and turgidity decreased with the increasing distance downwind from the lead edge and were directly proportional to the powers of the downwind dimension. The observed downwind variation of the mean evaporation over the period from 1020 hours to 1755 hours, March 21, 1961, is given in Figure 80. At the edge of the field, the clothesline effect caused a 10 per cent increase in the evaporation rate. Plants showed an early afternoon wilt during a mild autumn day although the soil moisture content was very high. Plant growth near the upwind edge was somewhat restricted. Millar concluded that, where advection is important, plant growth can be improved by having larger irrigated fields to minimize the clothesline effect.

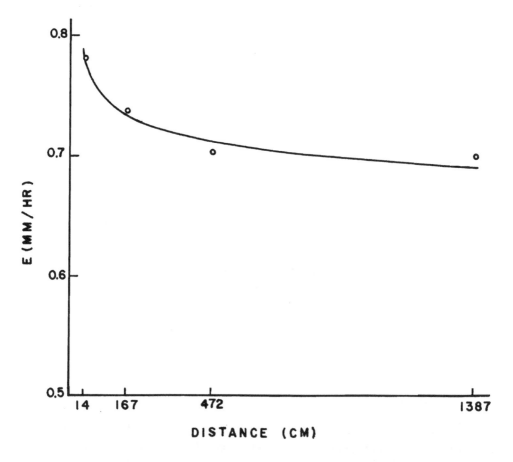

FIGURE 80. Variation of mean evaporation rate (E) with downwind distance over the period 1020-1755 hours, March 21, 1961 (after Millar).

In contrast to the clothesline effect, the oasis effect is often measurable many miles into an irrigated field in an arid climate. Lemon, Glaser, and Satterwhite (1957) reported a case in Texas where the oasis effect extended at least ten miles into the irrigated cotton field. The evapotranspiration rate was 1.65 times that attributable to the net radiation for a 24-hour period. Gal'tsov (1953) also observed appreciable oasis effect in the center of the large irrigated region in Kazakhstan, U.S.S.R. In humid climates, the oasis effect is weaker but by no means absent. Graham and King (1961a), working in Ontario, Canada, found that when the local surroundings of their irrigated corn were dry, the ratio between evapotranspiration and net radiation was 20 per cent higher than when the surroundings were moist. However, in a humid climate, the oasis effect rarely accounts for more than 30 per cent of the actual daily evapotranspiration according to my experience in Hawaii.

There is no simple way to evaluate the advected energy. The comparison between the lysimeter or evaporimeter readings and the net radiation gives only a rough estimate. Any attempt to quantitatively assess the advected energy requires a combination of both the turbulent transfer and energy budget approach. Such a study has been undertaken by de Vries (1959), who developed a theoretical model based on the assumption of homogeneous surfaces and constant diffusivities. However, his approach is too complicated for general use.

THE EFFECT OF PLANT HEIGHT ON THE RATE OF EVAPOTRANSPIRATION

In the definition of potential evapotranspiration, the vegetation is specified as "a short green crop of uniform height," which implies that the water need of a crop varies with its height. Isolated tall vegetation will intercept direct-beam solar radiation by the following proportion:

$$Q_T/Q = 1 + (h/w) \tan z$$

where Q_T is the radiation intercepted by an isolated tall plant; Q is the radiation received by a horizontal surface; h is the height of the vegetation above the surrounding; w is the width of the vegetation in the direction of the solar azimuth, and z is the solar zenith angle.

A tall crop covering a large area (e.g., 10 acres) will not intercept much more incident radiation, but may have a slightly higher net radiation and a higher roughness than a field of short crop. Because of the higher net radiation, the potential evapotranspiration of a tall crop in a humid climate may exceed that of a short crop, but the difference is very small. Because of the increased roughness, the potential maximum evapotranspiration of a tall crop could far exceed that of a short crop in areas of strong advection. Thus, in the humid Congo, Bernard (1954) observed that the rates of potential evapotranspiration are identical for both tall and short grass. In Hawaii, where a moderate amount of advection occurs, the potential maximum evapotranspiration of sugar cane is approximately 10 per cent higher than that of short grass. In the arid Sudan under advective conditions, El Nadi and Hudson (1965) reported that the evapotranspiration of tall vegetation is much accentuated.

Measurements of evapotranspiration rates were made under advective conditions in the Sudan from two lysimeters with similar exposures, but with a crop of lucerne clipped short in one case and unclipped in the other. As the difference in the height of the crops in the two lysimeters increased to 42 centimeters, the evapotranspiration rate of the tall crop was more than twice that of the short crop (Hudson, 1965) (Figure 81). Large differences in evapotranspiration rates between tall and short crops have also been noted for ryegrass and white clover in New Zealand. (Mitchell and Kerr, 1966).

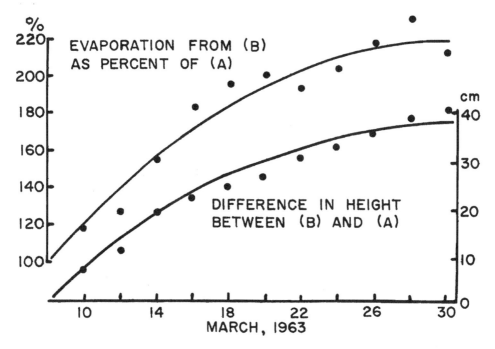

FIGURE 81. Evapotranspiration from lysimeters, with similar exposures, but where crop of lucerne was kept clipped short in one (A) but allowed to grow in height in other (B). Evaporation rates from (B) increased to more than double those from (A) as crop in (B) grew to 42 cm. higher than in (A). Both lysimeters were surrounded by areas of lucerne of similar heights to crops growing in them (after Hudson).

Generally, the boundary between the grassland and forest marks the transition from the humid to the arid climate. Various reasons have been advanced for the absence of trees in arid climates. One important reason that has yet to be mentioned is that clusters of trees in an arid climate are an effective sink for advected energy. As the aridity of the climate increases, the differences in water needs between a tall and a short plant are accentuated to such a degree that they render the survival of tall vegetation in arid climates most difficult.

CHAPTER 14

LYSIMETERS

DETERMINING POTENTIAL EVAPOTRANSPIRATION

The methods of determining or estimating potential evapotranspiration fall into five categories: (1) direct measurements by lysimeters, (2) empirical formulae using one or more common climatic factors, (3) the aerodynamic approach, (4) the energy budget approach, and (5) the use of evaporimeters. Both the empirical formulae and the simple energy budget approach do not make sufficient allowance for the influence of advection; hence, they must be modified when used to estimate the potential maximum evapotranspiration in arid climates.

Some of the methods are primarily research tools for the better understanding of the physical process of vapor transfer; others may be called "operational" in the sense that they can be used in the field as a guide to daily operation. The operational method uses either an inexpensive field instrument or readily available mean climatic data. Since the use of mean monthly or daily values depends upon a somewhat crude correlation between the instantaneous and mean values, the empirical formulae cannot be as accurate as the elaborate "research" methods. Tanner (1960b) maintains that for a day-to-day operation, agriculturists can use to advantage any method that is accurate to within 10 per cent except for very specific purposes.

INSTALLATION OF THE LYSIMETER

The term lysimeter was derived from the Greek word *lysis,* meaning loosening. Lysimeters are tanks buried in the ground to measure the percolation of water through the soils. Lysimeters are the most dependable means of directly measuring the evapotranspiration rate, but their installation must meet four requirements for the data to be representative of field conditions.

First, the lysimeter itself should be fairly large and deep to reduce the boundary effect and to avoid restricting root development. For short crops, the lysimeter should be at least one cubic meter in volume (McIlroy, 1957). For tall crops, especially row crops, the size of the lysimeter should be much larger. For example, I feel that lysimeters of four cubic meters may not be large enough for the tall sugar cane crop.

Secondly, the physical conditions within the lysimeter must be comparable to those outside. The soil should not be loosened to such a degree that the root ramification and water movement within the lysimeter are greatly facilitated. If the lysimeters are enclosed at the bottom, precautions must be taken to avoid the persistence of a water table and the presence of an abnormal thermal regime. To ensure proper drainage, the bottom of an isolated soil column will often require the artificial application of a moisture suction, equivalent to that present at the same depth in the natural soil (Coleman, 1946). In some cases, the soil temperature in the lysimeter is raised to such an extent that the air conditioning of the whole system becomes a necessity.

Thirdly, a lysimeter will not be representative of the surrounding area if the crop in the lysimeter is either taller, shorter, denser, or thinner, or if the lysimeter is on the periphery of non-cropped area. The effective area of a lysimeter is defined as the ratio of the lysimeter evapotranspiration per unit area to the average evapotranspiration per unit area of the surrounding field. The values of this ratio, other than unity, are caused by the inhomogeneity of the surface. The maintenance of uniform crop height and density is not an easy task in a tall crop, speced in rows. If the surface is indeed inhomogeneous, there is no adequate way to estimate the effective area from tank area overlap corrections or plant counts. For instance, Suomi (1953) used a lysimeter five feet in diameter, with a tank area of 19.6 square feet. Corn was transplanted in rows by hand with uniform row spacings and plant densities comparable to the surrounding areas. The effective lysimeter area, as determined by plant spacing, was 23.8 square feet. The lysimeter area, determined by heat budget calibration, was 16.6 square feet.

Fourthly, each lysimeter should have a "guard-ring" area around it maintained under the same crop and moisture conditions in order to minimize the clothesline effect. In an arid climate, Thornthwaite (1954) suggested that a guard-ring area of ten acres may not be large enough. Where several lysimeters are installed in the same field, Deacon, Priestley, and Swinbank (1958) indicated that the guard-ring radius may have to be about ten times the lysimeter separation.

Drainage lysimeter. Drainage lysimeters operate on the principle that evapotranspiration is equal to the amount of rainfall and irrigation water added to the system, minus leaching. Since percolation is a slow process, the drainage lysimeter is accurate only for a long period for which the water content at the beginning exactly equals that at the end. The length of such a period varies with the rainfall regime, frequency and amount of irrigation water application, depth of the lysimeter, water movement, and the like. Therefore, records of drainage lysimeters should be presented only in terms of long-period averages probably on a monthly basis.

The drainage lysimeter is useful only in determining the potential evapotranspiration rate and should be irrigated every four or five days, unless rainfall intervenes.

In their exhaustive survey of lysimetry, Kohnke, Dreibelbis, and Davidson (1940) counted a total of nearly 150 lysimeters throughout the world prior to 1940, the majority of which were the drainage type. This mass of data rarely has been used by agriculturists or climatologists for regional water resource planning, partly because the information is widely scattered, and partly because it lacks uniformity. The border effect, caused by a difference in the exposure and cultural treatment inside and outside the lysimeter is often a serious problem. Nevertheless, much valuable information may be salvaged if a determined effort is made to assemble all the records and adjust them to a comparable basis.

Weighing lysimeter. The first soil-block weighing lysimeter was built in 1937 by the U.S. Soil Conservation Service at Coshocton, Ohio (Harrold and Dreibelbis, 1951). The construction cost of weighing lysimeters was greatly reduced after the introduction of hydraulic methods of recording weights by Russian research workers (Fedrov, 1954), and King, Tanner, and Suomi (1956) in Wisconsin.

In addition to the Coshocton installation, there are at least six very elaborate weighing lysimeters: at Davis, California (Pruitt and Angus, 1960); Tempe, Arizona (van Bavel and Meyers, 1962); Valdai, U.S.S.R. (Popov, 1952); Wageningen, the Netherlands (Makkink, 1953); Aspendale, Australia (McIlroy and Angus, 1963), and Rothamsted, England (Morris, 1959). The lysimeter at Davis, California, has a twenty-foot diameter and a three-foot depth, and is installed near the center of a 480-foot by 1,165-foot field of grass. In spite of an average weight of over 90,000 pounds, the 50-ton lever-type scale is sensitive to within approximately two pounds, which is equivalent to about 0.0012 inch (0.0308 mm) of evaporation. The weight of the system is recorded every four minutes.

The installation at Tempe, Arizona, is equally impressive. It weighs 3,000 kilograms mass and has an accuracy of ten grams or a water-depth equivalent of 0.01 millimeter.

The weighing lysimeter is capable of measuring evapotranspiration for a period of as short as ten minutes. Thus, it can provide much more additional information than the drainage lysimeter does. Problems such as the diurnal pattern of evapotranspiration, the phenomenon of midday wilt, the short-term variation of energy partition, and the relationship between transpiration and soil moisture tension, can be answered only by studying the records obtained from the weighing lysimeter.

Lysimeters, especially the weighing type, are the most direct and accurate instrument for the determination of evapotranspiration. The high cost of installation and their immobility have, however, precluded their use as a routine field instrument. Lysimeters are used primarily as a research tool to check the accuracy of other methods of estimating evapotranspiration. The large weighing lysimeters at Davis, Tempe, and Coshocton should certainly be used as "reference" stations in any study of water balance in the United States.

CHAPTER 15

EMPIRICAL FORMULAE

THORNTHWAITE'S METHOD

Thornthwaite (1948) presented a formula for estimating potential evapotranspiration based on lysimeter and watershed observations of water loss in the central and eastern United States. His formula reads:

$$E = 1.6 \, (10T/I)^a$$

where E is the unadjusted (30 days — each a twelve-hour day) potential evapotranspiration in centimeters, T is the mean monthly temperature in degree C., a is a constant that varies from place to place, and I is the annual heat index.

The annual heat index, or I, is the sum of twelve monthly heat indices i, and:

$$i = (T/5)^{1.514}$$

To evaluate a, the following equation is used:

$$a = 0.000000675 \, I^3 - 0.0000771 \, I^2 + 0.01792 \, I + 0.49239$$

The unadjusted potential evapotranspiration is corrected by actual daylength of hours and days in a month to give the adjusted potential evapotranspiration.

Thus, Thornthwaite expressed the potential evapotranspiration as an exponential function of the mean monthly air temperature, and he applied a daylength adjustment to correct the relationship for season and latitude. He justified the omission of other meteorological elements by stressing the fact that they vary together with air temperature.

The Thornthwaite formula works well in the temperate, continental climate of North America, where the formula was derived, and where temperature and radiation are strongly correlated. This has been substantiated by Mather (1954) in New Jersey, Decker (1962) in Missouri, Baker (1958) in Minnesota, Burman and Partridge (1962) in Wyoming, and Sanderson (1950) in Canada. Good results have also been reported by Fitzgerald and Rickard (1960) in New Zealand.

However, in other parts of the world, the Thornthwaite approach has been less successful. The weaknesses of the method can be enumerated as follows:

The first, and the most serious, is that temperature is not a good indicator of the energy available for evapotranspiration. England (1963) observed at Waynesville, North Carolina, that the rate of water use of alfalfa in a weighing lysimeter was very high on June 24, 1961. The rate of water use then was reduced considerably the next day, although the air temperature remained the same. He concluded that radiation was by far the most important factor. At De Bilt in the Netherlands, the temperature is about the same during November (5.4° C.) and March (5.0° C.). The average solar radiation intensity in these two months is, however, 67 and 195 langleys per day, respectively, and the potential evapotranspiration calculated from the energy balance is nearly four times as high in March as it is in November (van Wijk and de Vries, 1954).

Secondly, because air temperature lags behind radiation, the Thornthwaite estimates also show a time lag in relation to the measured values of potential evapotranspiration. Van Bevel (1956) has noticed a lag of three to four weeks, while Lougee (1956) has cited a more specific example in Norway:

> In inner Sogn, the drought is worst in April, May and June, so irrigation is a regular part of spring's work. According to Thornthwaite's calculation, the greatest water deficit occurs late in the season.

Thirdly, according to the Thornthwaite formula, evapotranspiration will cease when the mean temperature is below 0° C. This is by no means true (Crowe, 1954), although the amount of evaporation may be small. Furthermore, the maximum temperature of various days during a cold winter month may rise well above freezing. Not surprisingly, therefore, the formula underestimates winter evapotranspiration in middle and high latitudes (Butler and Prescott, 1955), except in areas of strong advection of warm air. Mather (1965) has suggested the use of the maximum temperature rather than the mean.

Fourthly, in some areas, wind might be an important factor. Makkink (1955) has noticed that in the Netherlands, the Thornthwaite estimates agree with the observed potential evapotranspiration only after applying a correction factor for the wind velocity.

Fifthly, the formula does not take into consideration the effect of warm and cool air advection on the temperature. In regions of warm air advection, the air temperature may be markedly increased while the solar radiation and evapotranspiration are not. This has been observed in Ireland, particularly during the winter months (Guerrini, 1954). On the other hand, in areas of cold air advection, the estimated value may fall considerably below the observed value. At Davis, California, Pruitt (1960b) reported that the potential evapotranspiration of ladino clover in the weighing lysimeter averaged 1.72 times the Thornthwaite value.

In spite of its shortcomings, the Thornthwaite method has gained worldwide popularity, partly because it requires only temperature records, and partly because it is the foundation of a climatic classification that lends itself to the presentation of broad climatic patterns of the world. In an attempt to characterize the moisture regimes throughout the world, Thornthwaite assumed without the support of evidence that a surplus of six centimeters of moisture in one season may compensate for a deficit of ten centimeters in another. By adopting such an arbitrary adjustment, he was able to show that his classification displayed a general sympathy with the distributional patterns of soil and vegetation (Chang, 1959). However, this is no guarantee that the method is applicable in solving practical agricultural problems. The distribution of soil and vegetation is affected, to a large extent, by the long-term seasonal moisture regime of a place; whereas the agricultural operation demands more precise short-term estimates.

Numerous studies have shown that the Thornthwaite estimates differ significantly from short-term observed values of evapotranspiration. Garnier (1956) found that the measured potential evapotranspiration during the hot dry season in northern Nigeria amounted to almost twice the calculated Thornthwaite value. Even greater discrepancies were reported for Sibi (Baluchistan) and In-Salah (Sahara) (Sibbons, 1962). If sugar cane irrigation in Hawaii followed the Thornthwaite method, the yield would be less than two-thirds of the potential in many dry areas.

According to van der Bijl (1957), any formula solely dependent on air temperature to estimate short-term potential evapotranspiration is only accurate to within 10 to 20 per cent. The accuracy increases gradually with the lengthening of the period (King, 1956). Therefore, unless the Thornthwaite formula is verified by lysimeters or other more accurate methods, it should not be adopted as a guide to agricultural planning or operation.

THE BLANEY-CRIDDLE FORMULA

Blaney and Morin (1942) derived an empirical formula to relate evaporation to temperature, relative humidity, and length of daytime hours, based on measurements in New Mexico and Texas. The relation was of the form:

$$U = KTp\,(114 - h)$$

where U is the monthly consumptive use in inches, K is a crop coefficient, T is the mean monthly air temperature in °F., p is the monthly percentage of daytime hours in the year, and h is the mean monthly relative humidity.

The equation was later simplified by Blaney and Criddle (1950) by dropping the humidity term. Thus, it takes the form:

$$U = KTp$$

The value of p for latitude 24° to 50° N. are given in Table 7, and the values of crop coefficients, K, are given in Table 8. The crop coefficients were developed from actual measurements of consumptive use in tanks and soil moisture field studies.

TABLE 7

Monthly Percentages of Daytime Hours of the Year for Latitude 24° to 50° North of the Equator

Month	\multicolumn{14}{c}{Latitudes in Degrees North of Equator}
	24	26	28	30	32	34	36	38	40	42	44	46	48	50
January	7.58	7.49	7.40	7.30	7.20	7.10	6.99	6.87	6.76	6.62	6.49	6.33	6.17	5.98
February	7.17	7.12	7.07	7.03	6.97	6.91	6.86	6.79	6.73	6.65	6.58	6.50	6.42	6.32
March	8.40	8.40	8.39	8.38	8.37	8.36	8.35	8.34	8.33	8.31	8.30	8.29	8.27	8.25
April	8.60	8.64	8.68	8.72	8.75	8.80	8.85	8.90	8.95	9.00	9.05	9.12	9.18	9.25
May	9.30	9.38	9.46	9.53	9.63	9.72	9.81	9.92	10.02	10.14	10.26	10.39	10.53	10.69
June	9.20	9.30	9.38	9.49	9.60	9.70	9.83	9.95	10.08	10.21	10.38	10.54	10.71	10.93
July	9.41	9.49	9.58	9.67	9.77	9.88	9.99	10.10	10.22	10.35	10.49	10.64	10.80	10.99
August	9.05	9.10	9.16	9.22	9.28	9.33	9.40	9.47	9.54	9.62	9.70	9.79	9.89	10.00
September	8.31	8.31	8.32	8.34	8.34	8.36	8.36	8.38	8.38	8.40	8.41	8.42	8.44	8.44
October	8.09	8.06	8.02	7.99	7.93	7.90	7.85	7.80	7.75	7.70	7.63	7.58	7.51	7.43
November	7.43	7.36	7.27	7.19	7.11	7.02	6.92	6.82	6.72	6.62	6.49	6.36	6.22	6.07
December	7.46	7.35	7.27	7.14	7.05	6.92	6.79	6.66	6.52	6.38	6.22	6.04	5.86	5.65
Total	100.00	100.00	100.00	100.00	100.00	100.00	100.00	100.00	100.00	100.00	100.00	100.00	100.00	100.00

TABLE 8

CONSUMPTIVE USE COEFFICIENTS FOR THE MORE IMPORTANT IRRIGATED CROPS AND NATURAL VEGETATION IN WESTERN UNITED STATES

Irrigated land	Length of growing season	Consumptive use coefficient (K)
Alfalfa	frost-free	0.85
Beans	3 months	0.65
Corn	4 months	0.75
Cotton	7 months	0.62
Citrus orchard	7 months	0.55
Deciduous orchard	frost-free	0.65
Pasture, grass hay annuals	frost-free	0.75
Potatoes	3 months	0.70
Rice	3 to 4 months	1.00
Small grain	3 months	0.75
Sorghum	5 months	0.70
Sugar beets	5.5 months	0.70
Natural vegetation		
Very dense (large cottonwoods, willows)	frost-free	1.30
Dense (tamarisk, willows)	frost-free	1.20
Medium (small willows) tamarisk)	frost-free	1.00
Light (saltgrass, sacaton)	frost-free	0.80

K=1.00 for lake evaporation in arid areas and 0.90 for lake evaporation in coastal areas.

The Blaney and Criddle formula suffers from drawbacks similar to those of the Thornthwaite method, although it is the most widely used procedure for indirectly estimating evapotranspiration in the semiarid lands of the western United States. Tomlinson (1953), for instance, reported that the method checked very closely with the water needs of native hay in the Penedale area of Wyoming. Outside the semiarid climate of the western United States, the performance of the Blaney-Criddle method is usually very poor (Pruitt and Jensen, 1955; Stanhill, 1961; Jackson, 1960; and Dagg, 1965.)

One unique feature of the Blaney-Criddle method is the inclusion of a crop coefficient. They recognize that in an arid climate, the consumptive use of a crop may vary greatly with the condition of its physical stand. In general, the value of K increases with the height of the vegetation and the completeness of the ground cover. This agrees well with the theoretical consideration of surface roughness.

MAKKINK'S FORMULA

The Makkink formula (1957) differs from the two previous ones in that it is based on solar radiation measurements weighted according to air temperature. Makkink reasoned that the higher the temperature, the greater the proportion of solar energy used in evapotranspiration. His formula was derived in the Netherlands using the lysimeter measurements of potential evapotranspiration from short grass. The formula reads:

$$E = 0.61 \, Q \, \frac{\Delta}{\Delta + \gamma} - 0.12$$

where E is potential evapotranspiration, Q is incoming radiation expressed in millimeters per day (converted into the amount of water evaporated), Δ is the slope of the saturated vapor pressure-temperature curve at the mean air temperature, and γ is the psychrometric constant, 0.49 for degrees C. and millimeters of mercury, or 0.27 for degrees F. and millimeters of mercury.

The Makkink formula was derived in 1957 and has not been widely tested. Stanhill (1961) reported that, in comparison with the lysimeter readings at Gilat, Israel, it underestimated the potential evapotranspiration by a factor of 1.49, but showed high correlation. Again, the point is made: an empirical formula, derived in a humid climate, will underestimate the potential maximum evapotranspiration in an arid climate, because an insufficient allowance is made for advected energy.

TURC'S FORMULA

Turc (1954) derived an empirical formula expressing the evaporative power of the air as a function of mean temperature and solar radiation or hours of sunshine. Then he expressed the actual evaporation as a function of available moisture, including precipitation and irrigation water, and the evaporative power of the air. His formulae were based on the river-basin water balance and lysimetric data collected over a number of years at Rothamstead and Versailles. His expression for the evaporative power of the air is:

$$l = 0.0437 \, (T+2) \, \sqrt{Q}$$

where l is the evaporative power of the air, T is the mean air temperature in ° C., and Q is the incoming solar radiation in langleys per day.

Where actual radiation records are not available, it is then estimated by the formula given in Chapter 2:

$$Q = Q_A \, (0.18 + 0.55 \, n/N)$$

The evaporation from a calculated soil, E in millimeters, is then given by the following equation:

$$E = \frac{P + a + V}{\sqrt{1 + \left(\frac{P+a}{1} + \frac{V}{21}\right)^2}}$$

where P is the precipitation in millimeters, a is the soil moisture available for evaporation from bare soil, and V is additional soil moisture available for evaporation (through the vegetation) from cultivated soil.

In the absence of vegetation cover, $V = 0$, and the evaporation from bare soil is given by:

$$E = \frac{P + a}{\sqrt{1 + \left(\frac{P+a}{1}\right)^2}}$$

In the above equation, a and V represent the soil and the plant factors. The value of a, for a seven-day period, is determined from:

$$a(\text{mm/week}) = 35 - \Delta$$

where Δ is the soil moisture deficit from a reference value ($\Delta = 0$ for the soil at field capacity). The expression $a = 35 - \Delta$ indicates that the bare soil may not dry out until $\Delta = 35$ millimeters is reached.

The plant factor V may vary according to plant species and the state of the plant development. For a seven-day period, V (in millimeters), is given by:

$$V = 25 \sqrt{\frac{MK}{Z}}$$

where M is the total dry matter production in 100 kilograms per hectare, Z is the length of the growing season in days, and K is a crop factor. Values of K given by Turc are corn and beets, 0.67; potatoes, 0.83; cereal, flax, and carrots, 1.00; beans, clover, and other legumes, 1.17, and lucerne, meadow grass, and mustard, 1.33.

Turc's method is so complicated that it has rarely been used. However, the one study by Ahmad (1962) in using a modified version of Turc's method did give good estimates of the water balance of bare as well as cultivated soil (wheat crop) in the semiarid Quetta Valley of West

Pakistan. He concludes that the method is valuable in irrigation scheduling and water resources planning.

OTHER EMPIRICAL FORMULAE

Other less well known empirical formulae include the ones by Lowry and Johnson (1942), Baver (1937), and Haude (1952). The first two formulae use only temperature; whereas the latter expresses evapotranspiration as a function of the saturation deficit at 2:00 P.M.

All the empirical formulae are derived from a correlation between the measured evapotranspiration and one or more meteorological elements. Since such a relationship varies from region to region, the empirical formulae cannot be expected to have general validity. The formulae by Thornthwaite, Blaney and Criddle, Baver, and Lowry and Johnson—all based on temperature records from the United States—are different. Therefore, it would be far more rewarding to pursue an approach that has a firm physical foundation than to try to improve the empirical formulae by applying correction factors appropriate to a specific locality.

CHAPTER 16

THE AERODYNAMIC APPROACH

EVAPORATION AS A PROCESS OF DIFFUSION

In the very thin layer next to the evaporating surface, diffusion is entirely molecular. The molecular diffusivity of water vapor is known to be approximately 0.25 square centimeter per second. If the air flow were laminar, the evaporation from a saturated surface would be determined entirely by the molecular diffusivity. However, the unevenness of the surface and differential heating lead to turbulent motion. Above the laminar sublayer, the turbulent diffusivity far exceeds the molecular diffusivity. At a height of one meter, the former may be as much as 5,000 times the latter. Unlike the molecular diffusivity, the turbulent diffusivity is highly variable in time and space. The various aerodynamic methods are designed to measure or to estimate the rate of water vapor diffusion, particularly that caused by turbulence.

DALTON EQUATION

Probably the oldest equation used to estimate evaporation from a water surface E_o is the Dalton equation, which is a simplified aerodynamic method.

$$E_o = (e_s - e) f(u)$$

where e_s is the vapor pressure at the evaporating surface, e is the vapor pressure at some height above the surface, and $f(u)$ is a function of the horizontal wind velocity.

This equation provides an estimate of free water surface evaporation, but it has not been widely used for estimating evapotranspiration. One of the reasons for this is the difficulty in measuring value of e_s for a plant surface. The relative humidity at the surface of freely transpiring leaves, commonly assumed to be 100 per cent, may indeed be much lower (Inoue, 1963). Moreover, since the wind gradient varies with the roughness of the surface, wind speed at one height cannot give an adequate measurement of turbulence. Because the roughness of the water surface does not vary so greatly, Rohwer (1931) has evaluated the constants in the Dalton equation to give:

$$E_o = 0.40 \, (e_s - e_a) \, (1 + 0.17 \, u_2) \text{ mm/day}$$

where u_2 is the wind speed in miles per hour at a height of two meters.

PRINCIPLE OF SIMILARITY

The aerodynamic methods require that the wind speeds be measured at two heights or that a roughness parameter be included with the wind speed records at one height. The profile measurements make it possible to use the principle of similarity in the derivation of the aerodynamic approach. Near the ground surface, the transfer of momentum, water vapor, heat, and carbon dioxide may be expressed as follows:

Momentum $\quad\quad\quad\quad \tau = - \rho \, K_m \dfrac{du}{dz}$

Water vapor $\quad\quad\quad E = - K_e \dfrac{de}{dz}$

Heat $\quad\quad\quad\quad\quad\quad A = - \rho \, C_p K_h \dfrac{dT}{dz}$

Carbon dioxide $\quad\quad\, Q = - K_c \dfrac{dc}{dz}$

where τ is the momentum flux density or the shearing stress or the Reynold stress; E, A, and Q are flux densities for water vapor, heat, and carbon dioxide, respectively; ρ is air density; C_p is the specific heat of air at constant pressure, K_m, K_e, K_h, and K_c are eddy diffusivities for momentum, water vapor, heat, and carbon dioxide, respectively; u, e, T, and c are wind speed, vapor pressure, temperature, and carbon dioxide concentration measured at height z, respectively.

If the eddy diffusivities are identical, then the different quantities of u e, T, and c would have similar profiles in the air layer near the ground. The principle of similarity may be used to find any of the fluxes, provided that we know one flux and have other appropriate gradient measurements. For example, if the wind profile data provide a measurement of τ, the water vapor flux may be calculated from additional measurements of the water vapor gradient made over the same height interval.

Considerable controversy has arisen over the equality of eddy diffusivities. The relationship between the eddy diffusivities has been found to depend markedly on atmospheric stability. In neutral stability conditions, the eddy diffusivities are approximately equal, because momentum, water vapor, heat, and carbon dioxide are carried by essentially the same eddies. Under unstable conditions, where thermal convection (buoyancy) adds strongly to frictional turbulence, heat is being transferred upward by a generally larger eddy than is water vapor. Consequently, under moderately unstable conditions, K_h may be twice as great as K_m or K_e at a height of two meters (Swinbank, 1958). This ratio tends to increase with the instability of the atmosphere as well as the height above the ground. On the other hand, under inversion conditions, K_h may be smaller than both K_m and K_e.

Rider (1954b) has taken 51 profile readings and obtained the following mean ratios:

$$K_h/K_e = 1.14 \pm 0.06; K_h/K_m = 1.48 \pm 0.27; K_e/K_m = 1.23 \pm 0.17$$

The data indicate that the ratio between K_h and K_e is the closest to unity. The value 1.14 is the ratio of the molecular diffusivities for air and water vapor. The eddy diffusivities for water vapor averaged 23 per cent greater than the eddy diffusivity for momentum. Thus, the principle of similarity, on which the aerodynamic approach is based, is at best, a crude approximation. The fact that the ratio between eddy diffusivities is not constant with height is particularly disturbing. Van Bavel and Fritschen (1962) have cautioned that the vapor flux at some height above the ground is simply not equal to the vapor flux at the surface.

THE THORNTHWAITE-HOLZMAN EQUATION

Thornthwaite and Holzman (1939) derived the first aerodynamic equation for evapotranspiration over short vegetation. They give this expression:

$$E = \frac{\rho\, K^2\, (q_1 - q_2)\, (u_2 - u_1)}{\left(\ln \frac{z_2}{z_1}\right)^2}$$

where E is evaporation, ρ is air density, k is von Karman's constant or 0.40, u_1, u_2, q_1, and q_2 are the wind speeds and specific humidities at the height z_1 and z_2, respectively.

This equation was later expanded by Pasquill (1950) for tall crops in the form:

$$E = \frac{\rho\, K^2\, (q_1 - q_2)\, (u_2 - u_1)}{\left(\ln \frac{z_2 - d}{z_1 - d}\right)^2}$$

where d is the zero plane displacement.

The Thornthwaite-Holzman equation is dependent for its validity on the propositions: (1) that the principle of similarity is valid, and (2) that the wind profile near the ground can be described by the logarithmic equation. Since these two propositions are valid only under stable conditions, the equation cannot be very accurate under unstable conditions.

Although Rider (1954a) and House, Rider, and Tugwell (1960) have obtained reasonably accurate results with the aerodynamic equation (the latter under neutral conditions), other investigators have found the approach less encouraging. Extensive tests of the aerodynamic methods over Lake Hefner, the surface of which is smoother than that of the vegetation, did not give satisfactory results (Harbeck, 1958). Recently, Pruitt (1963) also showed that results obtained by the Thornthwaite-Holzman equation departed appreciably from the records of the weighing lysimeters at Davis, except during the periods when the wind speeds exceeded three to four meters per second.

Even if the aerodynamic equation were accurate, it cannot be adopted as a field method. While both humidity and wind can be readily measured at one level, the measurements of their gradients not only require that twice as many measurements be made, but that these measurements have nearly ten times greater accuracy (Thornthwaite and Halstead, 1942). A small error in the measurements may cause a very large difference in the final computation. Furthermore, the method requires instantaneous readings. The summarization of the data is extremely laborious without the aid of a computer.

Deacon, Priestley, and Swinbank (1958) pointed out that even though the aerodynamic equation may work well for short crops under certain conditions, it is not likely to be practicable with tall crops such as wheat, sugar cane, etc. The Pasquill equation has one more additional weakness than the Thornthwaite-Holzman equation in that the value of the zero plane displacement varies in a complicated manner with wind speed as discussed in Chapter 11. Another difficult problem with tall crops is the placement of instruments. If the readings are taken too close to the surface, they may not be representative because the surface is often uneven and patchy. On the other hand, if the readings are taken at considerable height above the surface, then the difference between eddy diffusivities for water vapor and momentum might be too large.

EDDY CORRELATION TECHNIQUE

Realizing the limitation of the Thornthwaite-Holzman type of approach, Swinbank (1951) was the first to attempt a direct measurement by the so-called eddy correlation technique. The method is based on the assumption that the vertical eddy flux can be determined by simultaneous measurements of the upward eddy velocity and the fluctuation in vapor pressure. Let a bar over a quantity denote its mean value in time and a prime indicate the difference between an instantaneous value of a quantity and its mean value; then the eddy flux is given by:

$$E = \overline{(\rho w)' q'}$$

where ρ, w, and q are simultaneous values of the air density, the vertical wind speed, and the specific humidity respectively.

In the equation, $(\rho w)'$ represents an instantaneous fluctuation in the rate of upward air flow, and q' is an associated fluctuation in water vapor. What is needed for the determination of the covariance is an anemometer to measure ρw, and a hygrometer to measure q, and devices to multiply the two results and integrate the product with respect to time. The same principle can also be used to measure the flux of heat and momentum.

The technique is theoretically sound. The difficulty lies in the design of the instruments. Munn (1961) has listed three requirements:

(1) The separation between instruments measuring q' and w' must be smaller than the smallest eddies contributing to the water vapor flux.

(2) The sensing instruments must have the same response time; otherwise spurious covariance will result.

(3) The responding time of the sensing elements must be sufficiently short so that the flux contributions of very small eddies will be recorded.

The response time required for accurate measurement depends on the nature of the turbulence. In general, the requirement decreases with the height above the ground, and increases with the wind speed and the roughness of the surface. McIlroy (1957) considers that, at a height of a meter or more, a response time of about a fifth of a second appears to be adequate under most circumstances. This would require extremely elaborate recording or computing systems. The machine analyzer, developed by Taylor and Webb (1955), is most helpful but it is by no means perfect. Observations indicated that the limited speed of the response of the recording equipment caused the fluxes to be underestimated sometimes by as much as 30 per cent (Priestley, 1959b).

Subsequently, Taylor and Dyer (1958) and Dyer and Maher (1965) developed portable equipment employing transistor circuitry, which they called "Evapotron." The Evapotron was tested by Pruitt (1962a) against results of the weighing lysimeter at Davis during the summer of 1961. He remarked:

> At first glance, the results are discouraging. However, for the May 15 test there was excellent agreement between lysimeter and the Evapotron. At this time there was little contrast in surface moisture conditions between the irrigated ryegrass field and the surroundings. As upwind surface conditions

in the fallowed or non-irrigated fields became dry, significant differences became apparent. The Evapotron, due to the extreme fast time response needed to pick up all of the up and down eddy motions, must be placed considerably above a surface where the eddy motion is less subject to fast variation. A 4-meter height appears to be sufficient over grass surfaces. However, with upwind fetches of grass of 300-600 feet, apparently the instrument was sometimes in a transition zone between the upper air which was representative of the upwind dry-field condition and the lower air which had been affected by the moist grass field.

Doubtless, with further improvement of instrumentation, the eddy correlation technique will be the ideal method for directly measuring evapotranspiration. However, even with the perfection of the instrument, much work remains to be done in determining how the instrument should be exposed, and how dense a network is needed in different climatic and surface conditions. This, together with the Evapotron's high cost may prevent its adoption as a field instrument for some time to come.

CHAPTER 17

THE ENERGY BUDGET APPROACH

THE ENERGY BUDGET EQUATION

Evapotranspiration is a process of turbulent transfer as well as of energy transformation. The complete energy budget equation may be given as follows:

Q_n + horizontal divergence of sensible and latent heat =
$S + A + E$ + heat storage in the crop + photosynthesis

where Q_n is net radiation, S is heat flux to the soil, A is heat flux to the air, and E is evapotranspiration.

As we have discussed in Chapter 6, most crops use less than 1 per cent of the solar radiation in photosynthesis during their life cycle. Even during a short period of a few hours, photosynthesis seldom accounts for more than 3 per cent of the solar radiation.

The heat storage in the crop is negligible during the day, but may be relatively important during sunrise and sunset when the temperature change is rapid, and when the values of Q_n, E, and A are small. However, for a period of a day or longer, Suomi and Tanner (1958) have found that even for a heavy stand of corn three meters high (which yields seventeen tons of silage to the acre), the storage term is at the most 1 per cent of the total heat budget. Therefore, for all practical purposes, both the photosynthesis and the heat storage in the crop may be neglected in the energy budget equation.

The horizontal divergence term represents the net gain of advected energy by the crop stand. The divergence term may be quite large in an arid climate, or in a small irrigated field in a humid climate. Suomi (1953) has shown that when measurements were made one to three meters above a crop of corn 2.5 meters high, the divergence term accounted for 40 per cent of the evapotranspiration. In arid climates, the divergence term may equal the net radiation. In spite of its importance, the divergence term usually is omitted, largely because no simple way has been devised to evaluate it. The omission is a serious drawback for the ordinary energy budget approach. However, the divergence term can be minimized by taking readings near the crop surface and at an appreciable distance downwind in a uniform cover.

By ignoring the advected energy, photosynthesis, and the heat storage in the crop, the energy budget equation is reduced to a form that accounts for only the simple vertical energy exchange. This equation

$$Q_n = S + A + E$$

states that the net radiation is disposed of in three ways: heat flux to the soil, heat flux to the air, and evapotranspiration. Of these three factors, heat flux to the soil is the smallest, especially under a dense vegetation cover. Decker (1959), for instance, reported that in Missouri the heat flux in a moist soil under corn was 15 per cent when the cover was incomplete and decreased to 4 per cent when the cover was fully developed. In England, the maximum heat flux to the soil is 10 per cent over a 24-hour period on a clear summer day, and less than a few percentage points for a period of several days (Monteith, 1958). Tanner and Pelton (1950) have assessed the magnitude of soil heat flux in these words:

> The authors found that (in Wisconsin) during the period July through September the average of the daily absolute (S/Q_n) was 0.05 when a good alfalfa-brome cover was present. When poor cover existed following cutting, $(S/Q_n) = 0.09$. When the daily losses and gains of soil heat were considered in an arithmetic average, $(S/Q_n) = -0.03$. The soil heat flux is least during the three summer months when the soil temperature is near maximum and during January through March when the soil temperature is minimum. However, since Q_n is least during the winter, (S/Q_n) will be greater in winter than in summer. Soil temperature records indicate that, in the higher latitudes where the diurnal variation in radiation is great, (S/Q_n) may be three to four times greater in the spring (May and early June) and fall (October through early December) than in September. The authors found the daily (S/Q_n) flux in September (0.09) to be about twice that in August, the month of minimum (S/Q_n). Thus the soil heat flux may be neglected during the important portion of the growing season (mid-June through September) but should be considered for daily and monthly estimates during the rest of the year. The yearly storage term will be small and may be neglected for annual estimates. Also, the storage term becomes less important at the lower latitudes where the amplitude of the annual radiation cycle is small.

When heat flux to the soil is neglected, the energy budget equation is reduced to its simplest form:

The Energy Budget Approach

$$Q_n = A + E$$

Now, all that is needed to solve the equation are measurements of the net radiation and a method for determining the relative magnitude of A and E.

THE BOWEN RATIO

Bowen (1926) proposed a method for partitioning the energy used in evaporation and in heating the air:

$$\beta = \frac{A}{E} = \gamma \frac{(K_h)}{(K_w)} \frac{(T_s - T_a)}{(e_s - e)}$$

where β is the Bowen ratio, γ is the psychrometric constant, 0.49 for degrees C. and millimeters of mercury, and 0.27 for degrees F. and millimeters of mercury, K_h and K_w are eddy diffusivities for heat and water vapor, respectively, T_s is the temperature of the surface, T_a is air temperature, e_s is vapor pressure of the surface, and e is vapor pressure of the air.

The Bowen ratio is negative when heat is transferred from air to crop, and positive when heat transfer is from crop to air.

Like the aerodynamic approach, the Bowen method utilizes the principle of similarity and assumes equality between eddy diffusivities for heat and water vapor. However, the energy budget method is less sensitive to incorrect assumptions concerning the eddy diffusivities than the Thornthwaite-Holzman type of equation. As long as the Bowen ratio is small, an error in its determination would have little effect on the estimated evaporation. Tanner (1960a) has shown that the Bowen ratio approach fails only when the value is less than —0.5. For well-watered crops, the value of the Bowen ratio is usually in the neighborhood of 0.1 during the day when the evaporation is large. At night, or during sunrise and sunset, when the evaporation may approach the sensible heat flux, there may be a large relative error in the calculated evaporation, but the absolute error is small. In arid regions, where the Bowen ratio is large, there is again a possibility for significant errors in the calculated evaporation.

Accurate determination of the Bowen ratio is not an easy task. There are no simple field instruments for measuring leaf temperature and the vapor pressure of an evaporating surface, which does not always have a relative humidity of 100 per cent. To determine the Bowen ratio accurately, one must take instantaneous readings. Mean daily values are misleading. The large temperature inversion and the small vapor pressure gradients during the night give undue weight to nighttime gradients, as compared with the weight given daytime gradients occurring during the high energy flux periods. Thus, the profile measurements for computing the Bowen ratio present problems not found in ordinary meteorological observations.

THE PENMAN EQUATION

Realizing the difficulty of gradient measurements, Penman (1948) combined the aerodynamic and the energy budget approaches to derive an approximate expression that eliminated the need for surface measurements. His equation is based upon reasonable physical principles, and hence, should not be regarded as an empirical formula. The equation reads:

$$E_o = \frac{\Delta Q_n + \gamma E_a}{\Delta + \gamma}$$

where E_o = evaporation from open water surface in millimeters per day;
Δ = slope of the saturation vapor pressure vs. temperature curve (de_a/dT) at the air temperature T in millibars per degrees C.;
e_a = saturation vapor pressure in millimeters of mercury at temperature T;
T = temperature in $°K$;
$Q_n = (1 - r) Q_A (0.18 + 0.55 \text{ n/N}) - \sigma T^4 (0.56 - 0.092 \sqrt{e_d}) (0.10 + 0.90 \text{ } n/N)$. Q_n is net radiation expressed in evaporation units;
r = reflection coefficient (for mean annual values, Penman used 0.05 for open water, 0.10 for wet, bare soil, 0.20 for fresh, green vegetation);
Q_A = Angot's value;
n/N = ratio between actual and possible hours of sunshine;
σ = the Stefan-Boltzman constant;
e_d = saturation vapor pressure in mm of mercury at the dew point temperature;
γ = the psychrometric constant or the ratio of the specific heat of air to the latent heat of evaporation of water;
$E_a = 0.35 (e_d - e_a) (1 + u_2/100)$. E_a is an aerodynamic component.
u_2 = wind speed in miles per day at height of two meters.

Penman uses the Brunt formula, a crude method, to estimate net radiation from sunshine duration, humidity, and temperature. Where net radiation is measured, or a relationship between incoming and net radiation is determined, such data should be used instead.

Subsequently, as a result of the Lake Hefner studies, Penman (1956a) changed the wind velocity term from $(1 + u_2/100)$ to $(0.5 + u_2/100)$. This affects the results only slightly. For example, in Hawaii, the Lake Hefner constant would lower the Penman estimates by 3.3 per cent (Chang, 1961).

The Energy Budget Approach

Penman's equation needs observations of radiation, temperature, humidity, and wind, which are usually available only in first-order weather stations. The computation of the equation is quite complicated. In order to expedite the computations, McCulloch (1965) has presented tables for the various components in the equation, whereas Rijkoort (1954) and Kohler, Nordenson, and Fox (1955) have prepared nomograms that enable the computations to be completed in several minutes. Recently a number of computer programs have been developed for calculating Penman's equation (Lamoureux, 1962; Young, 1963; Berry, 1964).

The Penman equation consists of two terms, namely the energy term Q_n and the aerodynamic term E_a. The relative importance of the two terms is dependent upon the ratio Δ/γ, which is a temperature-dependent, nondimensional weighing factor. The aerodynamic term is usually smaller than the energy term but may show greater variation over a smaller distance (Stanhill, 1962b). The aerodynamic term has been found to be highly correlated with evaporation from an evaporimeter as shown in Figure 82. Thus, where the net radiation is more or less the same within an area, a

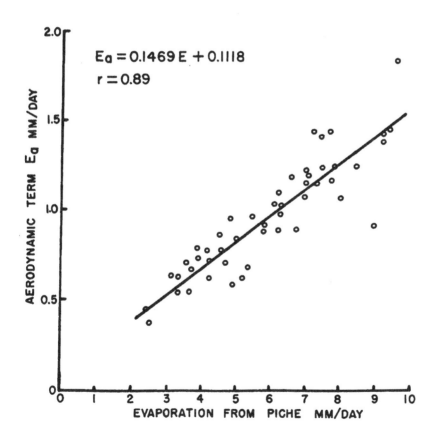

FIGURE 82. Comparison of observed Piche evaporation and the calculated aerodynamic term in Penman's equation for 49 mean weekly values at (Gilat (after Stanhill).

simplified procedure proposed by Stanhill (1962b) may be used to solve the Penman equation. He suggested that the net radiation be measured at a central station and the aerodynamic term be estimated from local measurements of the Piche evaporimeters. Then, the two terms are weighted according to mean air temperature as shown in Table 9. As the air temperature increases, the energy term becomes increasingly important.

TABLE 9

Weighting Factors for the Energy and Aerodynamic Terms in Penman's Formula

Mean air temperature (°C)	Energy term	Aerodynamic term
0	.40	.60
1	.42	.58
2	.44	.56
3	.46	.54
4	.47	.53
5	.48	.52
6	.50	.50
7	.52	.48
8	.53	.47
9	.54	.46
10	.56	.44
11	.58	.42
12	.59	.41
13	.60	.40
14	.62	.38
15	.63	.37
16	.64	.36
17	.66	.34
18	.67	.33
19	.68	.32
20	.69	.31
21	.70	.30
22	.71	.29
23	.72	.28
24	.73	.27
25	.74	.26
26	.75	.25
27	.76	.24
28	.77	.23
29	.78	.22
30	.79	.21
31	.80	.20
32	.81	.19
33	.81	.19
34	.82	.18
35	.83	.17
36	.83	.17
37	.84	.16
38	.85	.15
39	.85	.15
40	.86	.14

This simplified procedure has been tested at Beersheba, Israel. The results differ from the original Penman computation by 0.2 per cent for the annual value and by 7 per cent for the monthly values.

Where the variation of the aerodynamic term is small, it is also possible to simplify the Penman equation to estimate the evaporation from the energy term alone. For example, Wang and Wang (1962) have constructed monthly nomograms for the Penman estimates in Wisconsin in terms of temperature and sunshine hours. Figure 83 illustrates such a relationship for the month of July.

Experimental evidence so far lends support to the Penman equation as the best formula. Comparisons between the Penman and the Thornthwaite methods—Pruitt in California (1960a) and Washington (1960b), Stanhill in Israel (1961), Chapas and Rees in Rhodesia (1964), and Chang in

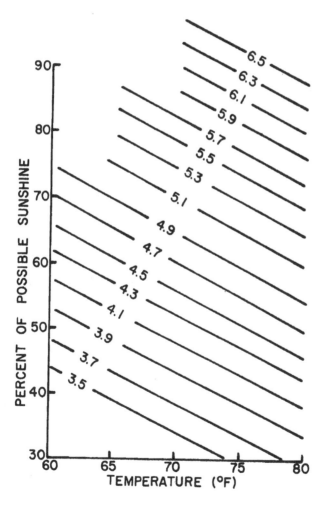

FIGURE 83. Nomogram for estimating of evaporation in Wisconsin and vicinities for July (after Wang and Wang).

Hawaii (1961)—indicate that the Penman equation is more accurate than the Thornthwaite formula. The only exception is found in the central United States, where the latter was derived. In Minnesota, Baker (1958) noted that the Thornthwaite formula was equal to or superior to the Penman method. In Missouri, the two methods worked equally well (Decker, 1962). However, in Wisconsin, King (1956) found that Penman's equation yielded better results than Thornthwaite's formula.

The Penman equation does not give accurate daily estimates and is usually accurate only for a period of five days or longer (Gilbert and van Bavel, 1954). For instance, in comparison with lysimeter readings, Businger (1956) obtained daily Penman estimates varying by as much as 25 per cent; whereas the total for more than 25 days was estimated to within 1 per cent.

The Penman equation only gives estimates of open water evaporation. The potential evapotranspiration from a vegetation surface is somewhat different. In southern England, the ratio between turf evapotranspiration and open water evaporation has been found to be 0.6 for November to February, 0.8 for May to August, and 0.7 for the other four months. The value is higher in the summer, presumably because of the greater amount of advected energy in that season.

The ratio between the potential evapotranspiration and Penman's estimates of open water evaporation, varies with the height of the vegetation and the climate, particularly the advected energy of the area. Penman (1952) argued, from a theoretical ground, that it is possible for tall vegetation to have a higher evapotranspiration rate. In areas of strong advection, the Penman equation may greatly underestimate potential maximum evapotranspiration. Abdel-Aziz (1962), for instance, found it necessary to add an advective term to the Penman equation in order to apply it to arid regions in the western United States. In the semiarid climate of central Washington, Pruitt (1960b) found that the water needs of ladino clover almost equal that of Penman's estimates of free water evaporation. In Hawaii, the potential maximum evapotranspiration of a tall sugar cane crop exceeded the Penman estimate by some 10 per cent (Chang, 1961).

THE FRACTION OF RADIATION USED IN EVAPOTRANSPIRATION

In the absence of advected energy, potential evapotranspiration is determined by the net radiation. Measurements of the energy budget in Ontario (Graham and King, 1961a), North Carolina (Harris and van Bavel, 1958), Missouri (Gerber and Decker, 1960), Hawaii (Chang, 1961), California (Halstead, 1954), the Netherlands (Scholte-Ubing, 1959), and England (House, Rider, and Tugwell, 1960), indicate that in the tropics, and during the warm seasons in the middle latitudes, 80 to 90 per cent of the net radiation is consumed in evapotranspiration. The lower values of 80 to 85 per cent are probably correct, because a small amount of advected energy exists even in humid climate.

During the winter season in the middle and high latitudes, the fraction of net radiation used in evapotranspiration is usually lower than 80 per cent. King (1961) cited one instance in which the fraction of net radiation used in evapotranspiration dropped from 0.85 to 0.52 after freezing.

In the presence of advected energy, the potential maximum evapotranspiration may well exceed the net radiation. Lemon, Glaser, and Satterwhite (1957) have found the daily evapotranspiration from a large field of irrigated cotton in Texas to be 1.75 times the net radiation. In Utah, the evapotranspiration of hay exceeded the net radiation nearly every day, sometimes by as much as 40 per cent (Abdel-Aziz, Taylor, and Ashcroft, 1964). Grable, Hanks, Whillhite, and Haise (1966) reported a maximum ratio between evapotranspiration and net radiation of 1.31 at Gunnison, Colorado. Perhaps, the most extraordinary case was that reported by Hudson (1965), in which a crop of lucerne transpired water at a rate of about an inch a day over a period of several days. Two-thirds of this transpiration was accounted for by advected energy. Therefore, in arid climates, the potential maximum evapotranspiration cannot be approximated from the net radiation alone.

DIURNAL VARIATION OF THE ENERGY BUDGET

In the absence of advected energy, a close relationship between the potential evapotranspiration and net radiation usually exists. In Figure 84, Pruitt (1962b) compared the potential evapotranspiration for ryegrass with the incoming and net radiation on a summer day in Davis, California. The quick response of evapotranspiration to radiation change is evident. In Figure 85, the evapotranspiration for a calm, clear day in spring is compared with the net radiation, air temperature, and saturation deficit. The evapotranspiration is considerably out of phase with the daily march of temperature and the saturation deficit; whereas it appears to be closely in phase with radiation. However, the fraction of net radiation used in the evapotranspiration varies during the course of the day. The plot of evapotranspiration versus net radiation, in Figure 86, indicates that the E/Q_n ratio is low during morning hours. The ratio increases throughout the day, reaching a value of 1.0 between 4 and 5 P.M. In the late afternoon, the evapotranspiration exceeds the net radiation.

ENERGY BUDGET THROUGHOUT THE CROP CYCLE

The fraction of solar radiation used in evapotranspiration varies with the development of the canopy and the physiological stage of the plant. Figure 87 illustrates the relationship for sorghum at two localities, Kansas

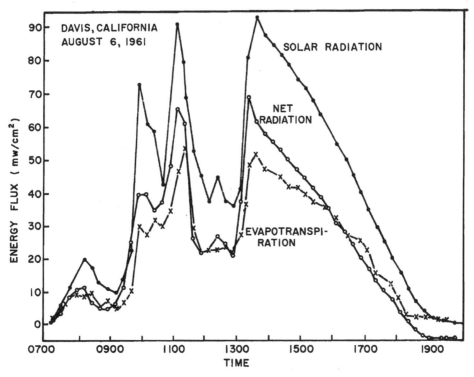

FIGURE 84. Evapotranspiration for ryegrass, incoming solar radiation, and net radiation for a day of variable cloudiness but clear after 1400 (after Pruitt).

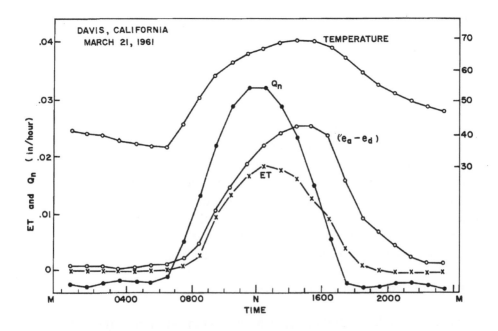

FIGURE 85. Evapotranspiration, net radiation, air temperature, and saturation deficit of air on March 21, 1961 (after Pruitt).

The Energy Budget Approach 173

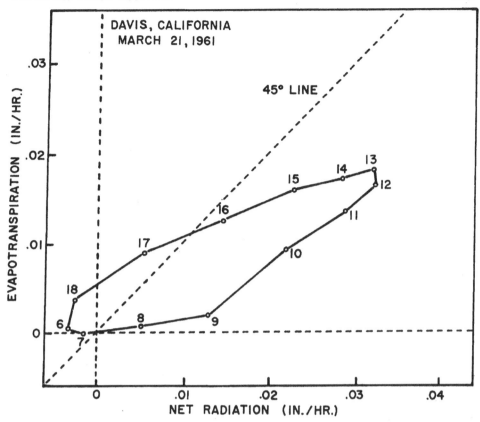

FIGURE 86. Evapotranspiration in inches for perennial ryegrass during the various hours of the day vs. net radiation expressed in equivalent evaporation terms (after Pruitt).

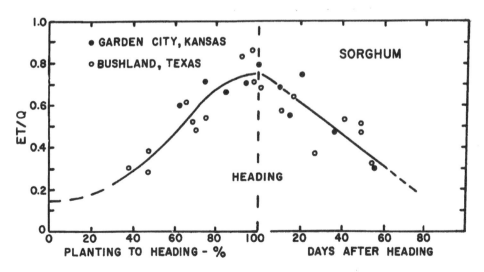

FIGURE 87. Variation in ET/Q ratio for grain sorghum in relation to stage of plant growth expressed as per cent of the period from planting to heading and days after heading (after Jensen and Haise).

and Texas (Jensen and Haise, 1963). The E/Q ratio increases as the vegetative cover develops. The low ratio in the early stage of the crop cycle is a result of the presence of a large proportion of bare ground that cannot be kept wet all the time by ordinary irrigation practices. The maximum ratio is reached just before heading and decreases almost linearly after heading because of the senescence of the crop.

In contrast, the E/Q ratio for alfalfa, as shown in Figure 88, is less variable throughout the year. The alfalfa develops a complete ground cover as soon as temperatures permit. The E/Q ratio remains fairly constant throughout the season, except during the periods immediately after cutting, when the net radiation decreases.

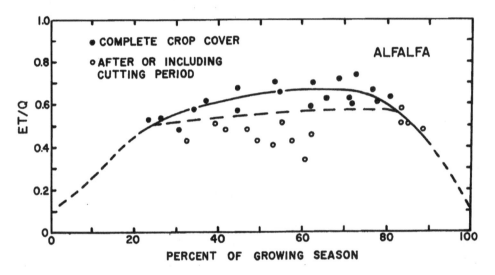

FIGURE 88. Variation in ET/Q ratio for alfalfa in relation to per cent of growing season at Prosser, Washington. Dashed line indicates weighted average, including cutting period (after Jensen and Haise).

RELATIVE MAGNITUDE OF EVAPORATION AND TRANSPIRATION

The relative magnitude of evaporation and transpiration from a vegetated surface is largely dependent upon the moisture condition of the soil, the radiation distribution within the canopy, and the completeness of the ground cover. In an attempt to separate evapotranspiration into its components, several investigators have compared natural, uncovered plots with plots with a plastic cover on the soil on the assumption that the plastic cover effectively prevents evaporation from the soil, so that any depletion of soil moisture could be attributed to transpiration. The results of these experiments on corn by Harrold, Peters, Dreibelbis, and McGuinness (1959) in Ohio; Shaw (1959) in Iowa; Letey and Peters (1957) and Peters and Russell (1959) in Illinois all indicate that for the growing season as a whole, the amount of evaporation is about the same as transpiration under normal cultural practices in the Midwest. The study by Peters and Johnson (1960) on soybean also suggests that evaporation accounts for half or more of the potential evapotranspiration when the soil is wet, but accounts for 25 to 50 per cent of the total moisture loss when the soil is dry.

However, the use of a plastic cover to determine the relative magnitude of evaporation and transpiration is subject to criticism. The plastic covers not only change the reflectivity and the temperature of the surface, but may also cause condensation in their lower surface. The net effect is often to reduce the net radiation in the plastic-covered plot below that of a natural plot (Peters, 1960a). Furthermore, soil moisture conditions may also be quite different in the two treatments.

A better method is to measure the net radiation at the soil surface and above the crop canopy. When the ground is wet, the ratio between the net radiation at the surface and above the canopy sets the upper limit for the fraction of soil evaporation that could occur. Thus:

$$\frac{Q_n \text{ at the soil surface}}{Q_n \text{ above the canopy}} = \frac{\text{Soil evaporation}}{\text{Potential evapotranspiration}}$$

It needs to be emphasized, however, that this equation is valid only when the soil is wet and the advected energy negligible.

Tanner, Peterson, and Love (1960) have made extensive net radiation measurements at Arlington, Wisconsin. They found that evaporation constituted, as a maximum, only about 30 per cent of the potential evapotranspiration from fully grown corn planted at 13 kiloplants per acre, and the percentage value was slightly lower when the population increased to 22 kiloplants.

At Ames, Iowa, Shaw (1959) reported that early in the season, net radiation at the ground level almost equaled that above the corn crop. As the corn developed, the amount of net radiation penetrating the crop cover decreased to 60 to 65 per cent when the crop was 60 inches tall, and to 14 per cent when the crop reached a maximum height of 90 inches.

Another interesting study was carried out by Mihara (1961) in Japan in a paddy field. With the growth of rice, evaporation gradually decreased, and transpiration increased as shown in Figure 89. The ratio between evaporation and evapotranspiration reached a minimum at the flowering stage in August, the value being about 0.2 in normal growth, and 0.1 in an extraordinarily luxuriant growth. This ratio is closely related to the completeness of the ground cover and, hence, the leaf area index (Figure 90). As the leaf area index increases from 1 to 3, the fraction of surface evaporation decreases from 0.5 to less than 0.2.

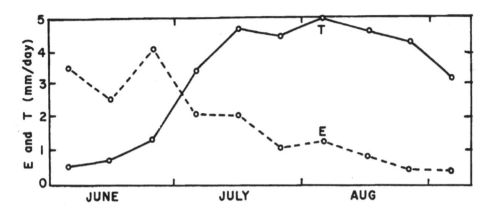

FIGURE 89. Seasonal march of evaporation and transpiration in a paddy field. Ten days mean, 1958, Tokyo (after Mihara).

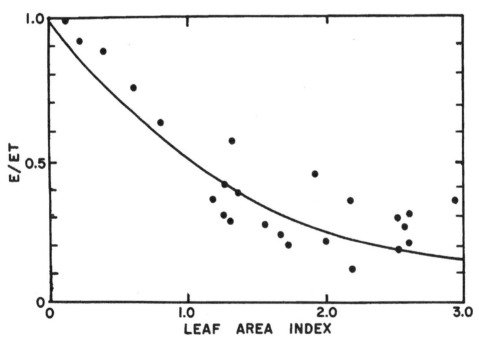

FIGURE 90. Relation between leaf area index and ratio of evaporation to evapotranspiration (after Mihara).

The separation of evaporation and transpiration has at least three practical applications.

First, if transpiration and dry-matter production are indeed linearly related, then precise determination of transpiration during successive stages of the crop cycle would improve our understanding of the relationship between water and yield.

Second, the practice of summer fallowing, to increase water for the following season, can be better evaluated if we know what the excess of transpiration is over soil evaporation. Holt and van Doren (1961) have pointed out that if evaporation does indeed account for nearly 50 per cent of the total evapotranspiration in a Midwest corn stand, then most of the summer rainfall would be lost through evaporation, and therefore summer fallow cannot be an efficient practice in that area. In North Dakota, Haas and Willis (1962) showed that a wheat crop averaging fifteen bushels per acre could be produced by adding only four more inches of water than a fallow field used during the same period.

Third, the selection of plant population and the width of the rows in semiarid or arid climates requires a knowledge of the relative magnitude of evaporation and transpiration. This point has been discussed by Aubertin and Peters (1961):

> A tremendous amount of energy can be absorbed in a high-density, narrow-rowed stand of corn. When soil moisture is limited, the plants under these conditions suffer greatly. They absorb much larger quantities of energy and transfer water vapor very rapidly. When water is not available for transpiration, the plants still absorb the same large amount of energy, heat up, and wilt badly. This factor was very evident within the test plots. High-population, 20″-row plots began to wilt daily about ten o'clock and remained badly wilted throughout the day, while high-population, 40″-rows showed very little or no noticeable wilting even during the hottest part of the day.

Obviously, planting practices that increase light absorption will decrease the ratio of evaporation to transpiration; hence, they are advisable only in the areas of high moisture. In arid areas, low population and wide row widths that decrease light absorption by the crop will conserve water and consequently result in a better yield. In desert regions, the spacing of shrubs is rigidly dictated by the moisture supply and ultimately only one shrub can be established for every one that dies. In cultivated lands, the best crop spacing is an equally delicate operation requiring a compromise between energy availability and moisture balance.

CHAPTER 18

EVAPORIMETERS

LIMITATIONS AND ADVANTAGES

Instruments for the measurement of evaporation are of two basic designs: open water evaporation pans and porous surface-type atmometers. Evaporimeters have been used for estimating the potential evapotranspiration of various crops. The process of evaporation is somewhat similar but not identical to that of transpiration. For even with fully opened stomata, there is still a diffusion resistance that is lacking in the evaporation from an open water surface. The energy budget and the aerodynamic roughness of the water surface may also differ from their counterparts on a vegetated surface. Furthermore, evaporation from the pan lags behind evapotranspiration for several hours because of the greater capacity for heat storage in the former. However, these differences are small, and a relationship—good enough for many agricultural purposes—usually exists between evaporation and evapotranspiration for a period of a day or longer.

Since the evaporation pan will incorporate the effects of all climatic factors, it is more accurate in estimating short-term fluctuations of evapotranspiration than empirical formulae that depend on fewer of the climatic factors involved. For instance, Pruitt and Jensen (1955), and Suzuki and Fukuda (1958) reported that pan evaporation rates gave much closer estimates of crop water use than either the Blaney-Criddle or the Thornthwaite procedures. Comparison between the evapotranspiration measured by a lysimeter from Bahia grass and results obtained by different methods in the Ruzizi Valley in the eastern Congo yielded the following correlation coefficient: pan evaporation, 0.977; the Penman equation, 0.79; the Blaney-Morin equation, 0.73; the Thornthwaite method, 0.72, and the Blaney-Criddle method, 0.59 (Brutsaert, 1965). Smith (1964) found that open water evaporation provided a better estimate of the monthly and seasonal evaporative power of the air than the Thornthwaite formula. Ward (1963) reported that evaporimeters give more accurate estimates than empirical formulae on the Thames floodplain. Weaver and Stephens (1963), Davis (1963), and Bowman and King (1965) found high correlations between evapotranspiration and evaporation from a U.S. Weather Bureau pan.

A number of studies (Hearn and Wood, 1964; Dagg, 1965, and Finkelstein, 1961) seem to indicate a good correlation between pan evaporation and the Penman estimates, or simply 85 per cent of the net radiation. However, in the presence of advected energy, the pan evaporation rate is in better agreement with evapotranspiration; whereas the simple energy budget approach will underestimate the evapotranspiration. Ekern (1965c), for instance, reported that at Wahiawa, Hawaii, during periods of strong positive advection, both the evapotranspiration of short grass and the pan evaporation approached a value of about 1.1 times the net radiation, which is 25 per cent higher than the simple energy budget estimate. Thus, evaporimeters have the advantage of giving sufficient allowance to the advected energy in an arid climate. Robins and Haise (1961) remarked:

> Perhaps the simplest and most successful [method] for use in the presence of advected energy are evaporimeters. These instruments reflect local and short-term climatic variations, and adjustment can be made for advected energy by control of the location or environment of the evaporimeter, provided necessary background on environmental effects is sufficiently understood and recognized.

Another advantage of evaporimeters over various formulae is that they can be used to estimate the evapotranspiration throughout the life cycle of a crop. Since all the formulae for estimating potential evapotranspiration presuppose a complete cover of actively growing vegetation, they will overestimate actual evapotranspiration under normal irrigation practice either when the crop is young or during the period of ripening. For instance, Denmead and Shaw (1959) have noted that the corn crop in Iowa approaches the condition of a green crop, actively growing and completely shading the ground, only for a period of two to three weeks during its growing season from May to September.

Evaporimeters are inexpensive and easy to handle in the field. Stanhill (1961) made a comparison among eight methods as to the accuracy, the cost, and the time needed for calculating potential evapotranspiration in Israel as shown in Tables 10 and 11. Clearly, when all factors are considered, the evaporation pan is the most satisfactory method for field use.

DESIGN AND INSTALLATION

The rate of evaporation from a pan varies with its size, color, material, depth, exposure, and the like. The relationship between the size of the evaporating surface and the rate of water loss has been schematically

TABLE 10

A Comparison of Eight Methods of Calculating Potential Evapotranspiration from Climatic Data

Method	Monthly periods			Weekly periods		
	Regression	r	cv	Regression	r	cv
Penman	$y = 0.97x + 0.96$	0.96	12	$y = 0.96x + 1.12$	0.76	36
Thornthwaite	$y = 1.48x + 1.85$	0.94	16	$y = 1.35x + 1.76$	0.73	38
Blaney-Criddle	$y = 1.22x + 0.72$	0.90	20	$y = 1.15x + 1.02$	0.70	40
Makkink	$y = 1.49x + 0.06$	0.95	15	$y = 1.45x + 0.15$	0.75	37
Evaporation tank	$y = 0.86x + 0.74$	0.94	16	$y = 0.84x + 0.73$	0.76	37
Evaporation pan	$y = 0.70x + 0.47$	0.95	15	$y = 0.72x + 0.36$	0.77	36
Piche evaporimeter	$y = 0.88x + 0.03$	0.69	30	$y = 0.94x + 0.35$	0.63	44
Solar radiation	$y = 0.72x - 1.04$	0.91	20	$y = 0.70x - 0.87$	0.77	29

y = measured value of potential evapotranspiration, millimeters per day
x = calculated values of potential evapotranspiration, millimeters per day
r = correlation coefficient
cv = coefficient of variation, per cent of mean y-value

TABLE 11

A Comparison of the Equipment and Time Needed for Eight Methods of Calcuating Potential Evapotranspiration

Method	Equipment (minimum requirements)	Cost of equipment (Israeli pounds)	Time needed for observations (minutes per day)	Time needed for each calculation (minutes)
Penman	Thermometer screen, thermometers, sunshine recorder, and totalizing anemometer	1,050	10	10
Thornthwaite	Thermometer screen and thermometers	550	5	5
Blaney-Criddle	Thermometer screen and thermometers	550	5	5
Makkink	Thermometer screen, thermometers, and sunshine recorder	850	5	5
Evaporation tank	Tank, stilling well, and hook gage	450	5	5
Evaporation pan	Pan, stilling well, hook gage, and wooden platform	100	5	5
Piche evaporimeter	Thermometer screen and evaporimeter	505	5	5
Solar radiation	Sunshine recorder	320	5	5
Gravimetric soil, moisture, sampling	Veihmeyer tube, hammer and jack, sampling tins, triple beam balance, and drying oven	650	180	30

presented by Mather (1959) in Figure 91. The value of evaporation is fairly independent from the size of the measuring pan when the humidity is high, but, as the air becomes drier, the size of the pan greatly influences the rate of evaporation. This fact emphasizes the importance of maintaining a guard area around the pan in an arid climate so that the humidity around the pan is relatively high.

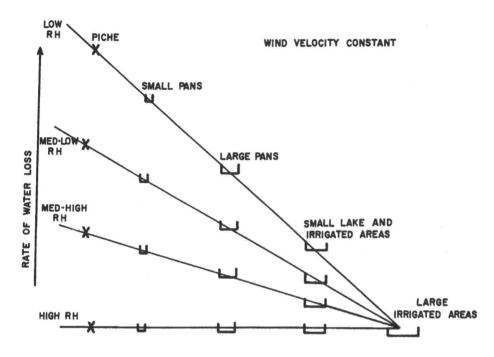

FIGURE 91. Schematic relation between size of evaporating surface and rate of water loss (after Mather).

As a result of the difference in reflectivity, a black pan may lose 23 per cent more moisture by evaporation than a white pan (Young, 1947). A copper pan may lose some 10 per cent more than an aluminum pan. The water level in the pan also affects the reading. Bonython (1950) has reported a difference of 15 per cent in an evaporation rate reading because of a change of five centimeters in a tank's water level. This is caused partly by the disturbing effect of the rim of the tank and the air flow, and partly by the excessive heating of the upper part of the pan walls. However, a change of water level probably has less effect on the evaporation rate of the raised pan than the sunken pan (Ventikeshwaran, Jagannathan, and Ramakishran, 1959).

Evaporation pans are sometimes buried in the ground, but these are less desirable. Their readings are greatly influenced by the temperature gradient between the soil and the pan. Moreover, the relationship between sunken and unsunken pans exhibits seasonal variations according to the moisture condition of the soil. McIlroy and Angus (1964) noted that a raised tank gives better performance than a buried pan, especially if it is installed inside the field of a crop.

The height at which the evaporation pan is installed is another important factor. Pruitt (1960a) compared the consumptive use by ladino clover with evaporation from six different pans. The data are presented by a double-mass type of plotting as shown in Figure 92. The relatively corresponding nature of data for all six pans indicates the close relationship between evaporation and evapotranspiration, regardless of the type of pan or its elevation above the ground. Of the three pans of two-feet diameter, the one elevated to a height of eighteen inches lost some 12 per cent more moisture by evaporation than the surface pan and 24 per cent more than the pan buried at ground level.

In order to eliminate the effect of pan design and exposure on evaporation rates, the World Meteorological Organization adopted the U.S.

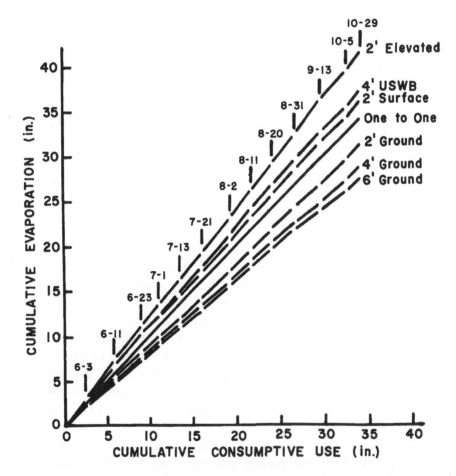

FIGURE 92. Comparison of consumptive use by ladino clover and evaporation from various pans at Prosser, Washington, 1955 (after Pruitt).

Weather Bureau Class A pan as the interim standard for the International Geophysical Year. A working group under the Commission for Instruments and Methods of Observation is examining the question of international standards, including the possibility of using an insulated pan (Nordenson and Baker, 1962). The insulated pan may have the advantage of minimizing the radiational and convective transfer of energy from the walls and bottoms.

The U.S. Weather Bureau Class A pan is of cylindrical design 10 inches deep and 47½ inches in diameter (inside dimension). It is constructed of galvanized iron or monel metal, preferably the latter in areas where the water contains large amounts of corrosive substances. The evaporation pan is placed on a wooden platform. The bottom of the pan is approximately four inches above the ground, so that air may circulate beneath the pan. The site should be fairly level, sodded, and free from obstructions. The grass should be frequently watered. The pan should be filled to a level of two inches below the rim, and refilled when the water has evaporated one inch.

In a humid climate, the size of the guard area or the location of the pan with respect to the crop field is not a serious problem. In an arid climate, the pans in a non-cropped environment of even a small sodded area will overestimate the water needs of the large field as a whole. Wartena (1959) reported an extreme case of a small irrigated rice field in central Iraq in which the daily evaporation from a Class A pan outside the field was sixteen millimeters and from an identical pan inside the field was five millimeters. In such a case, the potential maximum evapotranspiration will also decrease rapidly from the edge to the center of the field, and the evaporation pan gives a hint of this rapid change. However, this poses a problem as to where the pan should be located. Stanhill (1961) presented a relationship between relative evapotranspiration and distance downwind for four localities with different sizes of irrigated fields (Figure 93). The exponential relationship suggests that the pan should be located at least 300 meters downwind from a dry-wet border in order to minimize the clothesline effect. For tall crops, it would be advisable to elevate the pan to the canopy level.

In some areas, screening of the evaporation pan is necessary to prevent birds or stray animals from drinking the water. Various types of screens have been used, from chicken wire to heavy mesh. Stanhill (1962a) reported a 10.4 per cent reduction in the evaporation caused by the standard screen of the Israel Meteorological Service, which consists of wire netting of 0.8-centimeter hexagonal mesh supported by a light metal framework. The reduction was of the same magnitude throughout the year and at three different stations. In general, pans screened with chicken wire lose more moisture through evaporation than those screened with heavy mesh.

Mukammal (1961) is of the opinion that screens should be avoided

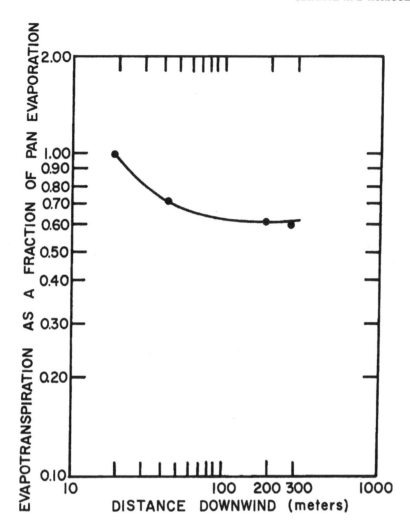

FIGURE 93. Relationship between relative evapotranspiration and distance downwind (after Stanhill).

whenever possible because they introduce an error in the catch of rainfall in the pan. His experiments at Accra showed a decrease in rainfall collected in the screened pans from that collected in the unscreened pans. The differences varied according to the intensity of the rainfall.

The effect of rainfall on the evaporation rate from the pan has been investigated by Bilham (1932a, 1932b). He concluded that two different effects may cause errors:

(1) Large rain drops cause a considerable splash-out from the tank, which retains less than the rain gauge, and thus apparent evaporation tends to be too large. The exaggerated rate is often aggravated by a simultaneous loss by wind.

(2) With small rain drops, the tank acts as a shielded rain gauge, and in very windy conditions, it tends to catch more than the normally exposed gauge. The apparent evaporation may thus be too small and even be negative in the winter.

The splashing losses from the pan in areas of high rainfall intensity can be mitigated by providing an overflow tank, such as the one designed by Richards and Stumpf (1966) or the one designed by Sumner (1963). The latter can be operated unattended for periods up to six months.

Various types of evaporation pans have been used throughout the world. In some cases, the evaporation rate of one type of pan can be converted into that of another with a high degree of precision. For instance, Brown and Hallstead (1952) have found the following relationships between the Class A pan and the Plant Industry pan, which is 72 inches in diameter and 24 inches deep and is set in the soil to a depth of 20 inches with 4 inches extending above the soil surface:

$$Y = 1.46 X + 0.031$$

where Y is the monthly evaporation in inches from the Class A pan and X is the monthly evaporation in inches from the Plant Industry pan. The equation was derived from fourteen years of records taken at Hays, Kansas. The correlation coefficient was as high as 0.986.

RATIOS BETWEEN EVAPOTRANSPIRATION AND PAN EVAPORATION

A number of studies have been conducted to determine the ratios between evapotranspiration and pan evaporation. The results of these are summarized in Table 12 (for the U.S. Weather Bureau Class A pan) and in Table 13 (for other types of pans).

The life cycle of most plants may be divided into three stages: vegetative, flowering, and fruiting. The vegetative stage includes two phases: the early stage when the cover is incomplete, and the late stage when the canopy is fully developed. The fruiting stage may also be subdivided into the wet fruiting stage and the dry fruiting stage. In Tables 12 and 13, the young crop is defined as that in the early vegetative stage. The mature crop encompasses both the late vegetative and the flowering stages, while the ripening stage is the same as the fruiting stage.

During the early stage of the crop cycle, the ratio is very low, varying from 0.2 to 0.5 as given in Table 12. In row crops, the greater part of the water loss during the early stage is by evaporation from the bare soil. In the absence of rain or irrigation, the evaporation rate is quickly reduced by

TABLE 12
RATIOS BETWEEN POTENTIAL EVAPOTRANSPIRATION AND EVAPORATION FROM U.S. WEATHER BUREAU CLASS A PAN FOR VARIOUS CROPS

Crop	Locality	Period	Young	Mature	Ripening	Average	Reference
Grass	Aspendale, Australia	1959-61				.84	McIlroy and Angus (1964)
Bermuda grass	Thorsby, Alabama	Apr.-Sept.	.45	.75			Doss, Bennett, and Ashley (1964)
Grass	Oahu, Hawaii	1958-59	.50	.80-1.00		.90	Ekern (1959)
Meadow grass	Gunnison, Colorado	May 22-Oct. 7, 1964				1.08	Grable, Hanks, Willhite, and Haise (1966)
Grass	Caesarea, Israel	June-Sept. 1959				0.94	Stanhill (1964)
Alfalfa	Thorsby, Alabama	Apr.-Sept. 1961-62	.50	1.05			Doss, Bennett, and Ashley (1964)
Alfalfa brome	Ithaca, New York	June-Oct. 1953-54				1.0	Gray, Levine, and Kennedy (1955)
Alfalfa clover	Lod, Israel	Mar. 1956-Dec. 1958				.82-.98	Lomas (1964)
Barley	North Logan, Utah	July-Aug. 1953	.25	.90	.30		Hansen (1963)
Sorghum	Thorsby, Alabama	Apr.-Aug.	.30	1.15			Doss, Bennett, and Ashley (1964)
Cotton	College Station, Texas	Aug. 4-5, 1954		1.00			Lemon, Glaser, and Satterwhite (1957)
Cotton	Gilat, Israel	Apr.-Oct. 1959	.20	.85	.10-.40		Stanhill (1962a)
Corn	Ames, Iowa	June-Sept. 1959	.27	.90	.40		Fritschen and Shaw (1961)
Sugar cane	Chaka's Kraal, South Africa	Oct. 1959-July 1962		1.00			Thompson, Pearson, and Cleasby (1963)
Sugar cane	Hawaii	Nov. 1958-Apr. 1959	.40	1.10	.98		Chang (1961)
Tule	California	Two years				.95	Blaney (1951)
Pineapple	Hawaii	1960-62		.35			Ekern (1965a)

TABLE 13

Ratios between Potential Evapotranspiration and Evaporation from Pans other than the Standard U. S. Weather Bureau Type for Various Crops

Crop	Locality	Period	Pan ratio			Type of evaporation pan	Reference
			Young	Mature	Average		
Grass	Aspendale, Australia	1959-61			1.05	Australian evaporation tank	McIlroy and Angus (1964)
Grass	Copenhagen, Denmark	1955-64			0.90	Sunken 12 square meters	Aslyng (1965b)
Alfalfa and grass	Kootenay River Valley, Canada	summer, 1957-59			0.83-1.00	buried four-foot evaporation pan	Krogman and Lutwick (1961)
Lucerne	Alice Spring, Australia	Sept. 1957-Aug. 1958	.40-.60		0.888		Jackson (1960)
Alfalfa	Prosser, Washington	July, 1949, 1952-54		1.38		U.S. Department of Agriculture BPI pan	Pruitt and Jensen (1955)
Timothy	Kapuskasing, Ontario, Canada	summer, 1954-58			1.24	buried four-foot Canadian pan	Chapman and Dermine (1961)
Pasture	Winchmore, New Zealand	Dec. to Feb., 1955-56		.65		New Zealand sunken pan	Finkelstein (1961)
Rice	Murrumbidgee, Australia	Oct. 1946-Mar. 1947			1.10	Australian evaporation tank	Butler and Prescott (1955)
Sugar cane	*Chaka's Kraal*, Tongaat, Illovo, South Africa	Sept. 1959-Oct. 1960	.35	1.40	1.09	Sunken Symon tank	Pearson, Cleasby, and Thompson (1961)
Coffee	Ruirn, Kenya	6 years		.80		Sunken insulated tank	Pereira (1957)
Pine, Cyprus, Bamboo	Kinal, Kenya	1948-61		.86		Raised British pan	Pereira and Hosegood (1962)

the formation of a surface mulch of dry soil. Thus, when the crop is young, the ratio will vary according to the frequency of rainfall and irrigation, soil characteristics, and the completeness of the vegetation cover.

The relatively low water requirements during the early stage of crop development is still not widely known. The experience reported by Laycock and Wood (1963b) of the misuse of water in tea nurseries is typical:

> The traditional method of watering tea nurseries heavily in the early stages and less heavily or not at all in the later stages . . . finds no support from these observations. Indications are that watering should be heavier and more frequent as the plant become older.

During the mature stage when the canopy is fully developed and the plants are actively growing, the ratio between potential evapotranspiration and evaporation from the Class A pan varies from 0.75 to 1.15 for various crops. The only exception is the pineapple with a ratio of 0.35. The extremely low water requirement of the pineapple is explained by the fact that its stomata are closed throughout the day. From the standpoint of water use, the pineapple is known as a nonconventional crop in the sense that its transpiration rate is regulated to a large extent by its peculiar physiology rather than by the energy available under the prevailing weather conditions.

The ratios given in Table 12 for mature crops are not strictly comparable. Some are the peak use values for a period of a few days; others are the monthly or other long-period averages. In a few cases the ground cover may not be complete. Furthermore, the pan site and its surroundings may not meet the desired specifications. In spite of these discrepancies, the data suggest that the ratio between potential evapotranspiration and evaporation from the Class A pan increases from about 0.8 for short crops to 1.0 or slightly higher for tall crops.

During the mature stage, the ratio may vary slightly with the season and the physical characteristics of the plants. Theoretically, the ratio may increase slightly during a season when the advected energy is strong. The smooth surface of water protected by the pan's rim is not as good an interceptor of warm air moving into the area as are crop surfaces that protrude and are rough. However, under the normal conditions of moderate advection, the variation in the pan ratio is rather small, as evidenced by the data for ryegrass at Davis, California (Pruitt and Angus, 1961) in Figure 94.

From the standpoint of energy budget, the pan ratio during the mature stage should not be affected by the soil types. Nor should the pan ratios for different varieties vary to an extent in excess of that suggested

Evaporimeters 189

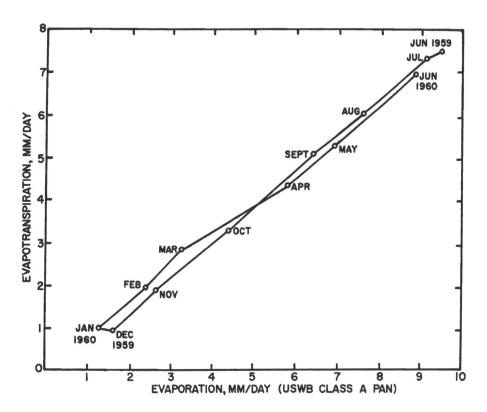

FIGURE 94. Relationship between average daily evapotranspiration for ryegrass and evaporation from U.S.W.B. Class A pan for monthly period. Data taken at Davis, California (after Pruitt and Angus).

by the small differences in albedo and surface roughness. For many practical purposes, the pan ratio during the mature stage may be considered as constant.

During the fruiting stage, evapotranspiration is usually reduced because of senescence. The reduction is relatively small, say about 10 to 20 per cent during the wet fruiting stage, but it is much greater during the dry fruiting stage. This explains the large variations of pan ratios during the ripening stage as shown in Table 12.

From planting to harvest, the shape of the curve relating evapotranspiration and pan evaporation will depend on the stage of growth during which the crop is harvested. A list of crops harvested during the different stages of growth is given by Hansen (1963) in Table 14. For a crop that is harvested during the fruiting stage, such as corn, the change of ratios throughout its life cycle is illustrated by Fritschen and Shaw (1961) in Figure 95. Such a curve permits a quick estimation of the short-term water needs from the pan evaporation data.

TABLE 14
Typical Crops Harvested during the Different Stages of Growth

Vegetative	Flowering	Wet fruiting	Dry fruiting
Lettuce	Flowers	Tomatoes	Dry peas
Alfalfa	Cauliflower	Green beans	Dry beans
Grasses	Broccoli	Sweet corn	Potatoes
Cabbage		Deciduous fruits	Mandioca
Asparagus		Sugar cane	Seeds and grains
Mint		Artichokes	Field corn
		Sugar beets	Rice
		Berries	Onions
		Bananas	Nuts
		Cantaloupe	
		Citrus	
		Watermelons	

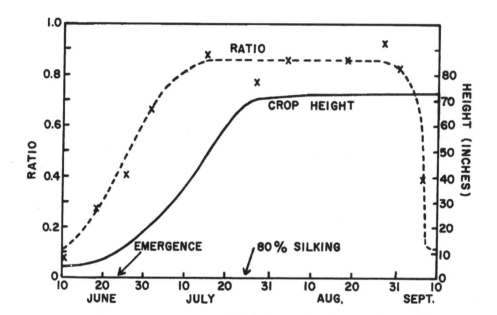

FIGURE 95. Ratio of evapotranspiration for corn to pan evaporation as function of crop height (after Fritschen and Shaw).

ATMOMETERS

Atmometers are either porous porcelain bodies or porous paper wick devices for measuring the evaporative power of the air. The former include the Livingston and the Bellani type of atmometer; whereas the Piche evaporimeter is an example of the latter. Atmometers are easy to install and operate. They require less water than the evaporation pan and do not have a splash problem. However, it is difficult to tell when errors are creeping into readings because of contamination. Pores in the porcelain are easily clogged by dust or salt in the water if it is not distilled. Atmometers are readily damaged by frost, and their breakage is high even under normal field conditions.

Probably the most serious shortcoming of atmometers is that they are unduly affected by wind and are not very responsive to radiation. Mukammal and Bruce (1960) derived regression equations showing the relative importance of three meteorological factors in affecting evaporation from the Class A pan and from the Bellani atmometer. For the pan, the relative importance of radiation, humidity, and wind are in the ratio of 80:6:14 respectively, while for the Bellani, these ratios are 41:7:52 respectively. The values for pan are more reasonable than those for the Bellani. Thus, Stanhill (1962b) used the Piche atmometer only for estimating the aerodynamic term in the Penman equation. The aerodynamic term is less important than the energy term, except at low temperatures.

As the wind speed near the ground changes rapidly with height, the height at which the atmometer is installed is of critical importance. Pruitt (1960c) reported a case in which the white atmometer placed at a 72-inch height lost 84 per cent more moisture by evaporation than one at a 3-inch height. For this reason, comparing the atmometer readings taken over surfaces with different roughness is difficult.

The profound effect of wind on the atmometer readings can be minimized by using the differences in evaporation between black and white atmometers instead of the absolute value of any one. Since the black atmometer absorbs and the white atmometer reflects nearly all the incoming radiation, the difference in evaporation between the two gives a good measure of the radiation intensity. Any attempt to estimate the potential evapotranspiration from the differences in evaporation between black and white atmometers becomes then a simplified energy budget approach.

In areas where the wind speeds do not vary greatly, a close correlation between atmometer and evaporation pan readings often exists. Observations in Lincoln, New Zealand, for the summer period from December 7, 1961, through February 5, 1962, gave the following linear relationship (Jessep, 1964):

$$Y = 0.0029 + 0.0025\,X$$

where Y is evaporation from a Class A pan in inches, and X is evaporation from an atmometer in cubic centimeters. The correlation coefficient was 0.987.

In Canada, Holmes and Robertson (1958) established a ratio of 0.0032 inch of open-pan evaporation and 0.0034 inch of evapotranspiration from irrigated fields for each cubic centimeter of water evaporated from a black Bellani plate atmometer based on observations from a large number of experimental farms. This relationship has been found to be so consistent that the Bellani plate has been used successfully as a guide to tomato, orchard, and pasture irrigation (Heeny, Miller, and Rutherford, 1961; Krogman and Hobbs, 1965; and Korven and Wilcox, 1965).

In California, Halkias, Veihmeyer, and Hendrickson (1955) used black and white atmometers to estimate crop consumptive use. They expressed the relationship in the following form:

$$U = KD$$

where U is the use of water by the crop in inches, K is the slope of the regression line, or a coefficient, and D is the difference in evaporation between black and white atmometers in cubic centimeters. The regression lines for various crops are shown in Figure 96, and the values of K are given in Table 15. The average correlation coefficient is as high as 0.98.

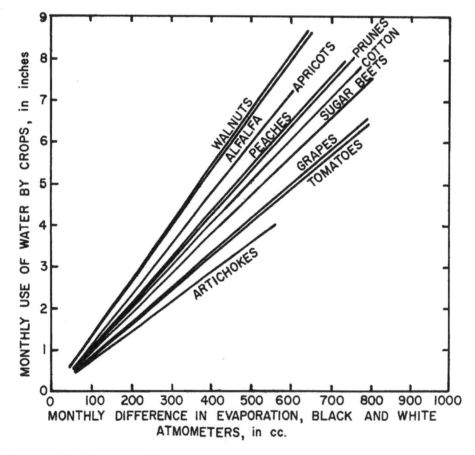

FIGURE 96. Mean monthly use of water by crops and mean monthly difference in evaporation between black and white atmometers (after Halkias, Veihmeyer, and Hendrickson).

TABLE 15

COEFFICIENTS IN THE FORMULA $U = KD$, AND THE CORRELATION COEFFICIENTS FOR WATER USE BY VARIOUS CROPS AND DIFFERENCE IN EVAPORATION BETWEEN BLACK AND WHITE ATMOMETERS

Crop	Coefficient K	Correlation coefficient
Alfalfa	0.0134	0.99
Walnuts	0.0135	0.97
Apricots	0.0120	0.95
Peaches	0.0110	0.98
Prunes	0.0108	0.98
Cotton	0.0105	0.97
Sugar beets	0.0096	0.99
Grapes	0.0086	0.98
Tomatoes	0.0082	0.98
Artichokes	0.0073	0.98

CHAPTER 19

WATER BALANCE

LIMITATIONS OF SOIL MOISTURE MEASUREMENTS

Methods of soil moisture measurements fall into two categories: measurements of moisture tension and measurements of water content. In the first group are included tensiometers and electric methods. In the second group, the oven dry method is still useful, though laborious; more recently the neutron scattering method has shown promise.

Tensiometers are the only instruments providing direct measurements of tension. A porous clay cell, filled with water and connected to a mercury manometer, is put into the soil. As the soil dries out, water moves from the cell into the soil, and the mercury in the manometer rises. As the soil becomes wet, water enters the cup, and mercury in the manometer falls. Tensiometers give accurate readings only over the wet end of the moisture range from 0 to 0.85 bar, which includes nearly 90 per cent of the available moisture from field capacity to the permanent wilting point for sandy soils, but only slightly over half for the finer textured soils. When the soil is dry, air enters the cup, and the tensiometer is no longer effective. The need for frequent refills of water is another shortcoming, which may hinder cultivation.

The electrical methods make use of the principle that heat conductivity in the soils can be used as an index of soil moisture. The sensing elements can be enclosed in resistance blocks made of different materials, such as gypsum, nylon, or fiber glass. Readings of nylon and fiber glass blocks are greatly affected by the salt content of the soil. The gypsum block is favored because its buffer capacity makes it less vulnerable than the other units to changes in the concentration of the soil solution. But it possesses properties that make it difficult to calibrate precisely and the calibration drifts with time. The gypsum blocks are more accurate than the tensiometer on the dry end of the moisture range. However, both tensiometers and electrical methods are subject to moisture hysteresis ambiguities.

The neutron scattering method, developed during the last fifteen years, is based on the principle that hydrogen is more effective in slowing down fast neutrons than any other elements and that most hydrogen nuclei present in the soil are found in water molecules. A source of fast neutrons and a detector of slow neutrons are housed in a probe that is lowered into a soil access tube. Fast neutrons, radiated into the soil, are absorbed by hydrogen and then are reflected back into the detector as slow neutrons. The count of slow neutrons is a measure of soil moisture content.

The neutron scattering method has distinct possibilities of being more accurate than any other method tried. Measurements can be taken repeatedly at the same site without disturbing the soil. However, some questions have arisen as to whether this type of equipment is suitable for measurements near the surface. In spite of this uncertainty, measurements of evapotranspiration by neutron scattering meters have been found to agree reasonably well with values obtained by weighing lysimeters (Bowman and King, 1965).

Nevertheless, none of the presently available methods of measuring soil moisture is satisfactory for routine field use by the farmer. Even when used in research work, these methods are expensive and time-consuming for operation and interpretation. Many more samples are needed for statistical reliability.

The practical limitations of the various methods of soil moisture measurements make it necessary to develop a simple method suitable for routine field operation. Recently, attention has been focused on micrometeorological methods. The rapid progress in the study of evapotranspiration has lead to the development of the water balance technique as a method of estimating soil moisture. This approach has much to recommend it: low cost of equipment, simplicity of the measurement, and reasonable accuracy of the results. For instance, Bartels (1965) has reported that over a number of seasons, evaporimeters are more accurate than gypsum blocks in estimating soil moisture in the root zone. Another advantage of the meteorological approach is that the estimation of the various components in the water balance equation lends itself to an intelligent planning of long-range water resources management; whereas the usefulness of soil moisture sampling is limited primarily to day-to-day operations.

THE WATER BALANCE EQUATION

The water balance in the soil can be stated by the following equation:

Rainfall + Irrigation water = Changes in soil moisture +
Evapotranspiration + Percolation + Runoff

If the moisture storage capacity of a soil is known, the water balance equation can be solved by comparing rainfall and irrigation water with evapotranspiration rate. The water balance can be computed on a daily, weekly, or monthly basis. For day-to-day agricultural operation, the daily water balance is much preferred, though weekly computation may give essentially the same results. For long-range water resources planning, the monthly computation may be adequate.

The bookkeeping form for water balance computation consists of six items, namely: precipitation, potential evapotranspiration, actual evapotranspiration, soil moisture storage, surplus, and deficit. Where irrigation water is applied, that item should also be included.

The bookkeeping method is simple and straightforward. Each day the amount of evapotranspiration is subtracted from soil moisture storage, while the precipitation is added. Rainfall in excess of the soil moisture storage capacity is regarded as a surplus, including both surface runoff and deep percolation. When the water balance reaches zero, drought occurs, and evapotranspiration ceases. On such a drought day, the difference between potential evapotranspiration and available moisture is registered as a deficit. For the entire period of water balance computation, the sum of the actual evapotranspiration and the deficit must equal the potential evapotranspiration, while the sum of the actual evapotranspiration and the surplus equals the precipitation.

Table 16 gives an example of daily water balance computation. The soil moisture storage capacity is assumed to be 2.00 inches. The running water balance can also be presented in graphical form as shown in Figure 97. However, the amount of surplus and deficit are not plotted in that graph.

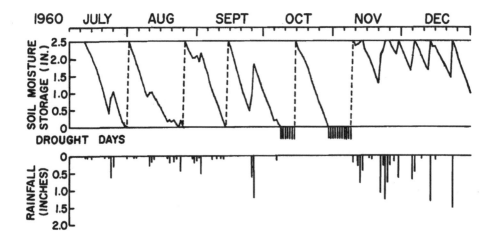

FIGURE 97. Running daily water balance at Waipio, Hawaii. Dash lines indicate irrigation.

TABLE 16 EXAMPLE OF DAILY WATER BALANCE COMPUTATION

STATION _____ YEAR _1963_ MONTH _July_ STORAGE CAPACITY _2.00_ inches

Day	1	2	3	4	5	6	7	8	9	10	11	12	13	14	15	16
PE	0.25	0.28	0.31	0.32	0.35	0.27	0.30	0.32	0.31	0.30	0.28	0.33	0.20	0.15	0.21	0.12
R	0.25	0	0	0.10	0	0.12	0	0	0	0	0.42	0	1.23	1.36	0.50	3.14
ST	2.00	1.72	1.41	1.19	0.84	0.69	0.39	0.07	0	0	0.14	0	1.03	2.00	2.00	2.00
AE	0.25	0.28	0.31	0.32	0.35	0.27	0.30	0.32	0.07	0	0.28	0.14	0.20	0.15	0.21	0.12
D	0	0	0	0	0	0	0	0	0.24	0.30	0	0.19	0	0	0	0
S	0	0	0	0	0	0	0	0	0	0	0	0	0	0.24	0.29	3.02

Day	17	18	19	20	21	22	23	24	25	26	27	28	29	30	31	Sum
PE	0.22	0.28	0.30	0.34	0.32	0.33	0.35	0.27	0.25	0.13	0.12	0.15	0.25	0.28	0.30	8.19
R	0	0	0	0.01	0	0	0	0	0.31	1.00	1.30	1.27	0.30	0	1.00	12.44
ST	1.78	1.50	1.20	0.87	0.55	0.22	0	0	0.06	0.93	2.00	2.00	2.00	1.72	2.00	
AE	0.22	0.28	0.30	0.34	0.32	0.33	0.22	0.13	0.25	0.13	0.12	0.15	0.25	0.28	0.30	7.19
D	0	0	0	0	0	0	0.13	0.14	0	0	0	0	0	0	0	1.00
S	0	0	0	0	0	0	0	0	0	0	0.11	1.12	0.05	0	0.42	5.25

PE - Potential Evapotranspiration ST - Storage D - Deficit
R - Rainfall AE - Actual Evapotranspiration S - Surplus

The water balance technique has been used to solve a number of problems such as irrigation interval control, water resources planning, yield forecasting, climatic classification, soil tractionability, stream flow, flood forecasting, worldwide fluctuations of sea level (van Hylckama, 1956), and forest fire forecasting (Nelson, 1959). The values of some items in the equation may vary somewhat according to the problems at hand. For example, the soil moisture storage capacity should be different for determining irrigation intervals of a crop with shallow rooting depth and for estimating stream flow. The limitations and the possible errors involved in the computation can be best understood by analyzing the various items individually.

MOISTURE STORAGE CAPACITY OF THE SOIL

The primary source of water available to plants is that held within the volume of soil invaded by roots. All soil moisture is not available to plants. Available water is commonly considered to be that portion held in the soil between the field capacity and the permanent wilting point. The field capacity is the amount of water held in the soil after excess water has drained away and the rate of downward movement has materially decreased, a state usually reached in two or three days after rain or irrigation. The field capacity is affected by so many factors, including soil texture, soil structure, organic matter content, and uniformity and depth, that it is not a precise constant. Yet it does serve as a practical measure of the upper limit of soil moisture in an unsaturated soil.

The concept of permanent wilting point was first introduced by Briggs and Shantz (1912). After conducting some 1,300 experiments in twenty soils, they concluded that all plants on a given soil reduce the moisture content of the soil to about the same extent when permanent wilting is attained. Thus, the permanent wilting point is a characteristic of the soil and not of the plant. Though several later studies (Burr, 1914; Alway, McDole, and Trumbull, 1919; and Batchelor and Reed, 1923) have demonstrated that some desert plants continue to deplete soil moisture to levels considerably drier than the permanent wilting point, the amounts withdrawn are too small to be of importance.

The permanent wilting point is usually reached at a tension of 15 atmospheres or at a *pF* of 4.2. However, the actual water content at the permanent wilting point varies with the soil texture. Fine-textured soils retain more water than coarse-textured soils at the wilting point. For example, the permanent wilting point for a sandy soil may be 3 per cent whereas that for a clay soil may be as high as 30 per cent when expressed on a weight basis.

In his study of global water balance, Thornthwaite (1948) assumed a storage capacity of ten centimeters for a normal soil, which he subsequently raised to 30 centimeters (Thornthwaite and Mather, 1955a). Such standard values are at best crude. Soils developed on recent lava may have such a small moisture storage capacity that even an annual rainfall of 100 inches can support only xerophytic plants. On the other hand, some deep alluvial soils may have a storage capacity well over 40 centimeters. Therefore, for practical agricultural purposes, the storage capacity of a soil should be determined on the spot. Where such measurements are not made, the table prepared by the U.S. Bureau of Reclamation (1951) for estimating available moisture per unit depth relative to soil texture may be used as a guide. In general, the total amount of usable water for plant growth is greater for clay than for coarse-textured soils.

Precipitation. The rain gauge is by no means a perfect instrument. During a period of strong wind, the turbulent eddies set up by the gauge itself may greatly reduce the catch of raindrops, especially if the gauge is elevated. On the other hand, rain splashing in may result in a gain. Another possible source of error is evaporation. Norum and Larson (1960) have reported that in Minnesota the daily evaporation loss from standard rain gauges may be as much as 0.010 inch. If observations are not taken daily, the evaporation loss may be significant during the summer season. Kurtyka (1953) has estimated that in Illinois the evaporation loss from the rain gauge is about 1 per cent.

The official weather bureau network of stations designed primarily for weather forecasting is not dense enough for planning irrigation and other agricultural activities. Brooks (1947), for example, has recommended that for forecasting purposes the number of permanent stations can be made as few as one for each zone of 1° or 2° of latitude and each zone of 500 to 2,000 feet altitude within a large coherent region, each serving some 2,000 square miles in rough or coastal terrain or some 2,000,000 square miles in interior plains country. Such a network cannot possibly give local variations of rainfall within a storm. Even for a network of one gauge per ten square miles, the ratio between the represented area and the sample area would exceed 1,000,000,000 for a gauge diameter of five inches. According to my experience in Hawaiian plantations, the minimum density needed for successful irrigation operation is one gauge per square mile. Thus, the density of rainfall stations needed for bringing out mesoscale variations is many times that for macroscale synoptic studies.

In the water balance computation, all light rains, no matter how small, should be credited. Light rains have often been neglected in water balance studies on the erroneous assumption that rain must get into the root zone to be fully effective. But in the first place, the plant may directly utilize water on the leaf surface. Moreover, Israelsen and Hansen (1962) have advanced the following argument from the standpoint of energy budget:

A light shower or rain remaining on the surface of leaves or on the ground can evaporate, utilizing the bulk of available energy. When this occurs, the transpiration of water through the plant will be reduced accordingly ...

For example, suppose a rainfall of only 0.1 inch occurred during the dry season and all the moisture was intercepted by leaves of the growing crop. Assume also that daily consumptive use was 0.25 inch per day. Experiments show that the bulk of consumptive use occurs during the day. Therefore, if consumptive use of 0.25 inch per day occurs over say 12 hours, the average use per hour would be 0.02 inch. During midday, the rate is likely to nearly double the average or 0.04 per hour. Hence, in 2.5 hours the rainfall of 0.1 inch would be evaporated and would have been fully utilized. Often overlooked is the fact that rainfall of 0.1 inch may represent [40%] of a normal daily consumptive use. Observations of the recovery of a wilted crop following a light shower will show that the rainfall is not without effect.

The benefit of light rainfall is usually accentuated for a plant community whose canopy is not closed. Isolated plants or clumps of plants can intercept more rainfall and concentrate it around their bases. The significant role played by light rainfall in the growth and survival of maize in arid East Africa has been demonstrated by Glover and Gwynne (1962). Owen and Watson (1956) reported that the beneficial effect of light rain on growth of sugar beets was especially pronounced after prolonged drought.

In areas where water balance computation has been carried out for only a short period, there is often a need for estimating long-period variations. Since rainfall is the most variable element in the water balance equation, an important first step would be to analyze the rainfall probabilites. The various statistical expressions of rainfall variability, discussed in standard textbooks (Conrad and Pollak, 1950, and Brooks and Carruthers, 1953), are designed primarily for single-station analysis. Recently, Landsberg (1951) devised a graphical method suitable for presenting rainfall probabilities over a large area. Figure 98 shows the probabilities (slanting lines) of having an annual rainfall of less than a given quantity (ordinate) as a function of the median annual rainfall (abscissae), based on the rainfall records of over sixty years in twenty sugar-plantation stations in Hawaii (Chang, 1963). Similar charts can also be constructed to show probabilities of monthly rainfall.

Another graphical method (Bierhuizen and de Vos, 1959) is to present probabilities of the differences between rainfall and potential evapotranspiration throughout the year for a single station as shown in Figure 99. Obviously, this can serve as a risk chart. However, this method, completely ignoring the soil factor, may grossly underestimate the surplus and overestimate the deficit. The results are accurate only in areas where the soil moisture storage capacity is large and heavy storms infrequent.

Water Balance 201

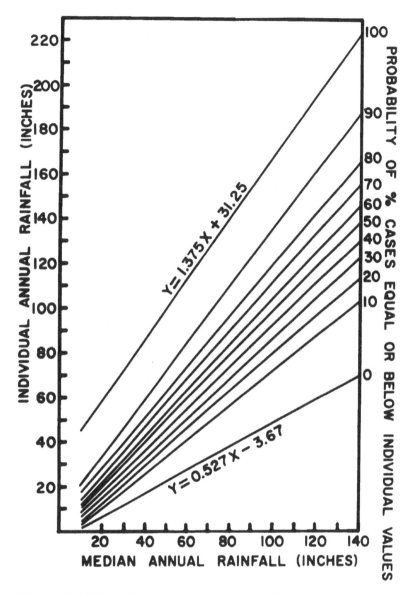

Figure 98. Probabilities of having individual rainfall amounts in Hawaii as function of median annual rainfall.

Evapotranspiration. The various methods of measuring or estimating potential evapotranspiration have been discussed in Chapters 14-18. The selection of the method will depend on the problem at hand and the data available. The lysimeter and the aerodynamic approach are primarily research tools for the better understanding of the physical process. Empirical formulae should not be adopted unless their accuracy in a given area is confirmed by direct measurements. The Penman equation or the simple energy budget approach is adequate only when the advected energy is small. For the determination of potential maximum evapotranspiration the

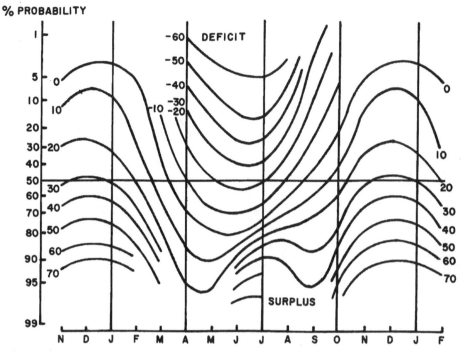

FIGURE 99. Probability of differences between rainfall and evapotranspiration during year. Calculated for twenty-day periods (after Bierhuizen and de Vos).

evaporation pan, when properly handled, is probably the best field instrument, although the apparatus for eddy correlation technique shows great promise.

In the simplest form of water balance computation, it is generally assumed that the rate of evapotranspiration does not change with the stage of crop development and with the change in soil moisture conditions between field capacity and permanent wilting point. These assumptions are unrealistic. Some of the methods for estimating potential evapotranspiration have serious limitations in that they cannot be applied to a crop with incomplete ground cover. In this respect, the use of the evaporation pan has decided advantage. The ratio between the evapotranspiration rate and the pan evaporation can vary according to the age of the crop. For a two-year sugar cane crop, the following ratios have been adopted in water balance computation as a guide to irrigation practice in some Hawaii plantations:

TABLE 17

RATIOS BETWEEN EVAPOTRANSPIRATION OF SUGAR CANE AND EVAPORATION FROM A U.S. WEATHER BUREAU CLASS A PAN

Age in months	0-1	1-2	2-3	3-4	4-5	5-6	6-18.5	18.5-22.5	22.5-24
Ratio	0.4	0.45	0.55	0.75	0.90	0.95	1.00	0.85	0

It is a common practice for sugar cane culture in Hawaii that irrigation water is withheld during the last six weeks in order to insure proper ripening. During the four months before ripening, the ratio is also slightly lower because of the senescence of the crop. Furthermore, this process of applying less water during the ripening period often increases the sugar content per ton of cane. Hendrickson and Veihmeyer (1950) and Weger (1953) likewise reported that irrigation applied late in the season had little effect upon grape yield.

The relationship between the evapotranspiration rate and the soil moisture tension has been discussed in Chapter 13. In the simple water balance computation, the model proposed by Veihmeyer and Hendrickson is adopted. This model holds true in a humid, cloudy climate, particularly if the soil is heavy and covered by dense vegetation. In an arid climate the depletion rate should decline with the increase of soil moisture tension, either as proposed by Pierce or by Thornthwaite and Mather. Such a procedure of adjusting the evapotranspiration rate progressively downward is known as a modulated water balance method. Holmes and Robertson (1959) found that in the arid climate of western Canada the modulated water balance computation gave results that coincided closely with the Coleman moisture block readings, and, hence, are related closely to dryland wheat yield (Figure 100).

The selection of the proper depletion curve is not simple. Ideally, it should be determined by soil moisture measurements on the spot. A good example is the work by Eagleman and Decker (1965), who derived a curvilinear relationship between relative evapotranspiration and soil moisture deficit for soybeans in Missouri of the form:

$$Y = 0.987 - 0.00262 X - 0.000028 X^2$$

where Y is the ratio of the measured evapotranspiration rate to the potential evapotranspiration and X is the percentage of soil moisture deficit.

In the absence of such direct measurements, the adjustment of the evapotranspiration rate can be no better than an educated guess. For some parts of the world the tables prepared by Thornthwaite and Mather (1957) and the nomograph by Wartena and Veldman (1961) are useful references.

Deficit. In the water balance computation, water deficit or drought occurs whenever the soil moisture is depleted to the permanent wilting point. The frequency of drought days will depend on the rainfall pattern, the moisture characteristics of the soil, the rooting depth, and the evapotranspiration rate. Drought, thus evaluated, is far more meaningful than direct interpretation of rainfall records, as in the work by Blumenstock

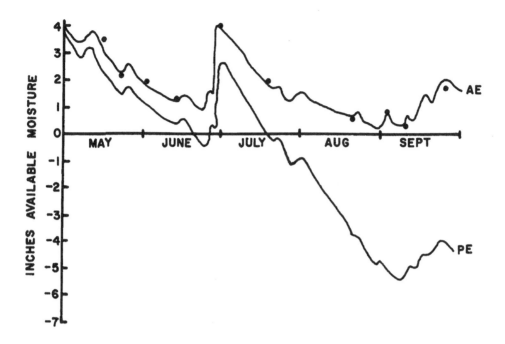

FIGURE 100. Soil moisture loss calculated by modulated water balance (AE) and simple water balance (PE). Circle indicates spot Coleman moisture block readings (after Holmes and Robertson).

(1942), or than the common definition that a drought is a period of at least fifteen consecutive days without a minimum of 0.04 inch of rainfall. The fallacy of the pure meteorological approach has been discussed by Rickard (1960):

> Fifteen consecutive days without rain in mid-winter would probably result in only a slight lowering of the percentage of moisture in the soil from field capacity, but the same rainless period in mid-summer may bring the soil from field capacity to close to the permanent wilting point. At this point a fall of rain of 0.01 inch would technically break the drought but would have no effect on the dry soil.

The amount of deficit is quite sensitive to a change in the moisture storage capacity of the soil. Van Bavel (1953b) investigated the dependence of drought days on the moisture storage capacity at Raleigh, North Carolina, for a period of 59 years, 1892-1951. He found that the number of drought days during the growing season, May 15 to August 27, was at least twenty days in any year for a storage capacity of one inch, but would be zero in the wettest year for a storage capacity of three inches or more (Figure 101). The average number of drought days decreased from 45

Water Balance   205

to 10 as the moisture storage capacity increases from one to four inches (Figure 102).

Surplus. In the water balance computation, surface runoff and deep percolation are lumped together as surplus. Surplus occurs whenever the rainfall exceeds the moisture storage capacity. This is only an approximation. In actuality, runoff is the excess of rainfall over infiltration rather than over storage. Furthermore, the disposal of excess water in the form of percolation does not take place immediately. The process takes a day or longer. While significant errors may occur in daily computation because of incorrect assumptions regarding runoff and percolation, they cannot be large when applied to a period of a month or longer, especially over a large area. Ligon, Benoit, and Elam (1964) compared the soil moisture surplus, computed by the water balance method, with the measured flow of Elkhorn Creek in Kentucky (Figure 103). In general, the agreement is quite good. They attributed part of the difference to the fact that only one station was used to represent precipitation over the entire watershed. There is also a tendency for stream flow to lag behind the computed surplus.

The estimation of surplus water is of considerable practical importance to farmers. It gives not only an indication of the possible extent of soil erosion, but also of the intensity of the leaching of chemical nutrients in

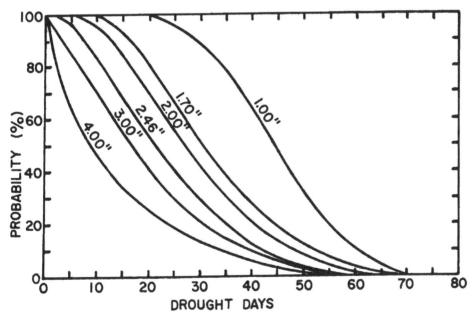

FIGURE 101. Curves showing empirical probability to have more than given number of drought days per season for varying storage capacity at Raleigh, North Carolina. Data of 59 years, 1892-1951, May 15-August 27 (after van Bavel).

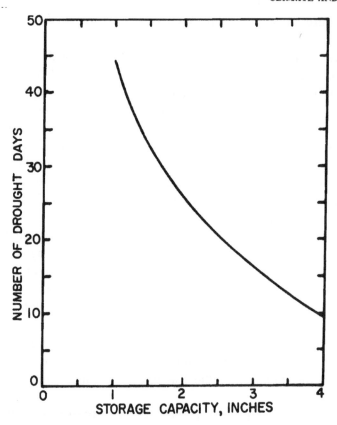

FIGURE 102. Dependence of average number of drought days per season on the soil moisture storage capacity at Raleigh, North Carolina. Data of 59 years, 1892-1951, May 15-August 27 (after van Bavel).

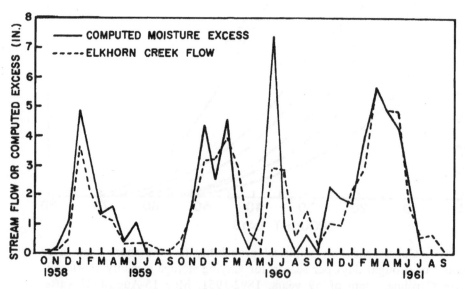

FIGURE 103. Comparison of computed monthly soil moisture excess with measured flow of Elkhorn Creek (after Ligon, Benoit, and Elam).

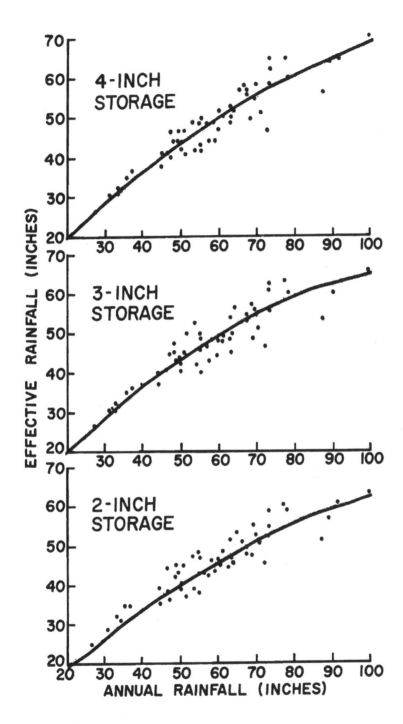

FIGURE 104. Relationships between annual rainfall and effective rainfall in Kohala, assuming varying moisture storage capacities of two, three, and four inches.

the soil. Nelson and Uhland (1955), for instance, have demonstrated that quantitative estimate of the surplus is required for judicious use of fertilizer.

The difference between rainfall and surplus is a measure of the effective rainfall. The effective rainfall is much less variable than rainfall and, when computed on a daily basis, it can be accurately determined with a short period of records, perhaps in three to five years. The effective rainfall is affected by the storage capacity, but the maximum difference caused by using two different values of storage capacity usually does not exceed the difference in storage capacity times the number of heavy rains exceeding the larger of the two. I have computed the effective rainfall at Kohala, Hawaii (Chang, 1963), by assuming varying storage capacities of two, three, and four inches. As shown in Figure 104, the effective rainfall increases with the storage capacity only slightly, especially when the annual rainfall is small. A few selected values of effective rainfall as a function of the annual rainfall at Kohala are given in Table 18.

TABLE 18

EFFECTIVE RAINFALL AS A FUNCTION OF ANNUAL RAINFALL AT KOHALA (ASSUMING VARYING SOIL MOISTURE STORAGE CAPACITIES OF 2, 3, AND 4 INCHES)

Annual rainfall (inches)	40	50	60	70	80	90	100
Effective rainfall for:							
two-inch storage	34	40	46	52	56	60	61
three-inch storage	37	43	49	55	58	62	65
four-inch storage	37	44	50	56	60	65	69

Hershfield (1962) has made a preliminary study of the effective rainfall in the United States assuming a soil moisture storage of three inches. His map indicates that the fraction of rainfall that is effective varies from 94 per cent in the arid West to 68 per cent in the humid South, where heavy storms are frequent. Detailed world maps of effective rainfall are urgently needed for agricultural planning.

CHAPTER 20

WATER AND YIELD RELATIONSHIP

The usefulness of water balance as a means of estimating soil moisture does not mean that the method alone is adequate for intelligent irrigation and water resources planning. Another key to the problem is the understanding of the relationship between water and yield. At what moisture status should irrigation be applied in order to obtain the maximum yield? What would be the yield loss if water application is inadequate?

IRRIGATION PRACTICE FOR MAXIMUM YIELD

Chapter 12 discussed de Wit's experimental evidence that for many if not all crops, dry matter production is proportionate to the amount of transpiration. Adequate application of water to meet the transpiration need will not only insure rapid leaf development when the crop is young but also maximum photosynthesis when the crop has reached the optimum leaf area index. Thus, for crops whose yield consists wholly, or in large part, of vegetative growth, maximum yield is usually obtained by applying water at the rate of potential evapotranspiration. Penman (1956b) and Dreibelbis and Harrold (1958) found that this expectation was fulfilled by grass. Pearson, Cleasby, and Thompson (1961) reported that the application of water at the rate of potential evapotranspiration produced a sugar cane crop in Natal, South Africa, far in excess of what was at one time thought possible for that area.

Therefore, Hagan and Vaadia (1961) have considered it a valid general principle to use the potential evapotranspiration as a guide to water application for the maximum production of a crop with a fully developed canopy. For a crop that does not completely cover the ground, the amount of water required for maximum production would be less. This is in fact one of the basic principles of dryland farming that seeks to conserve water by having a lower plant density. Of course, other things being equal, in areas of reasonably high radiation the maximum yield of a crop with an incomplete cover is usually less than that of a crop with a complete ground cover.

An interesting example is found in rice culture. For centuries, people held the opinion that both the water need and the yield potential of upland rice were substantially lower than those of paddy rice. Only recently have scientists realized that when either upland or paddy rice is adequately watered, the potential evapotranspiration is the same and when either completely covers the ground, the potential yields are also comparable. Under the ideal irrigation and cultural practices, the maximum yield is determined by environmental conditions as well as varietal characteristics, such as saturation light intensity, leaf arrangement, respiration rate, and the like.

One limitation of the use of potential evapotranspiration as a guide to irrigation is the fact that water deficit may have different effects on various aspects of plant growth (Hagan, Peterson, Upchurch, and Jones, 1957). Moisture deficit may severely retard vegetative growth but may have little effect upon the economical yield of some crops. This has been commonly noted for crops whose economical yield is a chemical constituent or a reproductive organ. The effect of moisture supply on the distribution of dry matter in the plant is a complicated subject (Brouwer, 1962a, 1962b).

The potential evapotranspiration specifies only the total amount of water required, but not the frequency of water replenishment. As I pointed out in Chapter 12, the photosynthetic rate for some crops falls quickly with a slight drop of moisture, while for others it may remain at a high level until the permanent wilting point is reached. Other things being equal, the former should be irrigated more often than the latter, but with a reduced amount of application per round.

Other factors should be considered as well. Some crops grow best under high soil moisture; others require a good aeration for maximum productivity. Paddy rice, for one, is grown customarily in waterlogged conditions, partly because the crop can take up oxygen through the roots (Goto and Tai, 1956, and Barber, 1962), and partly because the paddy serves as an effective means for weed control. In order for a plant to survive and grow under poor aeration, it must possess some or all of the following abilities (Bartlett, 1961): (1) a low oxygen requirement, (2) a structure of leaf and stem tissue that allows direct diffusion of atmospheric oxygen to the roots, (3) the use of oxygen obtained from the reduction of nitrate, (4) a specific tolerance to the toxic substance formed under waterlogging, and (5) an ability to carry out anaerobic respiration.

Experimental evidence has shown that plants such as lettuce, spinach, radishes (Bierhuizen, 1959), cauliflower (Salter, 1959), celery (Cannell,

Tyler, and Ashbell, 1959), grass (Doss, Bennett, Ashley, and Weaver, 1962), alfalfa (Lucey and Tesar, 1965), cotton (Sreenivasan, 1949), tobacco (van Bavel, 1953a), sweet potatoes (Jones, 1961), bananas (Shmueli, 1953), and peaches (West and Perkman, 1953) produce the maximum yield only when the soil mosture is kept at a high level of at least 50 per cent of the total moisture.

Other crops, such as tomatoes (Moore, Kattan, and Fleming, 1958), snapbeans (Kattan and Fleming, 1956), barley, sugarbeets (Penman, 1956b), Irish potatoes (Bonnen, McArthur, Magee, and Hughes, 1952; Bradley and Pratt, 1954; and Fulton and Murwin, 1955), peanuts (Mantell and Goldin, 1964), pears (Hendrickson and Veihmeyer, 1942), and apples (Magness, Degman, and Furr, 1935) do not respond with increase in yield from increases in soil moisture above the 50 per cent level. Once the range of optimum soil moisture is known, it is easy to determine at what point in the water balance computation that irrigation should be applied in order to assure maximum yield.

The development of irrigation agriculture in semiarid and arid lands has a long history in human civilization. However, the benefit of irrigation in many a subhumid and humid climate has not been realized fully until the recent progress in the study of potential evapotranspiration. In humid climates, the use of water balance as a guide to irrigation will not only bring about the maximum yield but will also keep the farmer from applying too much water. Among those to report that excessive water application may reduce the yield are Grainger, Sneddon, Chisholm, and Hastie (1955), for beans; Taylor (1961), for alfalfa; Jacob, Russell, Klute, Levine, and Grossman (1952), for potatoes; Chang, Campbell, and Robinson (1963), for sugar cane; Lovett (1953), for tobacco; Watson (1963), for wheat; Mantell and Goldin (1964), for peanuts; and Bourget, Finn, and Dow (1966) for flax and cereals. Apart from the leaching of soil nutrients, excessive water may restrict root development as reported by Aslyng and Kristensen (1958).

RELATIONSHIP BETWEEN ACTUAL EVAPOTRANSPORATION AND YIELD

When the actual evapotranspiration falls short of the potential, the actual yield will also be less than the maximum. However, the relationship between evapotranspiration and yield in the field may or may not be linear as it is between transpiration and dry matter production in container experiments. This is partly because the fraction of evaporation that does not contribute to plant growth varies throughout the crop life cycle.

Even when the dry matter production does increase linearly with evapotranspiration, the regression line seldom passes through the zero point as is often the case in the relationship between transpiration and yield (Viets, 1962). In other words, evapotranspiration in the field might be appreciable when the yield is still zero. Allison, Roller, and Raney (1958) analyzed the yields of a number of crops grown in the lysimeter near Columbia, South Carolina, for a period of more than five years. The data indicated that the first 18 inches of evapotranspired water was required to produce only enough for plant survival (Figure 105). The increase in dry matter was almost linear with increasing amounts of water

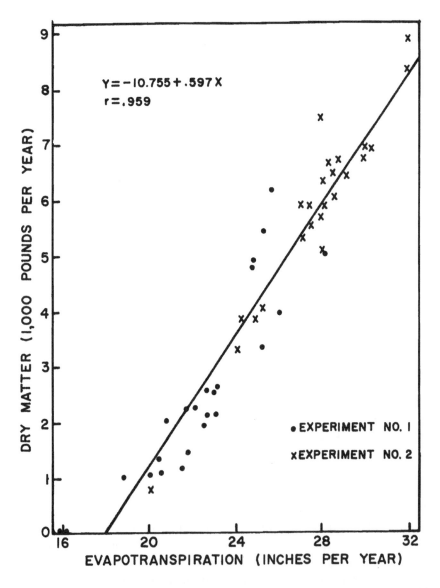

FIGURE 105. Relationship between crop yields and water use (after Allison, Roller, and Raney).

Water Yield Relationship 213

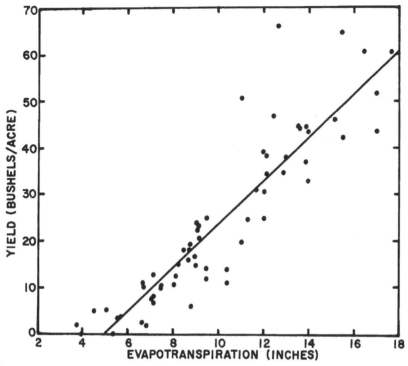

FIGURE 106. Relationship between wheat yield and evapotranspiration in tanks, 1922-1952 (after Staple and Lehane).

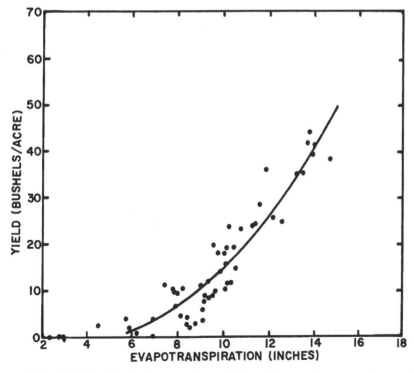

FIGURE 107. Relationship between wheat yield and evapotranspiration on field plots (after Staple and Lehane).

used from 18 to 32 inches. Staple and Lehane (1954) studied the use of water by spring wheat grown in tanks and in the open field in Swift Current, Canada. They reported that 4.9 inches of water for tanks and 5.64 inches for field were necessary to establish the plants (Figures 106 and 107). Beyond this, the yield in the tanks increased nearly linearly; whereas that in the field increased curvilinearly. In either case, the maximum production potential was not realized because of the shortage of water.

For the study of irrigation and water management, one must not only determine the water and yield relationship in its entire range, but also devise a method for comparing relationships obtained in different areas or different years. For the latter purpose, the use of the ratio between actual and potential evapotranspiration would be an improvement over the use of the absolute value of evapotranspiration. One example of the use of the ratio is found in the study of the relationship between water and su-

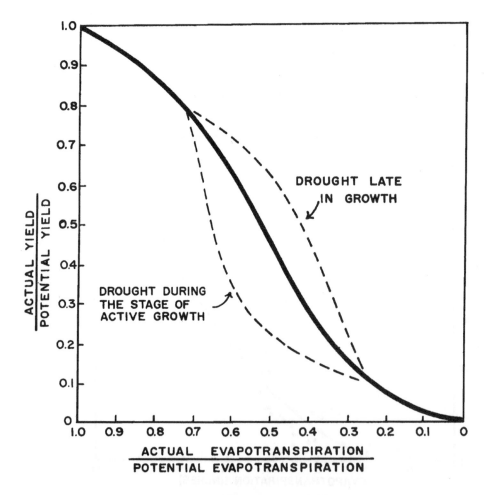

Figure 108. Generalized relationship between yield and adequacy of water application.

gar cane yield in Hawaii (Chang, Campbell, and Robinson, 1963). Figure 108 shows a generalized relationship between yield and adequacy of water application according to experimental and field data. The curve may be divided into three portions. The upper portion of the curve with a ratio between 0.75 and 1.00 represents conditions of relatively adequate water application. There is no drought severe enough to affect the metabolism of the plants.

The middle portion of the curve with a ratio between 0.40 and 0.75 is subject to large variations. Prolonged droughts are so frequent that their effect on yield will vary greatly according to the time of occurrence, plant response, and the like. If the drought occurs during the stage of active growth from twelve to fourteen months, the damage will be much more severe than if the drought sets in late in the growth. Most crops have "moisture-sensitive periods" during which a water deficit depresses the economic yield much more than at other periods. These critical periods for a number of crops are summarized in Table 19. Varieties may also respond differently under drought conditions. A drought-resistant variety may follow the upper broken curve in Figure 108, whereas a variety less resistant to drought may follow the lower broken curve.

The portion of curve with a ratio of less than 0.40 is not based on actual yield data. It is drawn for the sake of completeness and is not significant in irrigation practice.

For some perennial plants, the relationship between water and plant growth can be best understood only by analyzing the data over a long period. Zahner and Stage (1966) used a modulated water balance technique to compute the water deficit in a red pine stand in southern Arkansas. They found that the shoot growth of the red pine was closely related to the combined water deficits in the preceding summer and the current year as shown in Figure 109.

IRRIGATION EXPERIMENT

The relationship between water and yield can be established either from field yield data or through an irrigation experiment. The latter approach is preferred not only because it minimizes the number of variables, but also because it can be set up in such a way as to encompass a wide range of moisture treatments. The use of a water balance technique as a guide for an irrigation experiment has produced results far more meaningful than the simple counting of the number of irrigation rounds or than the measurement of the soil moisture tension alone.

If the evaporation pan is used for estimating the potential evapotranspiration of a crop, one method of conducting an irrigation experiment is to vary the rate of evapotranspiration in various treatments and to irrigate when the water balance reaches zero, according to the assumed evapotranspiration rate. For example, one may assume that the potential

TABLE 19
CRITICAL MOISTURE-SENSITIVE STAGE FOR SELECTED CROPS

Crop	Critical Moisture-Sensitive Stages	Reference
Cauliflower	No critical moisture-sensitive stage, frequent irrigation required from planting to harvest	Salter (1961)
Lettuce	Just before harvest when the ground cover is complete	Sale (1966)
Cabbage	During head formation and enlargement	Janes and Drinkwater (1959); Vittum, Alderfer, Janes, Reynolds, and Struchtemeyer (1963) Drew (1966)
Broccoli	During head formation and enlargement	Singh and Alderfer (1966)
Radishes and onions	During the period of root or bulb formation	Singh and Alderfer (1966)
Snap Beans	During flowering and pod development	Kattan and Fleming (1956)
Peas	At the start of flowering and when the pods are swelling	Salter (1962, 1963)
Turnips	From the time when the size of the edible root increases rapidly till harvest	Stanhill (1958)
Potatoes	After the formation of tubers	Winter (1960); Struchtemeyer (1960); Taylor and Rognerud (1959)
Potatoes (white rose)	From stolonization to the beginning of tuberization	de Lis, Ponce, and Tizio (1964)
Soybeans	Period of major vegetative growth and blooming	Runge and Odell (1960)
Oats	Commencement of ear emergence	van der Paauw (1949)
Wheat	During heading and filling	Lehane and Staple (1962)
Barley	The effects of water stress on grain yield and protein content were greater at the early boot stage than at the soft dough stage, and they were greater at the soft dough stage than at the onset of tillering or ripening stages	Wells and Dubetz (1966)
Corn	Period of silking and ear growth	Denmead and Shaw (1960); Robins and Domingo (1953); Runge and Odell (1958); Howe and Rhoades (1955)
Cotton	At the beginning of flowering	Marani and Horwitz (1963)
Apricots	Period of floral buds development	Uriu (1964)
Cherries and peaches	Period of rapid growth prior to maturity	Hildreth, Magness, and Mitchell (1941)
Olives	Later stages of fruit maturity	Crider (1922); Spiegel (1955)

Water Yield Relationship 217

evapotranspiration of a crop varies from 0.4, 0.6, 0.8, 1.0, and 1.2 of the pan evaporation in an experiment. These ratios may be called the control pan ratios. If the soil moisture storage capacity is 3.0 inches, and the daily pan evaporation 0.3 inch, then in the absence of rainfall, the irrigation interval would be 25, 16.7, 12.5, 10.0, and 8.3 days for the control ratio of 0.4, 0.6, 0.8, 1.0, and 1.2, respectively. The control ratio may be expressed as follows:

$$C = \frac{N \times M}{E - Re}$$

where C is the control pan ratio; N the number of irrigation rounds; M the soil moisture storage capacity; E the potential evapotranspiration; and Re the effective rainfall.

At the end of the experiment, the total water used, or the actual evapotranspiration (expressed as a fraction of the potential evaportanspiration), is known as the effective pan ratio. If the potential evapotranspiration of the crop in the experiment equals the pan evaporation, then both the plots with 1.0 and 1.2 control pan ratio should have an effective

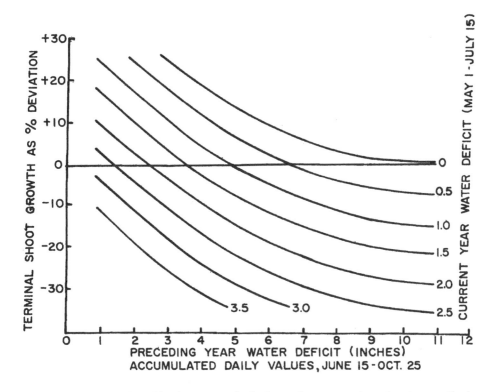

FIGURE 109. Relationship between deviations from average shoot growth in red pine and previous and current season's water deficits. Correlation coefficient = 0.85 (after Zahner and Stage).

pan ratio of 1.0. Of course the latter plot will have a greater amount of surplus water. The effective pan ratio, A, may be expressed as follows:

$$A = \frac{Re + (N \times M)}{E}$$

The effective pan ratio can be precisely determined only by computing a daily water balance throughout a crop cycle.

Control ratios are only predetermined values for running an irrigation experiment. The control ratio gives no indication of the total water used by the plant since the effective rainfall is excluded. In evaluating the relationship between water and yield, only the effective pan ratio should be considered. Any attempt to interpret irrigation yield results by the control pan ratio or, for that matter, the number of irrigation rounds alone would lead to faulty conclusions.

The effective pan ratio and the control pan ratio can be related as follows:

$$A = C + (1 - C)\frac{Re}{E}$$

In the absence of rainfall, the effective pan ratio clearly would be equal to the control pan ratio, except in the plots where the water application exceeds the potential evapotranspiration. In the presence of rainfall, the effective pan ratio will invariably exceed the control pan ratio for those plots whose water applications are less than the potential evapotranspiration. The relationship between effective and control pan ratios as a function of the ratio between effective rainfall and potential evapotranspiration can be presented in graphical forms as in Figure 110. By combining Figure 110 and the yield curve in Figure 108, one can construct a diagram to show the relative yields to be expected by applying water with different control ratios in different moisture regimes (Figure 111). The possible yield increase resulting from irrigation decreases gradually as the ratio between the effective rainfall and potential evapotranspiration increases. Such a diagram can aid in an analysis of costs and profits.

EFFICIENCY OF WATER USE IN DRY MATTER PRODUCTION

Once the water and yield curve is established, the efficiencies of water use for various irrigation treatments can be computed. Dreibelbis and Harrold (1958) analyzed the yield data for corn, wheat, and meadow crops in the lysimeter experiment at Coshocton, Ohio. They found that the efficiency of water use (pound of dry matter per inch of water) in-

Water Yield Relationship

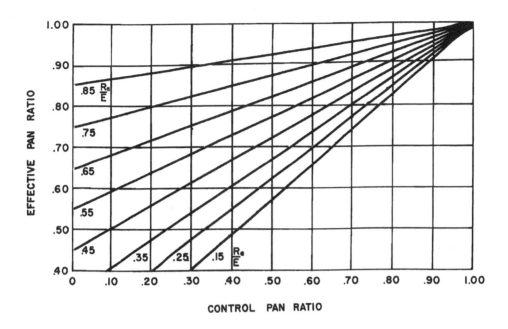

FIGURE 110. Relationship between effective and control pan ratios.

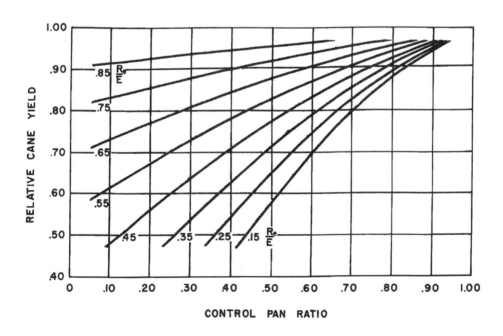

FIGURE 111. Relative yields to be expected by applying water with different control ratios in different moisture regimes.

creased with both the yield and water used up to the potential rate. The relationship for hay is presented in Figure 112. They explained the results for corn:

> With water supplied as needed, the efficiency of water use was greatly increased; 831 pounds of dry matter were produced per inch of water consumed, compared to an average of 552 pounds of dry matter per inch in the three years of no irrigation. With irrigation, only 273 pounds of water were used in producing one pound of dry corn and stalk; without irrigation, 438 pounds were used.

There are also instances in which the highest efficiency of water use is reached at a point considerably below the potential evapotranspiration. For example, the water-use efficiency of sugar cane, computed from the curve in Figure 108, increases with the effective pan ratio up to a value of 0.77 and then declines gradually (Figure 113). The efficiency at the ratio of 1.0 is about 8 per cent lower than at the ratio of 0.77. With corn (Howe and Rhoades, 1955, and Davis and Hagood, 1961) and grain sorghum (Musick and Grimes, 1961, and de Wit, 1958), the maximum efficiency of water use occurs at about 90 per cent of maximum yield.

The level of water application at which the maximum efficiency is reached can be artificially reduced by using a lower plant density. As was discussed in Chapter 18, the actual evapotranspiration rate is lower in an incomplete vegetation cover than in a reasonably well developed canopy. As the evaporation component increases, the evapotranspiration rate is often reduced at a rate faster than the reduction of dry matter production. Thus the efficiency of water use is usually higher in dryland farming than in irrigated agriculture.

Plants vary in their efficiency of water use. De Wit (1958) and de Wit and Alberda (1961) analyzed the yield data of a number of crops grown in containers. They found that sorghum and corn have much higher efficiency than wheat, oats, and barley. Alfalfa has the lowest efficiency among the crops tested. The high efficiency of sorghum and corn is a definite advantage for cultivation under conditions of limited water supply.

For the same plant species, the efficiency of water use may vary according to the climate. Stanhill (1960) has compared measurements of pasture growth and potential evapotranspiration at seven localities in different parts of the world. In Figure 114, the cumulative measured dry-weight yields are plotted against cumulative measured transpiration. A linear relationship exists at each site, but the slope of the line changes with latitude. In general, the growth rate per unit of water used is higher at high latitudes. This is a result of the increased respiration rate and the increased proportion of incoming radiation retained as net radiation in the tropics.

Water Yield Relationship

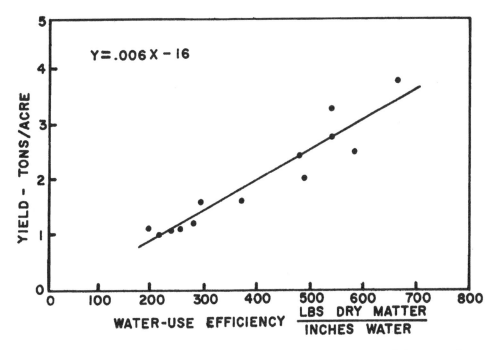

FIGURE 112. Relationship of yield of hay to water use efficiency (after Dreibelbis and Harrold).

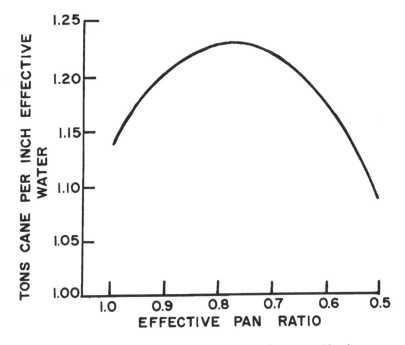

FIGURE 113. Efficiency of water use as a function of effective pan ratios (courtesy of Hawaiian Sugar Planter's Association).

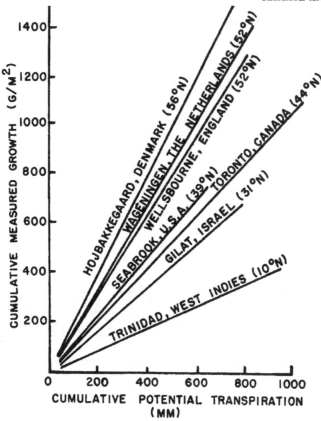

FIGURE 114. Measurements of potential evapotranspiration and dry matter production from pastures (after Stanhill).

Cultural practices, particularly fertilization, can alter significantly the efficiency of water use. A number of investigators[1] have reported that as long as the crop canopy is reasonably well developed, proper fertilization markedly increases crop yields without an appreciable change in water use. In well-watered fields, proper fertilization can bring about the full potential of crop growth. In times of water shortage, fertilized crops have been found to be more resistant to drought because of a better developed root system, which increases the capabilities of moisture extraction (Smith, 1953).

EFFECT OF IRRIGATION ON CROP QUALITY

The preceding sections have described only the relationship between water and dry matter production. The quality of an agricultural product, however, is not necessarily related to the yield. In analyzing the relation-

---

1. Allison, Roller, and Raney (1958); Weaver and Pearson (1956); Hanks and Tanner (1952); Kelley (1954); Gard, McKibben, and Jones (1961); Carlson, Alessi, and Mickelson (1959); Zubriski and Norum (1955).

TABLE 20

Effect of Irrigation on Crop Quality

Crop	Effect	Reference
Pasture	Irrigation increased the protein and decreased the fat contents of the herbage but had little effect on the crude fiber and ash content.	Turley, Webster, and Carson (1963)
Vegetable crops	Maintaining a low moisture stress during the whole growth period generally resulted in the highest yield and quality.	Bierhuizen and de Vos (1959)
Snap beans	Irrigation decreased the percentage of pods that were badly crooked or severely malformed. Fibrous content of beans was generally reduced.	Vittum, Alderfer, Janes, Reynolds, and Struchtemeyer (1963)
Sweet corn	Irrigation significantly increased the number of marketable ears per plants, the average weight per ear and the gross yield of unhusked ears and the percentage of usable corn cut from these ears for canning or freezing.	Vittum, Alderfer, Janes, Reynolds, and Struchtemeyer (1963)
Soybeans	Irrigated soybeans had slightly lower oil content and slightly higher protein content.	Schwab, Shrader, Nixon, and Shaw (1958)
Barley	Irrigation significantly increased the yield of grain and improved malting quality, mainly by increasing extract.	Bendelow (1958)
Potatoes	Irrigation that gave good increase in yield of potatoes very seldom reduced the specific gravity and was more likely to increase it.	Jacob, Russell, Klute, Levine, and Grossman (1952)
Tobacco	Irrigated tobacco had lower nicotine and protein, but higher carbohydrate content.	Anita (1949)
Tobacco	Adequate irrigation meant lower content in the cured leaves for nicotine, total nitrogen, CaO and MgO, but higher content for sugars, as well as better burning characteristics, high yield and higher market value.	van Bavel (1953a)
Tobacco	Irrigation produced higher yields of leaves of satisfactory quality.	Lovett (1953)
Fruits	Canned peaches that were tough and leathery in texture, pears that remained green and hard a week or more after the ripening season, prunes that were sunburned, and walnuts with partly filled shells were some of the results of a relatively long time without readily available moisture.	Veihmeyer and Hendrickson (1956)

TABLE 20 (Continued)

Crop	Effect	Reference
Cantaloupes	Heavier irrigation increased cull fruits associated with increased vine growth and succulence.	Flocker, Lingle, Davis, and Miller (1965)
Olives	The higher yield obtained by irrigation is due to an increase in fruit size, rather than in the number of fruits. Irrigated groves have a higher oil content than unirrigated ones.	Samish and Spiegel (1961)

ship between water and crop quality, one must differentiate between natural rainfall and controlled irrigation water. While rainfall usually is accompanied by high cloudiness and low radiation, the application of irrigation water is not complicated by a change of unfavorable weather conditions. The common misconception that irrigation exerts an adverse effect on crop quality is often deduced from a comparison of crop quality, which is usually better in a moderately dry than in a wet year.

The quality of a crop is affected by fertilizer application, radiation intensity, diurnal temperature range, rainfall during the ripening period, and the like. Irrigation as a cultural practice should not create a microclimatic or soil condition unfavorable for good crop quality. In fact, the radiation balance may even increase slightly after irrigation, a desirable change. Budyko and Pogosian (1959) attributed this to (1) an increase in absorption of short-wave incoming radiation as a result of decreased reflectivity, and (2) a decrease in outgoing radiation because of the decreased daytime temperature and increased humidity in the lower layer of the atmosphere.

The results of a number of experiments concerning the effects of irrigation on crop quality are summarized in Table 20. In general, adequate irrigation throughout periods of active vegetative growth results in an improvement in crop quality. For some crops, a detrimental effect on crop quality because of a moisture deficit at one stage of crop growth cannot be compensated for by later irrigation.

However, during the ripening period, moderate moisture stress has often been found to be desirable, especially in the case of certain compounds such as rubber, sugar, and tobacco. Wadleigh, Gauch, and Magistad (1946) reported that the rubber content of guayule is increased by slight moisture stress. In sugar cane culture, the withdrawal of irrigation water several weeks before harvest is a common practice. Late water stress has also been found to increase the sucrose concentration of sugar beets (Loomis and Worker, 1963). The aroma of Turkish tobacco is improved by water stress late in the crop cycle (Wolf, 1962). The flavor and taste of most fruits can also be enhanced by the same means (Aldrich, Lewis, and Work, 1940). For many crops, the regulation of water could be employed to some extent as a means of controlling the composition of the agricultural product.

CHAPTER 21

DEW, FOG, AND HUMIDITY

In the water balance computation, only daily rainfall equal to or in excess of 0.01 inch is credited. Dew, fog, and mist are neglected. These minor hydrometeors, though supplying only a small quantity of moisture, could be of considerable agricultural significance in some areas.

DEW

Dew has been defined by Long (1958) as "the deposition of water drops by *direct condensation* of water vapor from the adjacent *clear air,* in general upon surfaces cooled by nocturnal radiation." The instrument used to measure the amount of dew formed on a given surface is known as a drosometer. The early designs of drosometers were mostly of the weighing type: Aitken (1885) measured dewfall on a natural surface, and Lehmann and Schanderl (1942) made periodic weighing of a large pan containing soil covered with turf. More recently, Craddock (1951) constructed a continuous recording device that was further improved by Jennings and Monteith (1954). The continuous recording type of weighing drosometer provides the most accurate records necessary for the understanding of the physics of dew formation. However, it is too expensive for routine use by farmers. If the drosometer is weighed once a day, then the measured value could be lower than the actual dewfall because of the evaporative loss.

The most extensively used of the several forms of nonweighing drosometers is the one devised by Duvdevani (1947). It consists of a wooden block whose surface has been treated in such a manner that dew forms on it in characteristic patterns. By comparing these patterns with a set of standard photographs corresponding to a dewfall of from 0.1 to 0.45 millimeters, the intensity of dew can be estimated. However, these estimates give only a range of values lying between certain limits. They are fairly accurate only for monthly and annual totals.

Højendahl (1962) reported a method of using chemicals on paper for measuring dew:

> A piece of white cartridge paper is placed on a board lying flat on the ground and is kept in place with a small stone at each of the four corners. Some small crystals of $KMnO_4$ are left overnight on the paper. If dew falls during the night some $KMnO_4$ will dissolve in the water and seep into the paper where, due to the wood pulp present in the paper, $KMnO_4$ will be reduced to MnO_2. The following day the paper is changed, excess of $KMnO_4$ is eventually washed away with water and MnO_2 remains as a brown spot in the paper. The reaction is judged as follows: If the paper remains quite unstained, there has been no dew at all. If there are several tiny spots, there has presumably been no dew, but the humidity in the air has been sufficiently great that a small amount of water has been absorbed on the surface of $KMnO_4$ crystals. If there are several rather large brown spots on the paper, there has been actual dewfall during the night, or even a small drizzle or rain. If the paper is pale brown all over, a real shower has fallen.

Wallin and Polhemus (1954) made use of a lamb gut strip, which, when wet, allowed a pen to mark on a chart. When not wet, no mark was made. Yamamoto (1936) mounted a glass pen at one end of a balance that recorded changes in weight caused by dew at the other end of the balance on a soot-blackened paper.

Water condenses on a surface when the temperature of that surface drops below the dewpoint of the ambient air. Since radiation, the usual process of cooling, is greatly retarded by cloudiness, dust, and haze, dew can be deposited only on clear nights. Therefore, in spite of the high humidity in a humid climate, dew deposits are more frequent in an arid climate, where the sky is seldom overcast. Dusty areas, such as the Sahara, and regions with much haze, such as Los Angeles, are also unfavorable for dew formation.

Monteith (1957) found in southern England that the optimum wind speed for dew formation is between one and three meters per second, maintained throughout the night. In completely still air, the downward turbulent transfer of water vapor is negligible, being less than 1 per cent of the greatest possible deposition. On the other hand, when strong winds prevail, surface cooling will be retarded because of mixing. Thus, Lloyd (1961), for example, reported that dew has not been observed in northern Idaho on the exposed mountain slopes or under the forest canopy, where air movement is absent.

Within the plant community, the height of maximum dewfall may vary according to weather and surface conditions. Duvdevani (1953) measured a dew profile in Israel at different heights (100, 50, 30, 15, 7, 5, 3, and 2.5 centimeters) above the ground. In the dry summer months the maximum dewfall occurred at one meter, decreasing downward toward the surface. During the rainy season, when the ground was wet, the maximum value was recorded at three centimeters above the ground. The usual height of the dew instrument is at one meter above the soil surface.

Often in midlatitudes, on a windless, cloudy night when the soil is warm and wet, leaves have been observed to be profusely covered with water drops. Long (1955) and Monteith (1957) demonstrated that these water drops were either exuded from leaves by guttation or distilled upward from the soil. In either case, the accumulated water on the plants does not represent a gain from the atmosphere and, hence, should be carefully distinguished from dew. Measurements by Monteith indicate that in the British Isles distillation is more common than dewfall, and its rates could approach seven to eight mm/cm²/hr during the warm season.

Hofmann (1955) has presented a formula for estimating the rate of dew formation:

$$D = a(Q_b + A) - bu(1 - h)$$

where D is the rate of dew deposition; a and b are coefficients increasing with temperature at an order of 3 to 4 per cent per degree C.; Q_b is effective outgoing radiation; A is heat flux to the air; u is wind speed; and h is relative humidity.

The second term in the equation expresses evaporation; it increases with decreasing relative humidity and increasing wind velocity. It is zero when the relative humidity reaches saturation.

The first term represents the total energy loss from the ground surface. Since the sensible heat transfer is very small at night, the upper limit of dew formation can be approximated by the effective outgoing radiation. This has been estimated to be from 0.9 to 1.0 millimeters per night by Milthorpe (1961), Masson (1954), and Deacon, Priestley, and Swinbank (1958).

Most of the observed values are, however, much lower than the theoretical upper limit of 1.0 millimeters per night. Arvidsson (1958) reported a maximum value of 0.1 millimeters per night in Sweden and Egypt. The maximum dew deposit on the coastal plain of Israel is of the order of 30 millimeters over 200 nights, with a maximum of 0.2 millimeters on any one night (Duvdevani, 1957). In Germany, Hofmann (1958) gave the upper limit as 0.3 to 0.4 millimeters. The maximum dewfall on crops at Rothamstead, England, given by Monteith (1963) were:

Spring wheat	20 May 1957	0.26 millimeters per night
Sugar beet	8 August 1958	0.47 millimeters per night
Grass (60 centimeters)	1 August 1960	0.20 millimeters per night

However, two sets of figures in the United States are much higher than the maximum dewfall reported elsewhere. First, Thornthwaite and Holzman (1942) calculated a maximum dewfall of 0.8 to 1.2 millimeters per night by the aerodynamic approach of wind and vapor gradient measurements up to a height of 25 feet over grass. But their approach could seriously overestimate vapor transfer unless the measurements were made within a few feet of the ground. The second set of data came from the weighing lysimeters at Coshocton, Ohio, the maximum value being 0.5 millimeter per hour, 2 millimeters per night, 35 millimeters per month, and 250 millimeters per year for the period 1944 to 1949 (Harrold and Dreibelbis, 1951). These values, exceeding the outgoing radiation, have been questioned by most research workers. Either the surroundings or location provided readings unrepresentative of the natural surface, or the stress of wind on the lysimeter was not carefully distinguished from the deposition of dew.

Clearly, dew cannot be a significant item in the water budget, except in arid and semiarid tropical climates, where rainfall is scarce and dew formation is favored by excessive nocturnal radiative cooling. Dew has been repeatedly praised in both the Bible and the Talmud. There are Hebrew prayers for dew in summer as for rain in winter. Boyko (1955) has gathered evidence to show that in ancient days stones and gravels might have been used to trap dew:

> Since ancient times in arid areas, dew has been accumulated in dew ponds or gravel mounds for domestic use, and at some periods possibly also for agricultural purposes. Along the northern boundary of the Sahara, on the Sinai peninsula and in the Negev, we find hundreds of gravel mounds of about 1.70 m diameter and 70 cm high arranged in a grid, whose purpose can hardly have been anything but the accumulaion of dew for agricultural purposes. We examined the mounds in the Negev in 1949, opened them and studied the vegetation around them. The species between the stones are completely filled with loess dust, so that the wind can no longer blow through the mound and thus provide the necessary cooling effect. Nevertheless, in the spring, I constantly found many seedlings on the sides of these mounds; whereas in the hamada areas between the mounds not a single one was to be found. Since these fields of gravel mounds are always found in the vicinity of ancient dams and irrigation systems, it is assumed that plantations, such as vineyards, were at first watered by hand, by donkey or camel transport, until they were sufficiently well established to survive on the rain and on the supplementary dew.

In arid climates, dew plays a dual role in its contribution to plant growth. The passive role is to delay the rise in leaf temperature on the following morning, thereby reducing the rate of evapotranspiration. The active role is to provide water for direct plant use. Data for dew absorption show wide differences among species (Lehmann and Schanderl, 1942, and

Arvidsson, 1951). Wetzel (1924) investigated the water-absorbing ability of leaves with more than 100 species. Most of the leaves were able to absorb water, and the cuticle was not a barrier to the passage of water. Only when the leaf possessed a waxy bloom or a thick covering of hairs was water absorption prevented. Subsequently, Krause (1935) distinguished three groups of leaves on the basis of the rapidity of water absorption. The group with rapid absorption consisted mostly of mesophytes, but some xerophytes were included; the group with moderate absorption consisted mostly of leafy trees, although again some xerophytes were included; the group with very slow absorption consisted of plants with leathery leaves, orchids, succulents, and some of the ferns. In general, young leaves can absorb dew more readily than can the older ones (Went, 1955).

Although dew may be an important source of moisture for survival of some plants in arid climates as experimentally demonstrated by Stone (1957b), considerable controversy has arisen as to whether it can materially aid in plant growth. Some investigators argue that, as the maximum annual dewfall is no more than a few inches, its contribution is negligible. Baier (1966), for instance, pointed out that in the interior of South Africa the total annual dewfall of twelve millimeters could not have a significant bearing on the water economy of the soil. On the other hand, field experiments by Duvdevani (1964) in Israel showed that corn, squash, and cucumbers grew about twice as large when they received dew during the night.

The controversy can be reconciled at least in part by a consideration of the relationship between dewfall and plant spacing. Whereas the upper limit of dew formation is rigidly dictated by the energy balance in a completely covered vegetated surface, the single plant in a sparsely covered plant community can receive much more dew than its projected area would suggest. The leafier plants, with their low thermal diffusivity, will be cooler than the surrounding bare ground and consequently will attract dew. However, the process of dew attraction so brought about is one in which some parts gain at the expense of others. Angus (1956b), for instance, reported a case in which the shrub collected about 40 per cent more dew than its equivalent horizontal projected area of grass surface. He also noted that the lowest eight leaves of the shrub, though comprising more than half of the leaf area, collected only 10 to 15 per cent of the total dew. Deacon, Priestley, and Swinbank (1958) have reasoned that in desert areas, where dew supplies the greater part of available moisture, the habit of plants to grow in isolated clumps represents to some extent a natural regression toward optimum dew utilization.

Under certain circumstances, particularly in humid climates, the benefit of dew may be accompanied by some adverse effects. Dew may promote and spread plant diseases by providing a suitable environment for plant pathogen, according to Duvdevani, Reichert, and Palti (1946) for cucumbers, Kimmey (1945) for pine, and Stone (1957a) for a number of other crops. In the Mississippi Delta, cotton harvested with dew has been known to suffer from a loss of quality (Newton and Riley, 1964). A knowledge of dew there might be helpful in maintaining cotton quality by selecting the proper harvest time.

FOG

Fog may form either by the evaporation of warm water into the cold air or by the cooling of the air while in contact with the ground. The first process occurs predominantly over water bodies or during a period of warm rain. The fog thus formed is usually of little agricultural significance. The second process, or the cooling of air, is the more effective means of fog formation. The three causes of cooling are radiative cooling, advective cooling, and adiabatic cooling. The fogs formed by these three methods of cooling are known, respectively, as radiation ground fog, advection fog, and upslope fog.

The conditions favorable for the formation of ground fog are not much different from those for dew: a clear sky, a light wind, and a high humiity. Advection fog is prevalent along littorals having cold ocean currents over which the incoming warm, moist air is chilled. Upslope fog forms on mountain slopes where warm moist air becomes saturated through adiabatic cooling. The frequency of fog occurrence throughout the world is fairly well known.

The amount of moisture supplied by fog could exceed that of dew by an order of magnitude. Nagel (1956) reported that on Table Mountain, South Africa, a fairly constant rate of fog precipitation occurred throughout the year with an annual total of 3,294 millimeters. Measurements on the island of Lanai, Hawaii, showed 1,270 millimeters of annual fog drip (McGehee, 1963). In both cases, fog precipitation exceeded rainfall.

In some coastal deserts and dry mountain slopes, fog has been an important factor in determining the distribution of natural vegetation and agricultural crops. Went (1955) has related the following cases in California:

> An example is the coast of southern California. Immediately adjoining the ocean is a zone one-half to a few miles wide where tomatoes, peppers, beans, and other vegetables can be grown in summer. These crops develop well even without irrigation, although no rain falls during the growing season from May to October. The soil in the region may contain enough moisture for the last few months. The source can hardly be anything but dew or coastal fog.
>
> Vegetation in many places where fogs are frequent differs from the

vegetation in nearby localities where fogs do not occur. One of the best examples of this is the northern coast of California. Redwood occur there predominantly in a narrow belt and never range inland beyond the influence of sea fogs. In places where more water is available, redwoods grow very well outside the fog belt—an indication that fog is important in their water economy. In this belt, fogs occur almost daily. Anyone who walks in a redwood forest during a fog finds that the trees are dripping.

HUMIDITY

High atmospheric humidity has at least two possible beneficial effects on plant growth. First, many plants can directly absorb moisture from an unsaturated air of high humidity. This has been demonstrated experimentally by Haines (1952), Slatyer (1958), and Breazeale, McGeorge, and Breazeale (1950). Second, humidity may affect the photosynthesis of plant leaves. Baker (1965) has studied the effect of vapor pressure deficit on apparent photosynthesis of cotton leaves at 40° C. and at various high intensities and times of day. The photosynthetic rate increases with

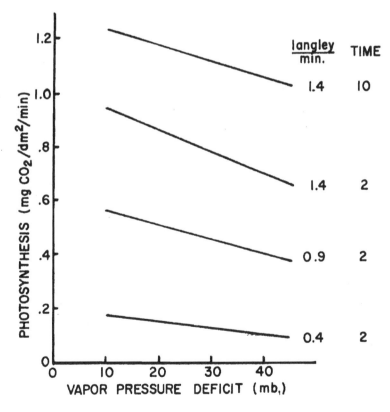

FIGURE 115. Effect of vapor pressure deficit on apparent photosynthesis at 40° C. and at various light intensities and times of day (after Baker).

humidity, only slightly at low light intensities but quite substantially at high intensities (Figure 115).

Most plants grow well under high atmospheric humidities, except when saturated air persists for weeks and completely stops transpiration. By varying the relative humidity from 50 to 90 per cent, Fortanier (1957) has found that the number of flowerings of peanuts increases with humidity. High humidity at night is especially beneficial. Breazeale and McGeorge (1953) reported that corn and tomato plants that were grown under high humidity not only gained more weight but also developed better root systems.

Other things being equal, the evapotranspiration rate decreases with the increase of humidity. Peters (1960b) compared the water use by corn under two different relative humidities. He noted a significantly higher rate of water use at 40 per cent relative humidity than at 95 per cent humidity in the early stages of growth. As time lapsed, the differences in water use controlled by relative humidity became almost negligible because of the faster rate of growth at 95 per cent relative humidity.

During the dry summer, crops often exhibit a distinctly different growth response to showers that are accompanied by a high humidity than they do to irrigation that is accompanied by a low humidity. Brierley (1934), for instance, noted an increase in size of raspberries after a very light shower, an increase that was not observed after irrigation.

In a recent study, Arkley (1963) examined the results of a number of container experiments on the relationship between transpiration and dry matter production. He found that differences in yield in different climates can be reconciled to a large extent by introducing a correction factor for humidity. He proposed the following equation:

$$D = a \frac{W}{(100 - h)}$$

where D is dry matter production; a is a constant for a given kind of plant when the soil is adequately fertilized; W is transpiration; and h is mean daily relative humidity in terms of percentage.

Thus, other things being equal, the efficiency of water use would increase with the humidity of the air. The equation is undoubtedly valid for a limited number of controlled experiments. However, much work needs to be done before the effect of humidity on crop growth in the field can be expressed in exact, quantitative terms.

CHAPTER 22

WIND

EFFECT OF WIND ON PLANT GROWTH

Wind affects plant growth in at least three significant ways: transpiration, CO_2 intake, and mechanical breakage of leaves and branches.

Controlled experiments show that transpiration increases with wind speed up to a certain point, beyond which either it does not increase (Stalfelt, 1932) or it decreases slightly at high wind speed (Hesse, 1954; Brown, 1910; Fibras, 1931; and Wilson, 1924). The exact relationship between wind and transpiration, however, varies greatly among plant species. Seybold (1929, 1931, 1932, 1933) found that wind exerts a much greater influence on cuticular transpiration than on stomatal transpiration. Therefore, only plants with a high cuticular transpiration, namely hydrophytes, show an appreciable increase in the transpiration resulting from the action of wind.

Under natural conditions, the effect of wind on transpiration will vary according to the roughness as determined by surface configuration. In general, the effect is greatest for an isolated tall plant. Copeland (1906), for instance, reported an increase in the transpiration rate of coconut palms in full sunlight of about 100 per cent with a wind speed estimated at five miles per hour. On the other hand, if the cover is complete and the canopy surface is more or less even and smooth, then the effect of wind on transpiration is usually small, certainly less important than solar radiation. From their studies of the correlation between environmental factors and transpiration rates for the growing season, Briggs and Shantz (1916a, 1916b) concluded that only 2 to 6 per cent of the water loss could be attributed to the wind.

Of course, the effect of wind on transpiration will vary with the temperature and humidity of the air sweeping over the plant surface. This determines largely the magnitude of the advective component in the energy budget equation. In arid climates, dry and hot winds often cause rapid wilting of plants. The damaging effect of the so-called *sukhovey* (Lydolph, 1964) in southern Russia offers a good example. In winter, when the soil is frozen, the damaging effect of increased transpiration resulting from wind is also accentuated as water lost cannot be readily replaced.

As discussed in Chapter 3, the rate of photosynthesis increases with the supply of CO_2, which in turn is favored by turbulence. Deneke (1931) concluded from wind tunnel experiments that the CO_2 uptake increases almost linearly up to a wind speed of 167 centimeters per second, above which no further increase was observed. The beneficial effect of increasing CO_2 exchange has also been reported by Lemon (1960), Hesketh (1961), Heinicke and Hoffman (1933), and Warren Wilson and Wadsworth (1958).

High wind speeds are harmful to plant growth. The peculiar configuration of trees in coastal or mountain area exposed to strong winds is familiar. Leaves mechanically damaged by wind have a reduced capacity for photosynthesis and translocation. Hartt (1963), in a thorough study for sugar cane, summarized:

> When only a midrib of a sugar cane blade (var. H. 50-7209) was broken, the laminae remaining uninjured, translocation was inhibited 34 to 38%; photosynthesis measured with an infrared analyzer above the injury in a similar blade was inhibited 30%. When both midrib and laminae were broken, translocation was inhibited 99 to 100%; photosynthesis above the injury in a similar blade was inhibited 84%. Translocation was measured 6 hours after inflicting the injury, and photosynthesis the day following the injury, using different blades. Moisture determinations of the leaves indicated that the inhibition of photosynthesis was not due to loss of water. When the blade is frayed (var. H. 49-5) by tearing the laminae lengthwise translocation was decreased (measured in 6 hrs.) by 54 to 60% and photosynthesis (measured immediately) 38 to 54%.

All plants, however, do not react in the same way to strong wind. Whitehead (1957) has classified plants into three groups:

EXPOSURE EVADER: These are usually short plants whose aerial parts do not grow above the layer of comparatively still air near the ground, and, as such, are least affected by strong wind. For example, when *Cerastium atrovirens* was exposed to wind velocities of 0.25, 9, 25, and 60 miles per hour, the plant height was not affected and the dry matter production showed only slight falling off with increase of wind velocity (Figure 104).

EXPOSURE TOLERANT: Plants in this group (e.g., barley) show a marked diminution of dry matter production with increase of wind velocity, but at a rate somewhat less than that of plants in the next group.

EXPOSURE SENSITIVE: These species, of which *Senecio nebrodensis* of the alpine desert is an example, are affected by strong wind to such a degree that they cannot survive in exposed areas. Both the plant height and the dry matter production decrease rapidly with an increase of wind speed (Figure 116).

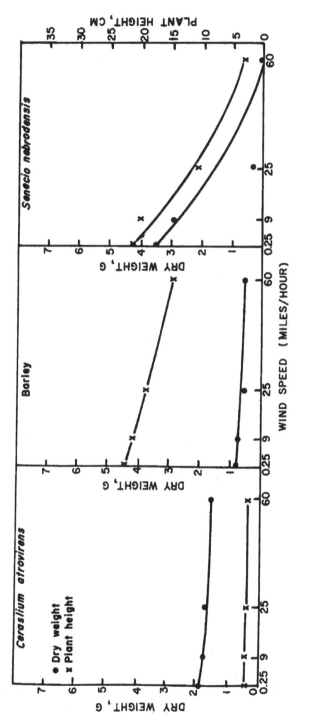

FIGURE 116. Growth in height and dry weight of three species at four wind speeds (after Whitehead).

Plants that have been growing in high wind conditions for a long period may develop certain physiological characteristics. In general, such plants have a greater proportion of root in comparison to shoot (Figure 117), and they often exhibit xerophytic characters, such as wider and thicker leaves, and a greater number of stomata per unit leaf area.

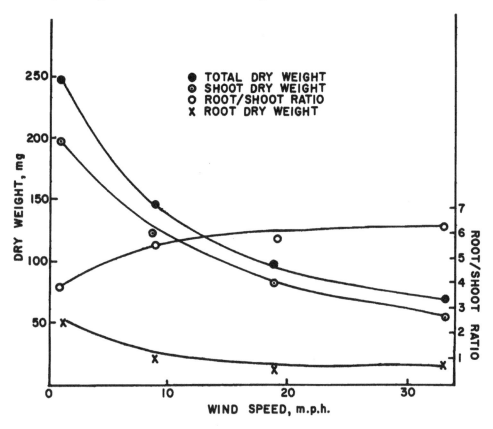

FIGURE 117. Graphs showing alterations of dry weight accumulation over 30 years in *Helianthus annuus* grown at four different wind speeds (after Whitehead).

In exposed mountain areas, wind is often a limiting factor in determining the distribution of plants, as illustrated by Whitehead (1963) in an ecological study in the central Apennines of Italy. He used the aerodynamic roughness, z_o, to measure the degree of shelter from wind affected by the surface irregularities. The average values of roughness were determined in the field for four plant communities, designated as A, E, G, and H in the order of increased roughness. A positive correlation existed between the degree of shelter and the number of plant species (Figure 118). In more exposed areas, both the plant height and the dry matter production of individual plants were greatly reduced.

Wind

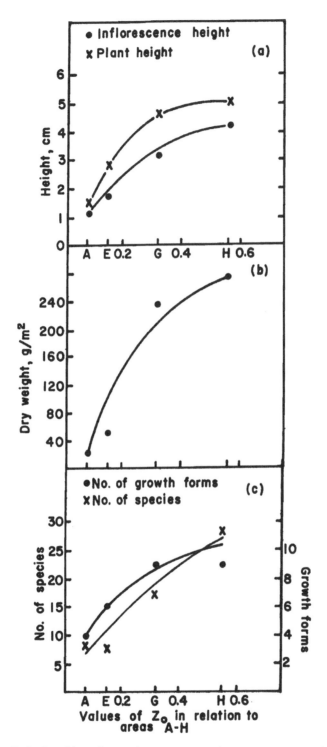

FIGURE 118. Relationship of roughness parameter, z_o, on mountain top to (a) heights of inflorescences and plants, (b) annual yield of dry matter, (c) number of growth forms and species (after Whitehead).

Warren Wilson (1959) studied the effect of strong winds on the net assimilation rates of *Oxyria digyna* leaves and the growth rate of *Brassica rapa* in Jan Mayen (71° N.). He found differences in growth rate of some 30 to 40 per cent between windswept summits and sheltered lowlands. The effect of wind is probably smaller in middle and low latitudes.

Since strong wind increases transpiration and decreases dry matter production, the efficiency of water use would be reduced. In a pot experiment using marigolds as a representative tender-foliage plant, Finnel (1928) found that the water use efficiency of the plant exposed to a constant wind speed of fifteen miles per hour was only about half that of a plant grown in still air. Exposure to wind also postponed maturity about ten days in a sixty-day growing period.

It may be concluded from the above consideration of transpiration, CO_2 uptake, and the detrimental effect of strong wind that moderate wind movements is best for plant growth. Of course, the optimal wind speed will vary with species and environment. In a wind tunnel study using stands of *Brassica napus,* Warren Wilson and Wadsworth (1959) observed that the growth rate increased with the wind speed up to an optimum near 30 centimeters per second and then fell off at higher speeds.

SHELTERBELTS

Shelterbelts have long been used to protect crops and to increase yields. By altering the surface roughness and, hence, the wind profile near the ground, massive planting of trees effectively changes both the energy budget and moisture balance of the crop field.

A shelterbelt oriented at a right angle to the prevailing wind reduces wind velocity on both the windward and leeward sides. To the windward, a small cushion of air is formed, and most of the wind is forced either up and over the barrier or around its sides. A small distance to the lee, the wind velocity is reduced to a minimum. At some distance farther downstream, wind speed again increases to its normal open velocity. Figure 119 summarizes the data obtained by Nageli (see Caborn, 1957) in Switzerland for a shelterbelt of medium penetrability. In his study, the wind velocity was reduced by over 10 per cent to a distance of twenty times the tree height to the lee, and three times the tree height to the windward. The minimum wind speed was observed at a distance of four times the tree height in the lee.

The exact percentage of reduction at any distance will vary with the permeability of the shelter as well as with the wind speed and direction. A very dense barrier will bodily uplift a wind current that descends abruptly in the lee. In contrast, a moderately dense stand acts as a filter rather than

Wind

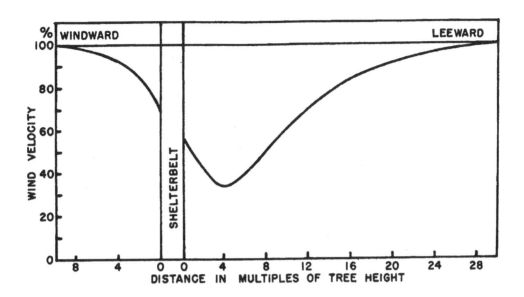

FIGURE 119. Mean wind speeds near shelterbelt of medium density. Data obtained by Nageli (after Staple).

as an obstruction. Thus, with denser belts the minimum wind velocity is lower but nearer to the belt; whereas with partially open stands, the minimum wind speed is higher but extends farther to the leeward. Gloyne (1955) reported the results of an experiment in which the wind speed was reduced to less than 80 per cent of its speed in the open to a distance of twelve times the tree height from a barrier of 30 per cent density, to a distance of 27 times the tree height from a barrier of 50 per cent density, but only to a distance of fifteen times the tree height from a barrier of 100 per cent density. Similar results have been reported by Lawrence (1955).

The area sheltered by the windbreak forms roughly the shape of a blunt triangle whose apex moves with the wind direction. The length of the shelterbelt should equal at least that of the field protected. Even for a small field, the length of barrier should not be less than twenty times the tree height in order to achieve maximum efficiency (Collins, 1962).

The reduced wind velocity in the sheltered area necessarily alleviates soil erosion and mechanical damage to the plants. For instance, Iizuka (1950) reported that a wind barrier that provided a 39 per cent reduction of the wind at a distance of ten times the tree height in the lee showed only 0.1 per cent sand movement in comparison with that in the open; at a distance of twenty times of tree height in the lee, movement was reduced 18 per cent.

Many other aspects of microclimate are affected as well. The temperature is increased during the day and decreased at night. But the change is usually small, especially in areas of low sunshine. Both Bates (1911) in the United States and Caborn (1957) in Europe cited daytime temperature increases of no more than 3° F. Greater increases in temperature up to 7° F. have been reported by Woodruff, Read, and Chepil (1959) in Kansas.

The effect of shelterbelts on wind speed and night temperature is such that radiation frost is often favored in the lee, particularly if the belts are dense. Belts that allow the penetration of a certain amount of wind prevent stable stratification of stilled air, and thereby minimize frost danger.

Where strong winds are frequent, shelterbelts tend to reduce evaporation in the protected area. Bates (1911) made extensive measurements of evaporation from piche evaporimeters near a shelterbelt of underplanted groves. His data, summarized in Figure 120, indicate that the reduction of evaporation increases with wind speed. At a high wind speed of over 15 miles per hour, evaporation is reduced to two-thirds of the open field value as far as ten times the tree height in the lee.

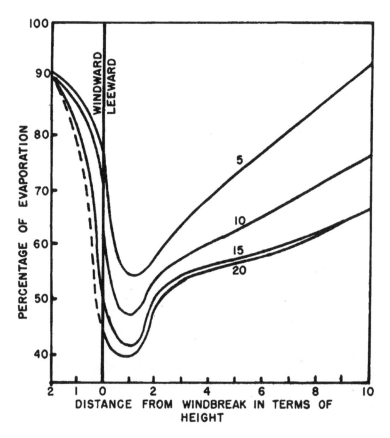

FIGURE 120. Evaporation near shelterbelt of underplanted groves, average height 75 feet. Wind speeds of 5, 10, 15, and 20 miles per hour (after Bates).

However, the data obtained by Bates are not representative of the actual field conditions, as the evaporation from the piche evaporimeter is greatly accentuated by wind (Chapter 18).

Measurements of tank evaporation, which is more representative of potential evapotranspiration that the piche evaporimeter, have been taken by Staple and Lehane (1955) near a three-row shelterbelt of 20 to 25 feet height at Conquest, Canada. They found for the period 1950 to 1954 that the reduction in evaporation averaged less than 5 per cent for a tank located some 80 feet in the lee.

When the soil is dry or when the wind speeds are low, evaporation and transpiration are only slightly affected by a shelterbelt. This has been found to be true by Alisov (1956) in the Soviet Union. In one extreme case, Kolasew (1941) even reported an increase in evaporation in the lee of a windbreak. He attributed this to the increased daytime temperature in the protected area.

In middle and high latitudes, the accumulation of snowdrifts near the shelterbelts often increases soil moisture far more than the amount of moisture saved because of the reduction of evaporation. If the main function of the shelterbelt is to trap snow and to spread it evenly over a field, it should be somewhat open near the ground level. The increase of soil moisture because of snow is usually greater near the trees and tapers off toward the center of the field between two parallel shelterbelts. This has been illustrated in a study at Aneroid, Canada, by Staple and Lehane (1955), who also found a close agreement between the stored moisture at seed time and the wheat yield (Figure 121). For the sheltered fields as a whole, the net increase in yield, taking into account the areas occupied by the trees, was reported to be 0.7 bushel for the period 1950 to 1954.

Although the effect of a shelterbelt on microclimate is varied and complicated, it is usually beneficial to plant growth, especially in areas of strong wind or in areas where snow moisture is an important factor. Of course, the over-all effect on crop yield will vary according to the prevalence of strong wind, rainfall distribution, temperature regime, snow drift, and the kind of crops, as well as the structure of the shelterbelt itself. The early study by Bates (1911) cited an increase of corn yield of about 45 per cent at four to five times the tree height in the lee, with beneficial effects extending to a distance of twelve times the tree height. Subsequently, he reported an increase of 60 to 70 per cent of alfalfa yield in the protected zone of a South Dakota windbreak (1944). Extensive measurements made near wide, dense shelterbelts in the Soviet Union, summarized by Rudolf and Gevorkiantz (1935), gave a yield increase of wheat up to 30 per cent. In Denmark, Jensen (1954) obtained a yield increase of 7 to 18 per cent for various crops. In England, Hogg and Carter (1962) showed that, when sheltered, lettuce crops increased the number of plants by 9 per cent, the average weight per plant by 16 per cent, and the

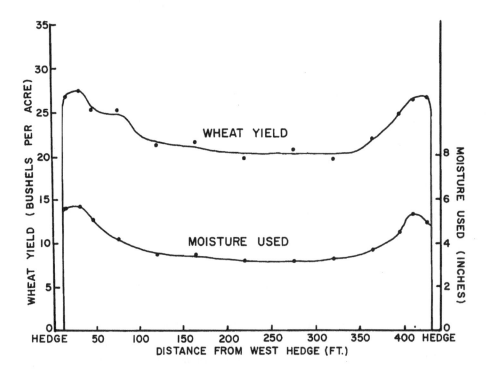

FIGURE 121. Wheat yield and stored moisture used between hedges, Aneroid, 1950-54 (after Staple and Lehane).

total yield by 27 per cent. These are only a few of the conflicting values reported by various authors, and they are meaningful only for local conditions.

In a recent review Stoeckeler (1965) classified the response of various crops to shelterbelts into three groups: comparatively low, medium, and high. Among the crops of comparatively low response to wind protection are the drought-hardy small grains and maize grown under dry farming conditions in subhumid to semiarid climates. Moderately responsive are rice and fodder crops such as alfalfa, lupine, clover, and seed of wheat grasses. Among the crops most responsive to shelterbelt protection are garden crops, including lentils, potatoes, tomatoes, cucumbers, beets, strawberries, watermelons, deciduous and citrus fruits, and other tender crops such as tobacco and tea. In another review, Fournier d'Albe (1958) considers 20 per cent as a reasonable average for a yield increase resulting from a shelterbelt in areas of strong wind. This would represent, as Andersen (1943) pointed out, a net gain of at least 15 per cent even when allowance is made for the land occupied by the belts themselves.

CHAPTER 23

CONCLUSION

The climatic elements and their effects on plant growth are far more complex than is apparent from their treatment in this book. In the field, the plant is never subjected only to a single variable at any given time, but to a constant interaction with an almost infinite number of combinations of the elements. We have not been able to isolate the various aspects of selective integration of the various climatic elements by the plants, except in a crude qualitative manner. In spite of the obvious limitations of studies on the effects of single climatic elements on plant growth, with our present knowledge we know what elements need to be measured and their relative importance in most cases. Furthermore, studies on the effect of single climatic elements are a necessary first step toward more complete understanding of our problem. For instance, under a known radiative and thermal regime, the estimation of potential yield when the water application is adequate and the yield loss caused by varying amounts of water deficit, opens up a new horizon in the assessment of various alternatives of water resources management. Starting from a different assumption, we can also estimate the potential photosynthesis as determined primarily by radiation and temperature. Should the two estimates agree with each other, we can proceed with confidence to assess their combined effects on productivity under less favorable conditions. The economic implications of this approach are many and varied.

In the last decade, the barrier separating the physical and biological sciences appears to have been breached. One reason for this progress is that climatologists are no longer content with the analysis of data designed only for weather forecasting. They have begun to realize that the study of climatology should not be limited to the atmosphere but should include both vegetation and soil surfaces as well. As a result of this reorientation, the traditional emphasis on temperature and rainfall is now being expanded to encompass the whole field of energy budget and water balance. This broadening of the sphere of research has rendered the seemingly noncomparable climatic elements amenable to precise and unifying physical interpretation. Without understanding concepts such as albedo, roughness, Bowen's ratio, and the like, studies on the relationship between climate and plant cannot advance beyond the level of statistical correlation. The combination of the approaches through micrometeorology and plant physiology holds great promise for the future, especially in the improved design of experiments.

Paralleling the progress of physiological studies, our climatological observations also must be expanded. Here the program may be broken down into three categories: (1) micrometeorological observations in sample areas for the understanding of the basic physical processes; (2) mesoscale climatic network designed for practicing farmers to improve their agricultural operation, and (3) macroscale regional network designed for weather forecasting and for gathering basic climatic information. The recommended climatic elements of observation in each of the three categories are listed in Table 21.

The climatic elements needed for understanding the microscale flux of heat, moisture, and momentum are numerous and difficult to assess. The requirements for instrumental precision are extremely high, and, in some cases, the lack of a satisfactory method of measurement has impeded progress. Equally essential is the need for more trained personnel. Fortunately, short observations often suffice to provide meaningful results, because the emphasis is on basic principles and the comparison of contrasting surface characteristics rather than on the long-term regional differences. Most of the micrometeorological studies concerning crop surfaces have been discussed in this book. Similar studies over other surfaces have been summarized by Sutton (1953), Sellers (1965), Miller, (1965), and Munn (1966).

The mesoscale climatic network of observation is intended to apply the knowledge gained in micrometeorological research to practical problems of a farm area of several hundred to several thousand acres. The climatic data will be used to explain the fluctuation of yields, to select the time of planting and harvest, to guide irrigation scheduling, and to improve other cultural practices. The instruments should be inexpensive, yet not fragile. The elements to be observed should be useful in making managerial decisions. For some elements, such as rainfall, which varies irregularly with time and space, relatively low grade "area measurements" can provide more meaningful records than individual point measurements of much higher precision. On the other hand, a number of poor quality measurements of solar radiation can be misleading and less useful than few accurate measurements made under skilled supervision. The ideal density of the network will vary not only with the climatic elements but also with the local microclimatic variations determined by terrain and ground

TABLE 21
Recommended Climatic Elements To Be Observed for Agricultural Purposes

GROUP	ELEMENT	SUBDIVISION	MACROSCALE		MESOSCALE	MICROSCALE
			First-order station	Second-order station		
Thermal elements and their associates	Radiation	Global radiation	x		x	x
		Diffuse radiation				x
		Reflectivity				x
		Sunshine hours	x			x
		Cloud form and cloudiness	x			x
		Terrestrial radiation				x
		Effective outgoing radiation				x
		Net radiation	x		x	x
		Emissivity*				x
		Heating coefficient*				x
		Radiation-distribution within the canopy				x
		Net radiation distribution within the canopy				x
	Temperature	Air temperature	x	x	x	x
		Soil temperature	x		x	x
		Leaf temperature				x
Moisture elements and their associates	Precipitation	Rain	x	x	x	x
		Snow	x	x	x	x
		Hail	x			
		Dew	x			x
		Fog	x			
	Humidity	Humidity	x	x	x	x
	Evaporation	Potential evapotranspiration				x
		Actual evapotranspiration				x
		Pan evaporation	x		x	x
	Soil moisture	Soil moisture storage capacity			x	x
		Depletion rate				x
Advective elements	Pressure	Pressure	x			
	Wind	Upper air wind	x			
		Surface wind	x	x	x	x
		Wind profile				x
		Roughness*				x

* derived elements

surfaces. This latter subject is treated only incidentally in this book. For a thorough discussion, the reader is referred to the work by Geiger (1965).

On the macroscale, the establishment and improvement of regional climatic stations has been the continuing concern of the World Meteorological Organization, as well as national weather services. The stations have been customarily separated into "first-class" and "second-class" according to the number of elements observed. The primary function of these stations has been weather forecasting, and because of this, the stations are usually spaced at least ten miles apart. Although their value for daily agricultural operation is limited, they do provide useful information for long-range planning. The existing network is adequate in most parts of North America, Europe, and Australia, but wide gaps exist in the tropics, and polar regions, and the arid lands. There is an urgent need for the development of inexpensive, automatic weather stations to be used in these sparsely inhabited areas. As the world becomes more densely populated and as our basic knowledge of agricultural climatology grows steadily, the climatic records of the world certainly will prove rewarding in the future planning of agricultural land use.

BIBLIOGRAPHY

ABD EL RAHMAN, A. A., and K. H. BATANOUNY. 1965. "Transpiration of desert plants under different environmental conditions." *Journal of Ecology*, 53: 267-72.

ABD EL RAHMAN, A. A., P. J. C. KUIPER, and J. F. BIERHUIZEN. 1959. "Preliminary observations on the effect of soil temperature on transpiration and growth of young tomato plants under controlled conditions." *Mededelingen van de Landbouwhogeschool te Wageningen*, 59(15): 1-12.

ABDEL-AZIZ, M. H. 1962. The influence of advective energy on evapotranspiration. Unpublished master's thesis, Utah State University.

ABDEL-AZIZ, M. H., S. A. TAYLOR, and G. L. ASHCROFT. 1964. "Influence of advected energy on transpiration." *Agronomy Journal*, 56: 139-42.

ADAMS, J. E., and R. J. HANKS. 1964. "Evaporation from soil shrinkage cracks." *Proceedings*, Soil Science Society of America, 28: 281-84.

AHMAD, M. S. 1962. "Water requirements of plants in the Quetta Valley, West Pakistan." In *Plant-water relationships*. UNESCO.

AIKMAN, J. M. 1941. "The effect of aspect of slope on climatic factors." Iowa State College Journal of Science, 15: 161-67.

AITKEN, J. 1885. "On dew." *Transactions*, Royal Society of Edinburgh, 33: 9-64.

ALBERDA, T. 1962. "Actual and potential production of agricultural crops." Netherlands Journal of Agricultural Science, 10: 325-33.

ALDRICH, W. W., M. R. LEWIS, and R. A. WORK. 1940. "Anjou pear responses to irrigation in a clay adobe soil." *Bulletin*, Oregon Agricultural Experiment Station, 374: 76.

ALISOV, B. P. 1956. "Metodika i resul'taty mikroklimaticheskikh nabliudenii 1949 g. v polezashchitnykh polosakh Voronezhskoi oblasti." [Method and results of Microclimatic investigations of 1949 in Shelter belt zones of the Voronezh region]. Voprosy Geografii [Geography], 23: 209-217.

ALLARD, H. A. 1938. "Complete or partial inhibition of flowering in certain plants when days are too short or too long." *Journal of Agricultural Research*, 57: 775-89.

ALLEN, L. H., and K. W. BROWN. 1965. "Short-wave radiation in a corn crop." *Agronomy Journal*, 57: 575-80.

ALLEN, L. H., C. S. YOCUM, and E. R. LEMON. 1964. "Photosynthesis under field conditions VII. Radiant energy exchanges within a corn crop canopy and implications in water use efficiency." *Agronomy Journal*, 56: 253-59

ALLISON, F. E., E. M. ROLLER, and W. A. RANEY. 1958. "Relationship between evapotranspiration and yields of crops grown in lysimeters receiving natural rainfall." *Agronomy Journal*, 50: 506-511.

ALLMARAS, R. R., W. C. BURROWS, and W. E. LARSON. 1964. "Early growth of corn as affected by soil temperature." *Proceedings*, Soil Science Society of America, 28: 271-75.

ALLMENDINGER, D. F., A. L. KENWORTHY, and E. L. OVERHOLSER. 1943. "The carbon dioxide intake of apple leaves as affected by reducing the available soil moisture to different levels." *Proceedings*, American Society for Horticultural Science, 42: 133-40.

ALTER, J. C. 1913. "Crop safety on mountain slopes." *Yearbook of U. S. Department of Agriculture.*

ALWAY, F. J., G. R. MCDOLE, and R. S. TRUMBULL. 1919. "Relation of minimum moisture content of subsoil of prairies to hygroscopic coefficient." *Botanical Gazette*, 67: 185-207.

ANDERSEN, P. C. 1943. *Laeplantnings-Bogen.* [*Book on shelterbelt.*] Viborg: Danish Heath Society.

ANDERSON, C. H., and D. W. L. READ. 1966. "Water use efficiency of some varieties of wheat, oats, barley, and flax grown in the greenhouse." *Canadian Journal of Plant Science*, 46: 375-78.

ANDERSON, D. T., S. DUBETZ, and G. C. RUSSELL. 1958. "Studies on transplanting sugar beets in southern Alberta." *Journal of American Society of Sugar Beet Technologists*, 10: 150-55.

ANDERSON, E. R. 1952. "Energy budget studies." In *Water loss investigation.* Lake Hefner Studies Technical Report, U.S. Geological Survey Circular 29.

ANDERSON, M. C. 1964. "Light relations of terrestrial plant communities and their measurement." *Biological Review*, 39: 425-85.

ÅNGSTRÖM, A. 1916. "Über die Geenstrahlung der Atmosphäre." *Meteorologische Zeitschrift*, 33: 529-38.

ÅNGSTRÖM, A. 1924. "Solar and terrestrial radiation." *Quarterly Journal*, Royal Meteorological Society, 50: 121-26.

ÅNGSTRÖM, A. 1925. "The albedo of various sufaces of ground." *Geografiska Annaler*, 7: 321-42.

ANGUS, D. E. 1955. "The use of heaters for frost prevention in a pineapple plantation." *Australian Journal of Agricultural Research*, 6: 186-95.

ANGUS, D. E. 1956a. "Radiation frosts, their causes and local prevention in agriculture." *Journal of the Australian Institute of Agricultural Science*, 22: 91-104.

ANGUS, D. E. 1956b. "Measurements of dew." In *Climatology and microclimatology*, UNESCO.

ANGUS, D. E. 1962. "Frost protection experiments using wind machines." Commonwealth Scientific and Industrial Research Organization, Australia. Technical Paper, 12: 44.

ANGUS, D. E., and H. BIELORAI. 1965. "Transpiration reduction by surface films." *"Australian Journal of Agricultural Research*, 16: 107-112.

ANITA, N. 1949. "The influence of soil moisture on the production and quality of tobacco." *Revue Internationale des Tabacs*, 24: 133:36.

ANONYMOUS. 1963. "Carbon dioxide, its influence on crop yield and maturity." *World Crops*, 15: 163-66.

ANSARI, A. Q., and W. E. LOOMIS. 1959. "Leaf temperatures." *American Journal of Botany*, 46: 713-17.

ANSTEY, T. H. 1966. "Prediction of full bloom date for apple, pear, cherry, peach, and apricot from air temperature data." *Proceedings*, American Society for Horticultural Science, 88: 57-66.

ARKLEY, R. J. 1963. "Relationships between plant growth and transpiration." *Hilgardia*, 34: 559-84.

ARTHUR, I. P. 1956. "An evapotranspirometer." *Australian Journal of Agricultural Research*, 7: 707-712.

ARTHUR, J. M., and W. D. STEWART. 1931. "Plant growth under shading cloth." *American Journal of Botany*, 18: 897.

ARVIDSSON, I. 1951. Austrocknungs-und Dürreresistenzverhältnisse einiger Repräsentanter Öländischer Pflanzenverine nebst Bemerkungen über Wasserabsorption durch Oberirdische Organ. Oikos. Supplement 1.

ARVIDSSON, I. 1958. "Plants as dew collectors." *International Union of Geodesy and Geophysics*. Association of Scientific Hydrology, General Assembly of Toronto, September 3-14, 1957. Transactions. 2: 481-84.

ASHLEY, D. A., B. D. DOSS, and O. L. BENNETT. 1963. "A method of determining leaf area in cotton." *Agronomy Journal*, 55: 584-85.

ASHLEY, D. A., B. D. DOSS, and O. L. BENNETT. 1965. "Relation of cotton leaf area index to plant growth and fruiting." *Agronomy Journal*, 57: 61-64.

ASHTON, F. M. 1956. "Effects of a series of cycles of alternating low and high soil water contents on the rate of apparent photosynthesis in sugar cane." *Plant Physiology*, 31: 266-74.

ASLYNG, H. C. 1965a. "Rain, snow, and dew measurements." *Acta Agriculturae Scandinavica*, 15: 275-83.

ASLYNG, H. C. 1965b. "Evaporation, evapotranspiration, and water balance investigations at Copenhagen 1955-64." *Acta Agriculturae Scandinavica*, 15: 284-300.

ASLYNG, H. C., and K. J. KRISTENSEN. 1958 "Investigation on the water balance in Danish agriculture II. 1953-1957." Yearbook, Royal Veterinary Agricultural College, Copenhagen, pp. 64-99.

ASLYNG, H. C., and B. F. NIELSEN. 1960 "The radiation balance at Copenhagen." *Archiv für Meteorologie, Geophysik, und Bioklimatologie*, B., 20: 342-58.

AUBERTIN, G. M., and D. B. PETERS. 1961. "Net radiation determinations in a cornfield." *Agronomy Journal*, 53: 269-72.

BAHRANI, B., and S. A. TAYLOR. 1961. "Influence of soil moisture potential and evaporative demand on the actual evapotranspiration from an alfalfa field." *Agronomy Journal*, 53: 233-37.

BAIER, W. 1965. "The interrelationship of meteorological factors, soil moisture and plant growth." *International Journal of Biometeorology,* 9: 5-20.

BAIER, W. 1966. "Studies on dew formation under semiarid conditions." *Agricultural Meteorology,* 3: 103-112.

BAIER, W., and G. W. ROBERTSON. 1965. "Estimation of latent evaporation from simple weather observations." *Canadian Journal of Plant Science,* 45: 276-84.

BAIER, W., and G. W. ROBERTSON. 1966. "A new versatile soil moisture budget." *Canadian Journal of Plant Science,* 46: 299-315.

BAINBRIDGE, R., G. C. EVANS, and O. RACKHAM. 1966. *Light as an ecological factor.* London: Blackwell Scientific Publications.

BAKER, D. G. 1958. A comparison of two evapotranspiration calculation methods and the application of one to determine some climatic differences between great soil groups of Minnesota. Unpublished doctoral dissertation, University of Minnesota.

BAKER, D. N. 1965. "Effects of certain environmental factors on net assimilation in cotton." *Crop Science,* 5: 53-56

BAKER, D. N. and R. B. MUSGRAVE. 1964. "Photosynthesis under field condition. V. Further plant chamber studies of the effects of light on corn (*Zea mays* L.)." *Crop Science,* 4: 127-36.

BAKER, F. S. 1929. "Effect of excessively high temperatures on coniferous reproduction." *Journal of Forestry,* 6: 949-75.

BALD, J. G. 1943. "Estimation of the leaf area of potato plants for pathological studies." *Phytopathology,* 33: 922-32.

BARBER, D. A. 1962. "The movement of ^{15}O through barley and rice plants." *Journal of Experimental Botany,* 13: 397-403.

BARNES, A. C. 1953. "Cane sugar in East Africa." *Sugar,* 48: 36-38.

BARRY, R. G., and R. E. CHAMBERS. 1966. "A preliminary map of summer albedo over England and Wales." *Quarterly Journal,* Royal Meteorological Society, 92: 543-48.

BARTELS, L. F. 1965. "A comparison of gypsum blocks and evaporimeters for irrigation control." *Australian Journal of Experimental Agriculture and Animal Husbandry,* 5: 453-57.

BARTLETT, R. J. 1961. "Iron oxidation proximation to plant roots." *Soil Science,* 92: 372-79.

BATCHELOR, L. D., and H. S. REED. 1923. "The seasonal variation of the soil moisture in a walnut grove in relation to the hygroscopic coefficient." California Agricultural Experiment Station. Technical paper No. 10.

BATES, C. G. 1911. "Windbreaks: their influence and value." USDA Forest Service Bulletin 86.

BATES, C. G. 1944. "The windbreak as a farm asset." USDA Farmer Bulletin 1405.

BAVER, L. D. 1937. "Rainfall characteristics of Missouri in relation to runoff and erosion." *Proceedings,* Soil Science Society of America, 2: 533-36.

BEGG, J. E., J. F. BIERHUIZEN, E. R. LEMON, D. K. MISRA, R. O. SLATYER, and W. R. STERN. 1964. "Diurnal energy and water exchanges in bulrush millet in an area of high solar radiation." *Agricultural Meteorology,* 1: 294-312.

BENDELOW, V. M. 1958. "The effect of irrigation on yield and malting quality of barley in southern Alberta." *Canadian Journal of Plant Science*, 38: 135-38.

BENEDICT, H. W. 1951. "Growth of some range grasses in reduced light." *Botanical Gazette*, 102: 582-89.

BERLIAND, T. G., and N. A. EFIMOVA. 1955. "Monthly charts of the total solar radiation and the radiation balance over the territory of the Soviet Union." Glavnaia Geofizicheskaia Observatoriia, Trudy [Main Geophysical Observatory, Transactions], 50: 48-82.

BERNARD, E. A. 1954. "Sur diverses conséquences de la méthode du bilan d'énergie pour l'évapotranspiration des cultures ou des couvertures végetables naturelles." *Assemblée générale de Rome*, International Association of Scientific Hydrology, 3: 161-67.

BERRY, G. 1964. "The evaluation of Penman's natural evaporation formula by electronic computer." *Australian Journal of Applied Science*, 15: 61-64.

BEST, R. 1962. "Production factors in the tropics." *Netherlands Journal of Agricultural Science*, 10: 347-53.

BIERHUIZEN, J. F. 1958. "Some observations on the relation between transpiration and soil moisture." *Netherlands Journal of Agricultural Science*, 6: 94-98.

BIERHUIZEN, J. F. 1959. "Plant growth and soil moisture relationships." In *Plant-water relationships in arid and semiarid conditions*. UNESCO.

BIERHUIZEN, J. F., and N. M. DE VOS. 1959. "The effect of soil moisture on the growth and yield of vegetable crops." Institute for Land and Water Management Research, Wageningen. Bulletin 11.

BILHAM, E. G. 1932a. "The splashing rain, with special reference to evaporation tanks." *Meteorological Magazine*, 67: 86-91.

BILHAM, E. G. 1932b. "On the measurements of evaporation at Valentia Observatory." In *British Rainfall* 1931. London: Her Majesty's Stationery Office.

BLACK, J. N. 1955. "The interaction of light and temperature in determining the growth rate of subterranean clover (*Trifolium subterraneum* L.)." *Australian Journal of Biological Science*, 8: 330-43.

BLACK, J. N. 1956. "The distribution of solar radiation over the earth's surface." In *Wind and solar energy*. UNESCO.

BLACK, J. N. 1960. "A contribution to the radiation climatology of northern Europe." *Archiv für Meteorologie Geophysik und Bioklimatolgie*, Series B_x, 10: 182-92.

BLACK, J. N. 1963. "The interrelationship of solar radiation and leaf area index in determining the rate of dry matter production of swards of subterranean clover (*Trifolium subterraneum* L.)." *Australian Journal of Agricultural Research*, 14: 20-38.

BLACK, J. N., C. W. BONYTHON, and J. A. PRESCOTT. 1954. "Solar radiation and the duration of sunshine." *Quarterly Journal*, Royal Meteorological Society, 80: 231-35.

BLACKWELL, J. H. 1954. "A transient-flow method for determination of thermal constants of insulating materials in bulk." *Journal of Applied Physics*, 25: 137-44.

BLANEY, H. F. 1951. "Definitions, methods, research data." *Transactions,* American Society of Civil Engineers, 117: 949-73.

BLANEY, H. F., and W. D. CRIDDLE. 1950. "Determining water requirements in irrigated areas from climatological data." Washington, D.C.: Soil Conservation Service, Technical Publications 96, USDA.

BLANEY, H. F., and K. V. MORIN. 1942. "Evaporation and consumptive use of water empirical formulas." *Transactions,* American Geophysical Union, 23: 76-83.

BLEKSLEY, A. E. H. 1956. "Solar and sky radiation in southern Africa." In *Wind and solar energy.* UNESCO.

BLUMENSTOCK, D. I. 1942. "Drought in the U. S. analyzed by means of the theory of probability." USDA Technical Bulletin 819.

BÖHNING, R. H., and C. A. BURNSIDE. 1956. "The effect of light intensity on rate of apparent photosynthesis in leaves of sun and shade plants." *American Journal of Botany,* 43: 557-61.

BOLAS, B. D. 1933. "The influence of light and temperature on the assimilation rate of seedling tomato plants variety E.S. 5." In *Report,* Experiment Research Station, Chestnut.

BONNEN, C. A., W. C. MCARTHUR, A. C. MAGEE, and W. F. HUGHES. 1952. "Use of irrigation water on the high plains." Texas Agricultural Station, College Station, Texas. Bulletin 756.

BONNER, J. 1962. "The upper limit of crop yield." *Science,* 137: 11-15.

BONYTHON, C. W. 1950. "Evaporation studies using some South Australian data." *Transactions,* Royal Society of South Australia, Part 2, 73: 198-219.

BORDEAU, P. 1954. "Oak seedling ecology determining segregation of species in Piedmont oak-hickory forests." *Ecological Monographs,* 24: 297-320.

BORTHWICK, H. A. 1957. "Light effects on tree growth and seed germination." *Ohio Journal of Science,* 57: 357-64.

BORTHWICK, H. A., and S. B. HENDRICKS. 1960. "Photoperiodism in plants." *Science,* 132: 1223-28.

BOURGET, S. J., B. J. FINN, and B. K. DOW. 1966. "Effects of different soil moisture tensions on flax and cereals." *Canadian Journal of Soil Science,* 46: 213-16.

BOUYOUCOS, G. J. 1913. "An investigation of soil temperature and some of the most important factors influencing it." Michigan Agricultural College Experiment Station. Technical Bulletin 17.

BOWEN, I. S. 1926. "The ratio of heat losses by conduction and by evaporation from any water surface." *Physical Review,* 27: 779-87.

BOWERS, S. A. and R. J. HANKS. 1965. "Reflection of radiant energy from soils." *Soil Science,* 100: 130-38.

BOWMAN, D. H. and K. M. KING. 1965. "Determination of evapotranspiration using the neutron scattering method." *Canadian Journal of Soil Science,* 45: 117-26.

BOYKO, B. H. 1955. "Climatic, ecoclimatic, and hydrological influences on vegetation." In *Plant ecology.* UNESCO. Proceedings of the Montpellier symposium.

BRADLEY, G. A., and A. J. PRATT. 1954. "Irrigation to make a crop—not to save it." *Farm Research,* 20(2): 10-11.

BREAZEALE, E. L., and W. T. MCGEORGE. 1953. "Influence of atmsopheric humidity on root growth." *Soil Science,* 76: 361-65.

BREAZEALE, E. L., W. T. MCGEORGE, and J. F. BREAZEALE. 1950. "Moisture absorption by plants from an atmosphere of high humidity." *Plant Physiology,* 25: 413-19.

BRIERLEY, W. G. 1934. "Absorption of water by the foliage of some common fruit species." *Proceedings,* American Society for Horticultural Science, 34: 277-83.

BRIGGS, L. J., and H. L. SHANTZ. 1912. "The wilting coefficient for different plants and its indirect determination." U.S. Bureau of Plant Industry, Bulletin 230.

BRIGGS, L. J., and H. L. SHANTZ. 1913. "The water requirements of plants: I investigations in the Great Plains in 1910 and 1911." U.S. Bureau of Plant Industry, Bulletin 284.

BRIGGS, L. J., and H. L. SHANTZ. 1916a. "Daily transpiration during the normal growth period and its correlation with the weather." *Journal of Agricultural Research,* 7: 155-212.

BRIGGS, L. J., and H. L. SHANTZ. 1916b. "Hourly transpiration rate on clear days as determined by cyclic environmental factors." *Journal of Agricultural Research,* 5: 583-650.

BRILLIANT, B. 1924. "Le teneur en eau dans les feuilles at l'energic assimilatrice." *Comptes Rendus,* Academie Sciences, 173: 2122-25.

BRIX, H. 1962. "The effect of water stress on the rates of photosynthesis and respiration in tomato plants and loblolly pine seedlings." *Physiologia Plantarum,* 15: 10-20.

BRODIE, H. W. 1964. "Instruments for measuring solar radiation: research and evaluation by the Hawaiian sugar industry, 1928-1962." *Hawaiian Planters' Record,* 57: 159-97.

BRODIE, H. W. 1965. "The wig-wag." *Solar Energy,* 9: 27-31.

BROOKS, C. E. P., and N. CARRUTHERS. 1953. *Handbook of statistical methods in meteorology.* London: Her Majesty's Stationery Office.

BROOKS, C. F. 1947. "Recommended climatological networks based on the representativeness of climatic stations for different elements." *Transactions,* American Geophysical Union, 28: 845-46.

BROOKS, F. A. 1955. "Storage of solar energy in the ground." In F. Daniel and J. A. Duffie, (Eds.), *Solar energy research.* Madison: University of Wisconsin Press.

BROOKS, F. A. 1956. "More food from solar energy." In *Proceedings,* World Symposium on Applied Solar Energy. Stanford Research Institute.

BROOKS, F. A. 1959. *An introduction to physical microclimatology.* Davis: University of California Press.

BROOKS, F. A., C. F. KELLEY, D. G. RHOADES, and H. B. SCHULTZ. 1952. "Heat transfers in citrus orchards using wind machines for frost protection." *Agriculture Engineering,* 23: 74-78, 143-47.

BROOKS, R. M. 1951. "Apricot harvest predictable." *California Agriculture,* 5: 3-14.

BROUGHAM, R. W. 1956. "Effect of intensity of defoliation on regrowth of pasture." *Australian Journal of Agricultural Research,* 7: 377-87.

BROUGHAM, R. W. 1958a. "Leaf development in swards of white clover (*Trifolium repens* L.)." *New Zealand Journal of Agricultural Research*, 1: 707-718.

BROUGHAM, R. W. 1958b. "Interception of light by the foliage of pure and mixed stands of pasture plants." *Australian Journal of Agricultural Research*, 9: 39-52.

BROUGHAM, R. W. 1960. "The relationship between the critical leaf area, total chlorophyll content, and maximum growth rate of some pasture and crop plants." *Annals of Botany* [N.S.], 24: 463-73.

BROUGHAM, R. W. 1962. "The leaf growth of *Trifolium repens* as influenced by seasonal changes in the light environment." *Journal of Ecology*, 50: 449-59.

BROUGHAM, R. W. 1966. "Aspects of light utilization, leaf development, senescence, and grazing on grass-legume balance and productivity of pastures." *Proceedings*, New Zealand Ecological Society, 13: 58-65.

BROUWER, R. 1956. "Radiation intensity and transpiration." *Netherlands Journal of Agricultural Science*, 4: 43-48.

BROUWER, R. 1962a. "Nutritive influences on the distribution of dry matter in the plant." *Netherlands Journal of Agricultural Science*, 10: 399-408.

BROUWER, R. 1962b. "Distribution of dry matter in the plant." *Netherlands Journal of Agricultural Science*, 10: 361-76.

BROWN, D. M. 1960. "Soybean ecology I. Development-temperature relationships from controlled environment studies." *Agronomy Journal*, 52: 493-96.

BROWN, M. A. 1910. "The influence of air currents on transpiration." *Proceedings*, Iowa Academy of Science, 17: 13-15.

BROWN, P. L., and A. L. HALLSTEAD. 1952. "Comparison of evaporation data from standard weather bureau and plant industry type of evaporation pans." *Agronomy Journal*, 44: 100-101.

BROWN, R. H., R. B. COOPER, and R. E. BLASER. 1966. "Effects of leaf age on efficiency." *Crop Science*, 6: 206-209.

BRUNT, D. 1934. *Physical and dynamical meteorology*. London: Cambridge University Press

BRUTSAERT, W. 1965. "Evaluation of some practical methods of estimating evapotranspiration in arid climates at low latitudes." *Water Resources Research*, 1: 187-91.

BUDYKO, M. I. 1956. *Teplovoi balans zemnoi poverkhnosti*. Gidrometerologicheskoe Izdatel'stvo, Leningrad. (Translated as: Stepanova. N. A. 1958. *The heat balance of the earth's surface*. Washington, D.C.: U.S. Department of Commerce, Office of Technical Services.)

BUDYKO, M. I., and KH. P. POGOSIAN. 1959. "Modification of climate in the surface layer of the atmosphere with the amelioration of drought areas." *Weekly Weather and Crop Bulletin*, U.S. Weather Bureau, November 30, 1959.

BUDYKO, M. I., N. A. YEFIMOVA, L. I. ZUBENOK, and L. A. STROKHINA. 1962. "The heat balance of the surface of the earth." *Soviet Geography*, 3(5): 3-16.

BUETTNER, K. 1955. "A small portable meter for soil heat conductivity and its use in the O'Neill test." *Transactions,* American Geophysical Union, 36: 827-30.

BURMAN, R. D., and J. R. PARTRIDGE. 1962. "Evapotranspiration of water by small grains, corn, and beans in northwestern Wyoming." Wyoming Agricultural Experiment Station, Circular 174.

BUROV, D. I. 1952. "Isparenie vody paruiushchei poshvoi i pochvoi pod rastitel'nym pokrovom v usloviiakh Zavolzhia." ["Evaporation of water from fallow and from soil under plant cover in the Trans-Volga."]. *Pochvovedenie* [*Soil Science.*] Leningrad, 1: 41-52.

BURR, G. O. 1961. "Factors determining the upper limit of yield in sugar cane." Presented before the western section of the American Society of Plant Physiology, Davis, California.

BURR, G. O., C. E. HARTT, W. H. BRODIE, T. TANIMOTO, H. P. KORTSCHAK, D. TAKAHASHI, F. M. ASHTON, and R. E. COLEMAN. 1957. "The sugar cane plants." *Annual Review of Plant Physiology,* 8: 275-308.

BURR, W. W. 1914. "The storage and use of soil moisture." Nebraska Agricultural Experiment Station. Research Bulletin No. 5.

BUSINGER, J. A. 1956. "Some remarks on Penman's equation for the evapotranspiration." *Netherlands Journal of Agricultural Sciences,* 4: 77-80.

BUSINGER, J. A. 1965. "Frost protection with irrigation." *Meteorological Monographs,* 6(28): 74-80.

BUTLER, P. F., and J. A. PRESCOTT. 1955. "Evapotranspiration from wheat and pasture in relation to available moisture." *Australian Journal of Agricultural Research,* 6: 52-61.

BUTLER, W. L., and R. J. DOWNS 1960. "Light and plant development." *Scientific American,* 203: 56-63.

CABORN, J. M. 1957. "Shelterbelts and microclimate." Forestry Commission Bulletin No. 29.

CALDER, K. L. 1939. "A note on the constancy of horizontal turbulent shearing stress in the lower layers of the atmosphere." *Quarterly Journal,* Royal Meteorological Society, 65: 537-41.

CALVERT, A. 1964. "The effects of air temperature on growth of young tomato plants in natural light conditions." *Journal of Horticultural Science,* 39: 194-211.

CANNELL, G. H., K. B. TYLER, and C. W. ASHBELL. 1959. "The effects of irrigation and fertilizer on yield, blackheart, and nutrient uptake of celery." *Proceedings,* American Society for Horticultural Science, 74: 539-45.

CARDER, A. C. 1960. "Atmometer assemblies: a comparison." *Canadian Journal of Plant Science,* 40: 700-706.

CARDER, A. C. 1961. "Rate of evaporation from a free-water surface as influenced by exposure." *Canadian Journal of Plant Science,* 41: 199-203.

CARDER, A. C. 1966. "Further data on the performance of Bellani-type atmometers." *Canadian Journal of Plant Science,* 46: 93-96.

CARLSON, C. W., J. ALESSI, and R. H. MICKELSON. 1959. "Evapotranspiration and yield of corn as influenced by moisture level, nitrogen fertilization, and plant density." *Proceedings,* Soil Science Society, 23: 242-45.

CAVRIELSEN, E. K. 1948. "Influence of light of different wavelengths on photosynthesis in foliage leaves." *Physiologia Plantarum,* 1: 113-23.

CHANDRARATHA, N. F. 1949. "The effect of daylength on heading time in rice." *Tropical Agriculture,* Ceylon, 104: 130-40.

CHANG, JEN-HU. 1957a. "Global distribution of the annual range in soil temperature." *Transactions,* American Geophysical Union, 38: 718-23.

CHANG, JEN-HU. 1957b. "World patterns of monthly soil temperature distribution." *Annals,* Association of American Geographers, 47: 241-49.

CHANG, JEN-HU. 1958. *Ground temperature, vol. II.* Blue Hill Observatory, Harvard University.

CHANG, JEN-HU. 1959. "An evaluation of the 1948 Thornthwaite classification." *Annals,* Association of American Geographers, 49: 24-30.

CHANG, JEN-HU. 1961. "Microclimate of sugar cane." *Hawaiian Planters' Record.* 56: 195-223.

CHANG, JEN-HU. 1963. "The role of climatology in the Hawaiian sugar cane industry: an example of applied agricultural climatology in the tropics." *Pacific Science,* 17: 379-97.

CHANG, JEN-HU, R. B. CAMBELL, and F. E. ROBINSON. 1963. "On the relationship between water and sugar cane yield in Hawaii." *Agronomy Journal,* 55: 450-53.

CHAPAS, L. C., and A. R. REES. 1964. "Evaporation and evapotranspiration in southern Nigeria." *Quarterly Journal,* Royal Meteorological Society, 90: 313-19.

CHAPMAN, H. W., and W. E. LOOMIS. 1953. "Photosynthesis in the potato under field conditions." *Plant Physiology,* 28: 703-716.

CHAPMAN, H. W., L. S. GLEASON, and W. E. LOOMIS. 1954. "The carbon dioxide content of field air." *Plant Physiology,* 29: 500-503.

CHAPMAN, L. J., and P. DERMINE. 1961. "Evapotranspiration at Kapuskasing, Ontario." *Canadian Journal of Plant Science,* 41: 563-67.

CHATTERJEE, B. N. 1961. "Analysis of ecotypic differences in tall fescue (*Festuca arundinacea* Schreb)." *Annals of Applied Biology,* 49: 560-63.

CHRELASHVILI, M. N. 1941. "The influence of water content and carbohydrate accumulation on the energy of photosynthesis and respiration." *Trudy Botanicheskii Institut Akademiia Nauk. U.S.S.R.,* Seriia 4, Eksperimental' naia Botanika, [Transactions. Botanical Institute. Academy of Science of the USSR. Series 4, Experimental Botany.] 5: 101-137.

CHRISTIDIS, B. D. 1962. "Growing cotton by transplantation." *Crop Science,* 2: 472-75.

CLOSS, R. L. 1958. "Transpiration from plants with a limited water supply." In *Climatology and microclimatology.* UNESCO.

COLEMAN, E. A. 1946. "A laboratory study of lysimetric drainage under controlled soil moisture tension." *Soil Science,* 62: 365-82.

COLEMAN, O. H., and B. A. BELCHER. 1952. "Some responses of sorgo to short photoperiods and variations in temperature." *Agronomy Journal,* 44: 35-39.

COLLINS, B. G., and R. J. TAYLOR. 1961. "Conditions governing the onset of dew on large leaves." *Australian Journal of Applied Science,* 12: 23-29.

COLLINS, P. E. 1962. "Shelterbelt influences." In *Proceedings*, International seminar on soil and water utilization, South Dakota State College, Bookings, South Dakota July 18-August 10, 1962.

COLLIS-GEORGE, N., and B. G. DAVEY. 1961. "The doubtful utility of present-day field experimentation and other determinations involving soil-plant interactions." *Indian Journal of Agronomy*, 5: 210-14.

COMMITTEE ON PLANT IRRADIATION OF THE NEDERLANDSE STICHTING VOOR VERLICHTINGSKUNDE. 1953. "Specification of radiant flux and radiant flux density in irradiation of plants with artificial light." *Journal of Horticultural Science*, 28: 177-84.

CONRAD, V. E., and L. W. POLLAK. 1950. *Methods in climatology*. Cambridge: Harvard University Press.

COPELAND, E. B. 1906. "On the water relations of the coconut palm." *Philippine Journal of Science*, 1: 6-57.

COX, H. J. 1910. "Frost and temperature conditions in the cranberry marshes of Wisconsin." U.S. Department of Agriculture, Weather Bureau Bulletin T.

CRADDOCK, J. M. 1951. "An apparatus for measuring dewfall." *Weather* 6: 300-308.

CRIDER, F. J. 1922. "The olive in Arizona." Arizona Agricultural Experiment Station. Bulletin 94.

CROWDER, L. V., R. RAMIREZ, and H. CHAVERRA. 1961. "Performance of temperate zone pasture species within tropical boundaries, as represented by Colombia." *Agronomy Journal*, 53: 229-32.

CROWE, P. R. 1954. "The effectiveness of precipitation: a graphical analysis of Thornthwaite's climatic classification." *Geographical Studies*, 1: 50.

CUNNINGHAM, R. K., and J. LAMB. 1959. "A cocoa shade and manurial experiment at the West African Cocoa Research Institute, Ghana: Part I, first year." *Journal of Horticultural Science*, 34: 14-22.

CUNNINGHAM, R. K., and J. LAMB. 1961. "A cocoa shade and manurial experiment at the West African Cocoa Research Institute, Ghana. Part II. Second and third years." *Journal of Horticultural Science*, 36: 116-25.

CURRY, J. R., and T. W. CHURCH. 1952. "Observations on winter drying of conifers in the Adirondacks." *Journal of Forestry*, 50: 114-16.

DAGG, M. 1965. "A rational approach to the selection of crops for areas of marginal rainfall in East Africa." *East African Agricultural and Forestry Journal*, 30: 286-95.

DALE, R. F. 1950. "Agricultural meteorology in the U.S.A." *Weather*, 5: 383-88.

DAS, U. K. 1931. "The problem of juice quality." *Hawaiian Planters' Record*, 35: 163-200.

DAVIDSON, J. L. 1965. "Some effects of leaf area control on the yield of wheat." *Australian Journal of Agricultural Research*, 16: 721-31.

DAVIDSON, J. L., and J. R. PHILIP. 1958. "Light and pasture growth." In *Climatology and microclimatology*. UNESCO.

DAVIES, J. A. 1966. "The assessment of evapotranspiration for Nigeria." *Geografiska Annaler*, 48A: 139-56.

DAVIS, J. R. 1955. "Frost protection with sprinkler irrigation." University of Michigan Extension Bulletin 327.

DAVIS, J. R. 1963. "Relationship of can evaporation to pan evaporation and evapotranspiration." *Journal of Geophysical Research*, 68: 5711-18.

DAVIS, J. R., and M. A. HAGOOD. 1961. "Efficient distribution of water in irrigating annual crops with limited supplies in drought years." *California Agriculture*, 15: 6-8.

DAY, G. J. 1961. "Distribution of total solar radiation on a horizontal surface over the British Isles and adjacent areas." *Meteorological Magazine*, 90: 269-84.

DEACON, E. L. 1949. "Vertical diffusion in the lowest layers of the atmosphere." *Quarterly Journal*, Royal Meteorological Society, 75: 89-103.

DEACON, E. L., C. H. B. PRIESTLEY, and W. C. SWINBANK. 1958. "Evaporation and water balance." In *Climatology*. UNESCO. p. 9-34.

DECKER, W. L. 1955. Determination of soil temperatures from meteorological data. Unpublished doctoral dissertation, Iowa State College.

DECKER, W. L. 1959. "Variations in the net exchange of radiation from vegetation of different heights." *Journal of Geophysical Research*, 64: 1617-19.

DECKER, W. L. 1962. "Precision of estimates of evapotranspiration in Missouri climate." *Agronomy Journal*, 54: 529-31.

DE LIS, B. R., I. PONCE, and R. TIZIO. 1964. "Studies on water requirement of horticultural crops: I. Influence of drought at different growth stages of potato on the tuber's yield." *Agronomy Journal*, 56: 377-81.

DENEKE, H. 1931. "Über den Einfluss bewegter Luft auf die Kohlensäureassimilation." *Jahrbuch für wissenschaftliche Botanik*, 74: 1-32.

DENMEAD, O. T., and R. H. SHAW. 1959. "Evapotranspiration in relation to the development of corn crop." *Agronomy Journal*, 51: 725-26.

DENMEAD, O. T., and R. H. SHAW. 1960. "The effects of soil moisture stress at different stages of growth on the development and yield of corn." *Agronomy Journal*, 52: 272-73.

DENMEAD, O. T., and R. H. SHAW. 1962. "Availability of soil water to plants as affected by soil moisture content and meteorological conditions." *Agronomy Journal*, 45: 385-90.

DENMEAD, O. T., L. J. FRITSCHEN, and R. H. SHAW. 1962. "Spatial distribution of net radiation in a corn field." *Agronomy Journal*, 54: 505-510.

DE VRIES, D. A. 1959. "The influence of irrigation on the energy balance and the climate near the ground." *Journal of Meteorology*, 16: 256-70.

DE VRIES, D. A. 1963. The physics of plant environments. In L. T. Evans (Ed.), *Environmental Control of Plant Growth*. Academic Press, New York and London.

DE VRIES, D. A., and J. W. BIRCH. 1961. "The modification of climate near the ground by irrigation for pastures on the riverine plain." *Australian Journal of Agricultural Research*, 12: 260-72.

DE VRIES, D. A., and A. J. PECK. 1957. "On the cylindrical probe method of measuring thermal conductivity with special reference to soils." *Australian Journal of Physics*, 11: 265-71.

DE WIT, C. T. 1958. "Transpiration and crop yields." Instituut voor Biologisch en Scheikundig Onderzoek van Landbouwgewassen, Wageningen, Mededeling. No. 59.

DE WIT, C. T. 1959. "Potential photosynthesis of crop surfaces." *Netherlands Journal of Agricultural Science*, 7: 141-49.
DE WIT, C. T. 1965. "Photosynthesis of leaf canopies." Instituut voor Biologisch en Scheikundig Onderzoek van Landbouwgewassen, Wageningen, Mededeling. No. 274.
DE WIT, C. T., and T. ALBERDA. 1961. "Transpiration coefficient and transpiration rate of three grain species in growth chambers." Instituut voor Biologisch en Scheikundig Onderzoek van Landbouwgewassen, Wageningen, Mededeling. 156: 73-81.
DONALD, C. M. 1961. "Competition for light in crops and pastures. In *Mechanisms in biological competition*, Symposia of the Society for Experimental Biology. New York: Academic Press.
DORE, J. 1959. "Response of rice to small differences in length of day." *Nature*, 183: 413-14.
DOSS, B. D., O. L. BENNETT, and D. A. ASHLEY. 1962. "Evapotranspiration by irrigated corn." *Agronomy Journal*, 54: 497-98.
DOSS, B. D., O. L. BENNETT, and D. A. ASHLEY. 1964. "Moisture use by forage species as related to pan evaporation and net radiation." *Soil Science*, 98: 322-27.
DOSS, B. D., O. L. BENNETT, D. A. ASHLEY, and H. A. WEAVER. 1962. "Soil moisture regime effect on yield and evapotranspiration from warm season perennial forage species." *Agronomy Journal*, 54: 239-42.
DREIBELBIS, F. R., and L. L. HARROLD. 1958. "Water-use efficiency of corn, wheat, and meadow crops." *Agronomy Journal*, 50: 500-03.
DREW, D. H. 1966. "Irrigation studies on summer cabbage." *Journal of Horticultural Science*, 41: 103-114.
DRUMMOND, A. J. 1958. "Radiation and the thermal balance. In *Climatology*. UNESCO.
DRUMMOND, A. J., and E. VOWINCKEL. 1957. "The distribution of solar radiation throughout southern Africa." *Journal of Meteorology*, 14: 343-53.
DUVDEVANI, S. 1947. "An optical method of dew estimation." *Quarterly Journal*, Royal Meteorological Society, 73: 282-96.
DUVDEVANI, S. 1953. "Dew gradients in relation to topography, soil and climate." In *Arid Research*, Proceedings of Desert Research Symposium, Jerusalem.
DUVDEVANI, S. 1957. "Dew research for arid agriculture." *Discovery*, 18: 330-34.
DUVDEVANI, S. 1964. "Dew in Israel and its effect on plants." *Soil Science*, 98: 14-21.
DUVDEVANI, S., I. REICHERT, and J. PALTI. 1946. "The development of downy and powdery mildew of cucumbers as related to dew and other environmental factors." *Palestine Journal of Botany*, Rehovost, 5: 127-51.
DYER, A. J., and T. V. CRAWFORD. 1965. "Observations of the modification of the microclimate at a leading edge." *Quarterly Journal*, Royal Meteorological Society, 91: 345-48.
DYER, A. J., and F. J. MAHER. 1965. "The evapotron: an instrument for the measurement of eddy fluxes in the lower atmosphere." Commonwealth

Scientific and Industrial Research Organization, Australia, Division of Meteorological Physics Technical Paper 15.

EAGLEMAN, J. R., and W. L. DECKER. 1965. "The role of soil moisture in evapotranspiration." *Agronomy Journal*, 57: 626-29.

EARLEY, E. B., R. J. MILLER, G. L. REICHERT, R. H. HAGEMAN, and R. D. SEIF. 1966. "Effects of shade on maize production under field conditions." *Crop Science*, 6: 1-7.

EHRLER, W. L., F. S. NAKAYAMA, and C. H. M. VAN BAVEL. 1965. "Cyclic changes in water balance and transpiration of cotton leaves in a steady environment." *Physiologia Plantarum*, 18: 766-75.

EHRLER, W. L., C. H. M. VAN BAVEL, and F. S. NAKAYAMA. 1966. "Transpiration, water absorption, and internal water balance of cotton plants as affected by light and changes in saturation deficit." *Plant Physiology*, 41: 71-74.

EIK, K., and J. J. HANWAY. 1966. "Leaf area in relation to yield of corn grain." *Agronomy Journal*, 58: 16-18.

EKERN, P. C. 1959. "Evapotranspiration patterns under trade wind weather regime on central Oahu, Hawaii." *Agronomy Abstracts*, 6: 4-5.

EKERN, P. C. 1965a. "Evapotranspiration of pineapple in Hawaii." *Plant Physiology*, 40: 736-39.

EKERN, P. C. 1965b. "The fraction of sunlight retained as net radiation in Hawaii." *Journal of Geophysical Research*, 70: 785-93.

EKERN, P. C. 1965c. "Disposition of net radiation by a free water surface in Hawaii." *Journal of Geophysical Research*, 70: 795-800.

EL NADI, A. H., and J. P. HUDSON. 1965. "Effects of crop height on evaporation from lucerne and wheat grown in lysimeters under advective conditions in the Sudan." *Experimental Agriculture*, 1: 289-97.

ELSASSER, W. M. 1942. "Heat transfer by infrared radiation in the atmosphere." Harvard Meteorological Studies. No. 6.

ELSASSER, W. M., and M. F. CULBERTSON. 1960. *Atmospheric Radiation Tables*. Meteorological Monograph, 4(23).

ENGER, I. 1959. "Optimum length of record for climatological estimates of temperature." *Journal of Geophysical Research*, 64: 779-87.

ENGLAND, C. B. 1963. "Water use by several crops in a weighing lysimeter." *Agronomy Journal*, 55: 239-42.

ERTEL, H. 1933. "Beweis der Wilh. Schmidtschen Konjugierten Potenzformeln für Austausch und Windgeschwindigkeit in den bodennahen Luftschichten." *Meteorologische Zeitschrift*, 50: 386-88.

EVERSON, J. N., and J. B. WEAVER. 1949. "Effect of carbon black on properties of soils." *Industrial and Engineering Chemistry*, 41: 1798-1801.

FEDROV, C. F. 1954. "Papers of the government hydrological institute." Soviet State Publishing, Hydrometeorology Section, No. 6.

FERGUSON, W. S. 1963. "Effect of intensity of cropping on the efficiency of water use." *Canadian Journal of Soil Science*, 43: 156-65.

FIBRAS, F. 1931. "Die Wirkung des Windes auf die Transpiration." *Bericht der Deutschen Botanischen Gesellschaft*, 49: 443-52.

FINKELSTEIN, J. 1961. "Estimation of open water evaporation in New Zealand." *New Zealand Journal of Science*, 4: 506-522.

FINNEL, H. H. 1928. "Effect of wind on plant growth." *Journal of the American Society of Agronomy*, 20: 1206-1210.

FISCHER, R. A., and R. M. HAGAN. 1965. "Plant water relations, irrigation management, and crop yield." *Experimental Agriculture*, 1: 161-77.

FISCHER, R. A., and G. D. KOHN. 1966. "The relationship between evapotranspiration and growth in the wheat crop." *Australian Journal of Agricultural Research*, 17: 255-67.

FITZGERALD, P. D., and D. S. RICKARD. 1960. "A comparison of Penman's and Thornthwaite's method of determining soil moisture deficits." *New Zealand Journal of Agricultural Research*, 4:106-112.

FITZPATRICK, E. A., and W. R. STERN. 1966. "Estimates of potential evaporation using alternative data in Penman's formula." *Agricultural Meteorology*, 3: 255-39.

FLOCKER, W. J., J. C. LINGLE, R. M. DAVIS, and R. J. MILLER. 1965. "Influence of irrigation and nitrogen fertilization on yield, quality and size of cantaloupes." *Proceedings*, American Society for Horticultural Science, 86: 424-32.

FOLEY, J. C. 1945. "Frost in the Australian region." Commonwealth Meteorological Bureau Bulletin 32.

FORTANIER, E. J. 1957. "Control of flowering in *Arachis Hypogaea* L." *Mededelingen van de Landbouwhogeschool te Wageninger*, 57: 1-116.

FOURNIER D'ALBE, E. M. 1958. "The modification of microclimates." In *Climatology*. UNESCO.

FREAR, D. E. H. 1935. "Photoelectric apparatus for measuring leaf areas." *Plant Physiology*, 10: 569-74.

FRITH, H. J. 1955. "Trials of a wind machine for frost protection in citrus." *Australian Journal of Agricultural Research*, 6: 903-912.

FRITSCHEN, L. J. 1963. "Construction and evaluation of a miniature net radiometer." *Journal of Applied Meteorology*, 2: 165-72.

FRITSCHEN, L. J. 1965a. "Miniature net radiometer improvements." *Journal of Applied Meteorology*, 4: 528-32.

FRITSCHEN, L. J. 1965b. "Accuracy of evapotranspiration determinations by the Bowen ratio method." *Bulletin*, International Association of Scientific Hydrology, 2: 38-48.

FRITSCHEN, L. J. 1966. "Evapotranspiration rates of field crops determined by the Bowen ratio method." *Agronomy Journal*, 58: 339-42.

FRITSCHEN, L. J., and R. H. SHAW. 1961. "Evapotranspiration for corn as related to pan evaporation." *Agronomy Journal*, 53: 149-50.

FRITSCHEN, L. J., and C. H. M. VAN BAVEL. 1962. "Energy balance components of evaporating surfaces in arid lands." *Journal of Geophysical Research*, 67: 5179-85.

FRITSCHEN, L. J., and C. H. M. VAN BAVEL. 1964. "Energy balance as affected by height and maturity of sudangrass." *Agronomy Journal*, 56: 201-204.

FRITZ, S. 1957. "Solar energy on clear and cloudy days." *Scientific Monthly*, 84: 55-65.

FRITZ, S., and T. H. MACDONALD. 1949. "Average solar radiation in the United States." *Heating and Ventilating*, 46: 61-64.

FULLER, H. J. 1948. "Carbon dioxide concentrations of the atmosphere above Illinois forest and grassland." *American Midland Naturalist,* 39: 247-49.

FULTON, J. M. 1966. "Evaporation from bare soil compared with evapotranspiration from a potato crop." *Canadian Journal of Soil Science,* 46: 199-204.

FULTON, J. M., and H. F. MURWIN. 1955. "The relationship between available soil moisture levels and potato yields." *Canadian Journal of Agricultural Science,* 35: 552-56.

FUNK, J. P. 1959. "Improved polyethylene shielded net radiometer." *Journal of Scientific Instruments,* 36: 267-70.

FUNK, J. P. 1962. "A net radiometer designed for optimum sensitivity and a ribbon thermopile used in a miniaturized version." *Journal of Geophysical Research,* 67: 2753-60.

FUNK, J. P. 1963. "Radiation observations at Aspendale, Australia, and their comparison with other data." *Archiv für Meteorologie, Geophysik, und Bioklimatologie,* Series B, 3: 52-70.

FUQUAY, D., and K. BUETTNER. 1957. "Laboratory investigation of some characteristics of the Eppley pyrheliometer." *Transactions,* American Geophysical Union, 38: 38-43.

GAASTRA, P. 1958. "Light energy conversion in field crops in comparison with the photosynthetic efficiency under laboratory conditions." *Mededelingen van de Landbouwhogeschool te Wageningen,* 58:(4): 1-12.

GAASTRA, P. 1959. "Photosynthesis of crop plants as influenced by light, carbon dioxide, temperature, and stomatal diffusion resistance." *Mededelingen van de Landbouwhogeschool te Wageningen,* 59: 1-68.

GAASTRA, P. 1962. "Photosynthesis of leaves and field crops." *Netherlands Journal of Agricultural Science,* 10: 311-24.

GAASTRA, P. 1963. "Climatic control of photosynthesis and respiration. In L. T. Evans (Ed.), *Environment control of plant growth.* New York: Academic Press.

GABITES, J. F. 1951. "A preliminary estimate of the mean daily insolation in New Zealand." New Zealand Meteorological Service Circular 71.

GAL'TSOV, A. P. 1953. *O Klimatishekom vzaimodeistvii oroshaemykh i neoroshaemykh ploshchadei.* [*Climatic interaction of the irrigated and nonirrigated areas*]. Akademiia Nauk SSSR, Izvestiia. Seriia Geograficheskaia. [Academy of Sciences of the USSR. Bulletin; Series in Geography,] 3: 11-20.

GARD, L. E., G. E. MCKIBBEN, and B. A. JONES. 1961. "Moisture loss and corn yields on a silt-pan soil as affected by three levels of water supply." *Proceedings,* Soil Science Society, 25: 154-57.

GARNER, W. W., and H. A. ALLARD. 1920. "Effect of the relative length of day and night and other factors of the environment on growth and reproduction in plants." *Journal of Agricultural Research,* 18: 553-606.

GARNER, W. W., and H. A. ALLARD. 1923. "Further studies in photoperiodism, the response of the plant to relative length of day and night." *Journal of Agricultural Research,* 23: 871-920.

GARNIER, B. J. 1956. "A method of computing potential evapotranspiration in West Africa." *Bulletin de l'Institut Francais d'Afrique Noire, Dakar,* 18: 665-76.

GATES, D. M. 1962. *Energy exchange in the biosphere.* New York: Harper and Row.
GATES, D. M. 1964. "Leaf temperature and transpiration." *Agronomy Journal,* 56: 273-77.
GATES, D. M. 1965a. "Energy, plants, and ecology." *Ecology,* 46: 1-13.
GATES, D. M. 1965b. "Heat transfer in plants." *Scientific American,* 213: 76-84.
GATES, D. M., H. J. KEEGAN, J. C. SCHLETER, and V. R. WEIDNER. 1965. "Spectral properties of plants." *Applied Optics,* 4: 11-20.
GEIGER, R. 1965. *The climate near the ground.* Cambridge: Harvard University Press.
GERBER, J. F., and W. L. DECKER. 1960. "A comparison of evapotranspiration as estimated by the heat budget and measured by the water balance from a corn field." Final Report USWB Contract Cwb-956. University of Missouri.
GIER, J. T., and R. V. DUNKLE. 1951. "Total hemispherical radiometers." *Transactions,* American Institute of Electrical Engineers, 70: 339-45.
GILBERT, M. J., and C. H. M. VAN BAVEL. 1954. A simple field method for measuring maximum evapotranspiration. *Transactions,* American Geophysical Union, 35: 937-42.
GLOVER, J., and J. FORSGATE. 1964. "Transpiration from short grass." *Quarterly Journal,* Royal Meteorological Society, 90: 320-24.
GLOVER, J., and M. D. GWYNNE. 1962. "Light rainfall and plant survival in East Africa. I. Maize." *Journal of Ecology,* 50: 111-18.
GLOVER, J., and J. S. G. McCULLOCH. 1958. "The empirical relation between solar radiation and hours of sunshine." *Quarterly Journal,* Royal Meteorological Society, 84: 172-75.
GLOYNE, R. W. 1955. "Some effects of shelterbelts and windbreaks." *Meteorological Magazine,* 84: 272-81.
GOODALL, G. E., D. E. ANGUS, A. S. LEONARD, and F. A. BROOKS. 1957. "Effectiveness of wind machine." *California Agriculture,* 2: 7-9.
GOSS, J. R., and F. A. BROOKS. 1956. "Constants for empirical expressions for downcoming atmospheric radiation under cloudless sky." *Journal of Meteorology,* 13: 482-88.
GOTO, Y., and K. TAI. 1956. "Studies on oxidizing power of roots of paddy rice plants." *Journal of the Science of Soil and Manure,* 26: 403-404. (In Japanese)
GRABLE, A. R., R. J. HANKS, F. M. WHILLHITE, and H. R. HAISE. 1966. "Influence of fertilization and altitude on energy budgets for native meadows." *Agronomy Journal,* 58: 234-37.
GRAHAM, W. G. and K. M. KING. 1961a. "Fraction of net radiation utilized in evapotranspiration from a corn crop." *Proceedings,* Soil Science Society of America, 25: 158-60.
GRAHAM, W. G., and K. M. KING. 1961b. "Short-wave reflection coefficient for a field of maize." *Quarterly Journal,* Royal Meteorological Society, 87: 425-28.
GRAINGER, J., J. L. SNEDDON, E. DE C. CHISHOLM, and A. HASTIE. 1955. "Climate and the yield of cereal crops." *Quarterly Journal,* Royal Meteorological Society, 81: 108-111.

GRAY, H. E., G. LEVINE, and W. K. KENNEDY. 1955. "Use of water by pasture crops." *Agricultural Engineering*, 36: 529-31.

GREGORY, F. G. 1926. "The effect of climatic conditions on the growth of barley." *Annals of Botany*, 40: 1-26.

GUERRINI, V. H. 1954. "A year of evapotranspiration in Ireland." *Publications in Climatology*, Laboratory of Climatology, 7(1): 90-111.

GUNNESS, C. I. 1941. "Blowers for frost protection." *Agriculture Engineering*, 22: 252.

HAAS, H. J., and W. O. WILLIS. 1962. "Moisture storage and use by dryland spring wheat cropping systems." *Proceedings*, Soil Science Society of America, 26: 506-509.

HAGAN, H. R. 1933. "Hawaiian pineapple field soil temperatures in relation to the nematodes *Heterodera radicicola* (Greef) Müller." *Soil Science*, 36: 83-95.

HAGAN, R. M., M. L. PETERSON, R. P. UPCHURCH, and L. G. JONES. 1957. "Relationships of soil moisture stress to different aspects of growth of ladino clover." *Proceedings*, Soil Science Society of America, 21: 360-65.

HAGAN, R. M., and Y. VAADIA. 1961. "Principles of irrigated cropping." In *Plant-water relationships in arid and semiarid conditions*. UNESCO.

HAINES, F. M. 1952. "The absorption of water by leaves in an atmosphere of high humidity." *Journal of Experimental Botany*, 3: 95-98.

HALKIAS, N. A., F. J. VEIHMEYER, and A. H. HENDRICKSON. 1955. "Determining water needs for crops from climatic data." *Hilgardia*, 24: 207-33.

HALSTEAD, M. H. 1954. "The heat flux of momentum, heat, and water vapor in microclimatology." *Laboratory of Climatology Publication*, 7(2): 326-61.

HALSTEAD, M. H., and W. COVEY. 1957. "Some meteorological aspects of evapotranspiration." *Proceedings*, Soil Science Society of America, 21: 461-64.

HAND, D. W. 1964. "Advective effects on evaporating conditions as hot day air crosses irrigated lucerne." *Empire Journal of Experimental Agriculture*, 32: 262-73.

HANKS, R. J., S. A. BOWERS, and L. D. BARK. 1961. "Influence of soil surface conditions on net radiation, soil temperature, and evaporation." *Soil Science*, 91: 233-38.

HANKS, R. J., and C. B. TANNER. 1952. "Water consumption by plants as influenced by soil fertility." *Agronomy Journal*, 44: 98-100.

HANSEN, E., and G. F. WALDO. 1944. "Ascorbic acid content of small fruits in relation to genetic and environment factors." *Food Research*, 9: 453-61.

HANSEN, V. E. 1963. "Unique consumptive use curve related to irrigation practice." *Journal of the Irrigation and Drainage Division*, Proceedings of the American Society of Civil Engineers, 89: 43-50.

HARBECK, G. E. 1958. "The Lake Hefner water-loss investigations." *International Union of Geodesy and Geophysics*. International Association of Scientific Hydrology, General Assembly of Toronto, Gentbrugge, 3: 437-43.

HARRIS, D. G., and C. H. M. VAN BAVEL. 1958. "A comparison of measured and computed evapotranspiration from Bermuda grass and sweet corn." *Agronomy Abstracts*, 5: 37.

HARROLD, L. L., and F. R. DREIBELBIS. 1951. "Agricultural hydrology as evaluated by monolith lysimeters." USDA Technical Bulletin 1050.

HARROLD, L. L., D. B. PETERS, F. R. DREIBELBIS, and J. L. MCGUINNESS. 1959. "Transpiration evaluation of corn grown on a plastic-covered lysimeter." *Proceedings*, Soil Science Society of America, 23: 174-78.

HART, S. A., and F. W. ZINK. 1957. "Brushing and brushing material for frost protection." *Proceedings*, American Society for Horticultural Science, 69: 475-79.

HARTT, C. E. 1963. "Translocation as a factor in photosynthesis." *Sonderdruck aus die Naturwissenschaften*, 21: 1-2.

HARTT, C. E. 1965. "Light and translocation of C^{14} in detached blade of sugar cane." *Plant Physiology*, 40: 718-24.

HASSELKUS, E. R., and G. E. BECK. 1963. "Plant responses to light transmitted into a fiber glass reinforced plastic greenhouse." *Proceedings*, American Society for Horticultural Science, 82: 637-44.

HAUDE, W. 1952. "Verdunstungsmenge und Evaporationskraft eines Klimas." *Berichtes des Deutschen Wetterdienstes*, U.S. Zone, 42: 225-29.

HAWAIIAN SUGAR PLANTERS' ASSOCIATION, Experiment Station. 1963. *Annual Report*.

HEARN, A. B., and R. A. WOOD. 1964. "Irrigation control experiments on dry-season crops in Nyasaland." *Empire Journal of Experimental Agriculture*, 32: 1-17.

HEENEY, H. B., S. R. MILLER, and W. M. RUTHERFORD. 1961. "A meteorological method of calculating the irrigation requirement of the canning tomato crop." *Canadian Journal of Plant Science*, 41: 31-41.

HEINICKE, A. J., and M. B. HOFFMAN. 1933. "The rate of photosynthesis of apple leaves under natural conditions." Cornell Agricultural Experiment Station Bulletin 577.

HEINICKE, A. J., and N. F. CHILDERS. 1937. "The daily rate of photosynthesis during the growing season of 1935 of a young apple tree of bearing age." Cornell University Agricultural Experiment Station Memoir 201.

HEINICKE, D. R. 1963. "The microclimate of fruit trees: I. light measurements with uranyl oxalate actinometers." Canadian Journal of Plant Science, 43: 561-68.

HENDRICKSON, A. H., and F. J. VEIHMEYER. 1942. "Irrigation experiments with pears and apples." University of California, Berkeley, Bulletin 667.

HENDRICKSON, A. H., and F. J. VEIHMEYER. 1950. "Irrigation experiments with grapes." California Agricultural Experimental Station Bulletin 728.

HERSHFIELD, D. M. 1962. "Effective rainfall." *Weekly Weather and Crop Bulletin*. USDA (November 12), pp. 7-8.

HESKETH, J. D. 1961. Photosynthesis: leaf chamber studies with corn. Unpublished doctoral dissertation, Cornell University.

HESKETH, J. D. 1963. "Limitations to photosynthesis responsible for differences among species." *Crop Science*, 3: 493-96.

HESKETH, J. D., and R. B. MUSGRAVE. 1962. "Photosynthesis under field conditions: IV. light studies with individual corn leaves." *Crop Science*, 2: 311-15.

HESSE, W. 1954. "Der Einfluss meteorologischer Faktoren auf die Transpiration der Pfefferminze (*Mentha piperita* L.)." *Angewandte Meteorologie*, 2: 14-18.

HILDRETH, A. C., J. R. MAGNESS and J. W. MITCHELL. 1941. "Effects of climatic factors on growing plants." In *Climate and man*. USDA Yearbook.

HOFMANN, G. 1955. "Die Thermodynamik der Taubildung." Berichte deutsche Wetterdienstes No. 3.

HOFMANN, G. 1958. "Dew measurements by thermodynamical means." International Union of Geodesy and Geophysics, International Association of Scientific Hydrology, General Assembly, Toronto, Gentbrugge, 2: 443-45.

HOGG, W. H., and A. R. CARTER. 1962. "Shelter screens at Luddington, 1957-59." *Experimental Horticulture*, 7: 47-51.

HØJENDAHL, K. 1962. "Synopsis of 11 years of measurements of light and microclimate in fertile soils with common farm crops." In *Royal Veterinary and Agricultural College Yearbook 1962. Copenhagen, Denmark.*

HOLMES, R. M. 1961. "Estimation of soil moisture content using evaporation data." In *Proceedings of Hydrology Symposium*: No. 2, evaporation. Toronto: Department of Northern Affairs and National Resources, Water Resources Branch.

HOLMES, R. M., and G. W. ROBERTSON. 1958. "Conversion of latent evaporation to potential evapotranspiration." *Canadian Journal of Plant Science*, 38: 164-72.

HOLMES, R. M., and G. W. ROBERTSON. 1959. "A modulated soil moisture budget." *Monthly Weather Review*, 67: 101-106.

HOLMES, R. M., and G. W. ROBERTSON. 1963. "Application of the relationship between actual and potential evapotranspiration in dry land agriculture." *Transactions*, American Society of Agricultural Engineers, 6: 65-67.

HOLT, R. F., and C. A. VAN DOREN. 1961. "Water utilization by field corn in western Minnesota." *Agronomy Journal*, 53: 43-45.

HOOVER, W. H., E. S. JOHNSON, and F. S. BRACKETT. 1933. "Carbon dioxide assimilation in a higher plant." *Smithsonian Miscellaneous Publication*, 87(16): 1-19.

HOPEN, H. J., and S. K. RIES. 1962. "Atmospheric carbon dioxide levels over mineral and muck soils." *Proceedings*, American Society for Horticultural Science, 81: 365-67.

HOPKINS, J. W. 1939. "Estimation of leaf area in wheat from linear dimensions." *Canadian Journal of Research*, Section C, 17: 300-304.

HORD, H. H. V., and D. P. SPELL. 1962. "Temperature as a basis for forecasting banana production." *Tropical Agriculture*, 39: 219-23.

HOUNAM, C. E. 1958. "Evaporation pan coefficients in Australia. In *Climatology and microclimatology*. UNESCO.

HOUNAM, C. E. 1965. "Comparison of evaporation from U. S. Class "A" pan and Australian standard tank and selected estimates." *Australian Meteorological Magazine*, 49: 1-13.

HOUSE, G. J., N. E. RIDER, and C. P. TUGWELL. 1960. "A surface energy-balance computer." *Quarterly Journal*, Royal Meteorological Society, 86: 215-31.

HOWE, O. W., and H. F. RHOADES. 1955. "Irrigation practice for corn production in relation to stage of plant development." *Proceedings*, Soil Science Society of America, 19: 94-98.

HUDSON, J. P. 1965. "Evaporation from lucerne under advective conditions in the Sudan: I. factors affecting water losses and their measurement." *Experimental Agriculture*, 1: 23-32.

HUGHES, R. 1965. "Climatic factors in relation to growth and survival of pasture plants." *Journal of the British Grassland Society*, 20: 263-72.

HUMPHRIES, E. C., and S. A. W. FRENCH. 1964. "Determination of leaf area by rating in comparison with geometric shapes." *Annals of Applied Biology*, 54: 281-84.

HURD, R. G., and A. R. REES. 1966. "Transmission error in the photometric estimation of leaf area." *Plant Physiology*, 41: 905-906.

HUXLEY, P. A. 1965. "Climate and agricultural production in Uganda." *Experimental Agriculture*, 1: 81-97.

IIZUKA, H. 1950. "Wind erosion prevention by windbreak." In Bulletin 45, Experiment Station, Neguro, Tokyo.

ILJIN, W. S. 1923. "Der Einfluss des Wassermangels auf die Kohlenstoffassimilation duren die Pflanzen." *Flora*, 116: 360-78.

INOUE, E. 1963. The environment of plant surfaces. In L. T. Evans (Ed.), *Environment control of plant growth*. New York: Academic Press.

ISRAELSEN, O. W., and V. E. HANSEN. 1962. *Irrigation principles and practices*. New York: John Wiley.

JACKSON, E. A. 1960. "Water consumption by lucerne in central Australia." *Australian Journal of Agricultural Research*, 11: 715-22.

JACOB, W. C., M. B. RUSSELL, A. KLUTE, G. LEVINE, and R. GROSSMAN. 1952. "The influence of irrigation on the yield and quality of potatoes on Long Island." *American Potato Journal*, 29: 292-96.

JANES, B. E., and W. O. DRINKWATER. 1959. "Irrigation studies on vegetables in Connecticut." Connecticut Agricultural Experiment Station Bulletin 338.

JENKINS, H. V. 1959. "An airflow planimeter for measuring the area of detached leaves." *Plant Physiology*, 34: 532-36.

JENNINGS, E. G., and J. L. MONTEITH. 1954. "A sensitive recording dew balance." *Quarterly Journal*, Royal Meteorological Society, 80: 222-26.

JENSEN, M. 1954. *Shelterbelt effect*. Copenhagen: Danish Technical Press.

JENSEN, M. E., and H. R. HAISE. 1963. "Estimating evapotranspiration from solar radiation." *Journal of the Irrigation and Drainage Division*, Proceedings of the American Society of Civil Engineers, 89(1R4): 15-41.

JESSEP, C. T. 1964. "The development, testing, and calibration of a portable atmometer suitable for field use." *New Zealand Journal of Agricultural Research*, 7: 205-218.

JONES, S. T. 1961. "Effect of irrigation at different levels of soil moisture on yield and evapotranspiration rate of sweet potatoes." *Proceedings*, American Society for Horticultural Science, 77: 458-62.

JOHNSON, F. A. 1954. "The solar constant." *Journal of Meteorology*, 11: 431-39.

KARSTEN, H. 1912. "Untersuchungen über die Wärmeleitungsfähigkeit einiger Bodenarten." *Internationale Mitteilung für Bodenkunde*, 2: 45.

KASANAGA, H., and M. MONSI. 1954. "On the light transmission of leaves and its meaning for the production of matter in plant community." *Japanese Journal of Botany,* 14: 304-324.

KATSCHON, R. 1949. "Üntersuchungen über die physiologische Variabilität von Förenkeimlingen autochtoner Populationen. Mitteilungen der Schweizer. *Anstalt für das forstliche Versuchwesen,* 26: 205-44.

KATTAN, A. A., and J. W. FLEMING. 1956. "Effect of irrigation at specific stages of development on yield, quality, growth, and composition of snapbeans." *Proceedings,* American Society for Horticultural Science, 68: 329-42.

KAWABATA, V., and M. FUJITO. 1955. "Distribution of total horizontal insolation in Japan and its neighborhood." *Geophysical Magazine,* 22: 143-56.

KEEN, B. A. 1932. "Soil physics in relation to meteorology." *Quarterly Journal,* Royal Meteorological Society, 58: 229-50.

KELLEY, O. J. 1954. "Requirements and availability of soil water." *Advances in Agronomy,* 6: 67-94.

KEMP, C. D. 1960. "Methods of estimating the leaf of grasses from linear measurements." *Annals of Botany, N. S,* 24: 491-99.

KHALIL, M. S. H. 1956. The interaction between growth and development of wheat as influenced by temperature, light, and nitrogen. Unpublished thesis, Wageningen Agriculture University.

KIHLMAN, A. O. 1890. "Pflanzenbiologische Studien aus Russisch Lapland." *Acta Societas pro Fanna et Flora fennica,* 6: 1-263.

KIMMEY, J. W. 1945. "The seasonal development and the defoliating effect of *Cronartium ribicola* on naturally infected *Ribes roezli* and *R. nevadense.*" *Phytopathology,* 35: 406-416.

KINBACKER, E. J. 1963. "Relative high temperature resistance of winter oats of different relative humidities." *Crop Science,* 3: 466-68.

KING, F. H. 1899. *Irrigation and Drainage, Principles and Practice of Their Cultural Phases.* New York: Macmillan.

KING, K. M. 1956. Pasture irrigation control according to soil and meteorological measurements. Unpublished doctoral dissertation, University of Wisconsin.

KING, K. M. 1961. "Evaporation from land surfaces." In *Proceedings of Hydrology Symposium No. 2, evaporation.* Toronto: National Research Council of Canada.

KING, K. M., C. B. TANNER, and V. E. SUOMI. 1956. "A floating lysimeter and its evaporation recorder." *Transactions,* American Geophysical Union, 37: 738-42.

KMOCH, H. G., R. E. RAMING, R. L. FOX, and F. E. KOEHLER. 1957. "Root development of winter wheat as influenced by soil moisture and nitrogen fertilization." *Agronomy Journal,* 49: 20-35.

KNOERR, K. R., and L. W. GAY. 1965. "Tree leaf energy balances." *Ecology,* 46: 17-24.

KOHLER, M. A., T. J. NORDENSON, and W. E. FOX. 1955. "Evaporation from pans and lakes." U.S. Weather Bureau Research Paper 38.

KOHNKE, H., F. R. DREIBELBIS, and J. M. DAVIDSON. 1940. "A survey and discussion of lysimeters and a bibliography on their construction and performance." USDA Miscellaneous publication.

KOLASEW, F. E. 1941. "Ways of suppressing evaporation of soil moisture." *Sbornik rabot po agronomicheskoi fizike*, [Collection of papers on physical aspects of agronomy], 3: 67.

KONIS, E. 1949. "The resistance of maquis plants to supermaximal temperatures." *Ecology*, 30: 425-29.

KORVEN, H. C., and J. C. WILCOX. 1965. "Correlation between evaporation from Bellani plates and evapotranspiration from orchards." *Canadian Journal of Plant Science*, 45: 132-38.

KRAMER, P. J. 1963. "Water stress and plant growth." *Agronomy Journal*, 5: 31-35.

KRAUSE, H. 1935. "Beiträge zur Kenntnis der Wasseraufnahme durch oberirdische Pflanzenorgane." *Österreichische Botanische Zeitschrift*, 84: 241-70.

KROGMAN K. K., and E. H. HOBBS. 1965. "Evapotranspiration by irrigated alfalfa as related to season and growth stage." *Canadian Journal of Plant Science*,45: 309-313.

KROGMAN, K. K., and L. E. LUTWICK. 1961. "Consumptive use of water by forage crops in the upper Kootenay River Valley." *Canadian Journal of Soil Science*, 41: 1-4.

KUNG, E. C., R. A. BRYSON, and D. H. LENSCHOW. 1964. "Study of a continental surface albedo on the basis of flight measurements and structure of the earth's surface cover over North America." *Monthly Weather Review*, 92: 543-64.

KURTYKA, J. C. 1953. "Precipitation measurements study." Illinois State Water Survey Division Investigation Report 20.

LACHENBRUCH, A. H. 1957. "Probe for measurement of thermal conductivity of frozen soils in place." *Transactions*, American Geophysical Union, 38: 691-97.

LAKE, J. V. 1956. "The temperature profile above bare soil on clear nights." *Quarterly Journal*, Royal Meteorological Society, 82: 187-97.

LAL, K. N., and M. S. SUBBA RAO. 1951. "A rapid method of leaf area determination." *Nature*, 167: 72.

LAMOUREUX, W. W. 1962. "Modern evaporation formulae adapted to computer use." *Monthly Weather Review*, 90: 26-28.

LANDSBERG, H. E. 1951. "Statistical investigations into the climatology of rainfall on Oahu." *Meteorological Monographs*, 1(3): 7-23.

LANDSBERG, H. E. 1961. "Solar radiation at the earth's surface." *Solar Energy*, 5: 95-98.

LANDSBERG, H. E., and M. L. BLANC. 1958. "Interaction of soil and weather." *Proceedings*, Soil Science Society of America, 22: 491-95.

LANDSBERG, H. E., and W. C. JACOBS. 1951. "Applied climatology." In *Compendium of meteorology*. Boston: American Meteorological Society.

LARSEN, A. 1960. "Experiments on the net assimilation rate of flax (*Linum usitatissimum*)." *Acta Agriculturae Scandinavica*, 10: 226-36.

LARSSON, P. 1963. "The distribution of albedo over arctic surfaces." *Geographical Review*, 53: 572-79.

LAWRENCE, E. N. 1955. "Effects of a windbreak on the speed and direction of wind." *Meteorological Magazine,* 83: 244-51.

LAYCOCK, D. H., and R. A. WOOD. 1963a. "Some observations on soil moisture use under tea in Nyasaland." *Tropical Agriculture,* 40: 35-48.

LAYCOCK, D. H., and R. A. WOOD. 1963b. "Some observations on soil moisture use under tea in Nyasaland (III)." *Tropical Agriculture,* Trinidad, 40: 121-28.

LEE, D. H. K. 1957. *Climate and economic development in the tropics.* New York: Harper.

LEHANE, J. J., and W. J. STAPLE. 1962. "Effects of soil moisture tensions on growth of wheat." *Canadian Journal of Soil Science,* 42: 180-88.

LEHENBAUER, P. A. 1914. "Growth of maize seedling in relation to temperature." *Physiological Researches,* 1: 247-88.

LEHMANN, P., and H. SCHANDERL. 1942. Tau und Reif: Pflanzenwetterkundliche Untersuchungen. Reichsamt Wetterdienstes, Wissenschaft Abhandlung, No. 9.

LEMON, E. R. 1960. "Photosynthesis under field conditions: II. an aerodynamic method for determining the turbulence carbon dioxide exchange between the atmosphere and a corn field." *Agronomy Journal,* 52: 697-703.

LEMON, E. R. 1963. "Energy and water balance of plant communities." In L. T. Evans (Ed.), *Environment Control of Plant Growth.* New York: Academfic Press.

LEMON, E. R., A. H. GLASER, and L. E. SATTERWHITE. 1957. "Some aspects of the relationship of soil, plant, and meteorological factors to evapotranspiration." *Proceedings,* Soil Science Society, 21: 464-68.

LETEY, J., and D. B. PETERS. 1957. "Influence of soil moisture levels and seasonal weather on efficiency of water use by corn." *Agronomy Journal,* 49: 362-65.

LETTAU, H. 1952. "Synthetische Klimatologie." *Berichte des Wetterdienstein Bad Kissingen,* 38: 127-36.

LEVITT, J. 1956. *The hardiness of plants.* New York: Academic Press.

LIGON, J. T., G. R. BENOIT, and A. B. ELAM. 1964. "A procedure for determining the probability of soil moisture deficiency and excess." *Transactions,* American Society of Agricultural Engineers, 8(2): 219, 220-22.

LINACRE, E. T. 1964. "A note on a feature of leaf and air temperatures." *Agricultural Meteorology,* 1: 66-72.

LINDSTROM, R. S. 1965. "Carbon dioxide and its effect on the growth of roses." *Proceedings,* American Society for Horticultural Science, 87: 521-24.

LINDSTROM, R. S. 1966a. "Carbon dioxide as a factor in total environment control." *Florist Review,* 138(3586): 25-26, 62-64.

LINDSTROM, R. S. 1966b. "The use of carbon dioxide in European horticulture." *Florist Review,* 138(3583): 56-57, 73.

LIST, R. J. 1966. *Smithsonian Meteorological Tables.* Smithsonian Miscellaneous Collections, Vol. 114.

LIU, B. Y. H., and R. C. JORDAN. 1960. "The interrelationship and characteristic distribution of direct, diffuse, and total solar radiation." *Solar Energy,* 4: 1-19.

LLOYD, M. G. 1961. "The contribution of dew to summer water budget of northern Idaho." *Bulletin,* American Meteorological Society, 42: 572-80.

LOMAS, J. 1964. "A simple method of assessing relative irrigation requirements." *Agricultural Meteorology,* 1: 142-48.

LONG, I. F. 1955. "Dew and guttation." *Weather,* 10: 128.

LONG, I. F. 1958. "Some observations on dew." *Meteorological Magazine,* 87: 161-68.

LÖNNQVIST, O. 1954. "Synthetic formulae for estimating effective radiation to a cloudless sky and their usefulness in comparing various estimation procedures." *Arkiv für geofysik,* 2: 245-94.

LOOMIS, R. S., and G. F. WORKER. 1963. "Responses of the sugar beet to low soil moisture at two levels of nitrogen nutrition." *Agronomy Journal,* 55: 509-515.

LOOMIS, W. E. 1965. "Absorption of radiation energy by leaves." *Ecology,* 46: 14-16.

LOUGEE, C. R. 1956. Climatic classification and the practice of irrigation in Norway. Unpublished doctoral dissertation, Clark University.

LOUSTALOT, A. J. 1945. "Influence of soil moisture conditions on apparent photosynthesis and transpiration of pecan leaves." *Journal of Agricultural Research,* 71: 519-32.

LOVETT, W. J. 1953. "Water requirements of tobacco grown under irrigation at Claire, North Queensland." *Australian Journal of Agricultural Research,* 4: 168-76.

LOWRY, R. L., and A. F. JOHNSON. 1942. "Consumptive use of water for agriculture." *Transactions,* American Society of Civil Engineering, 107: 1243-1302.

LOWRY, W. P. 1956. "Evaporation from forest soils near Donner Summit, California, and a proposed field method for estimating evaporation." *Ecology,* 37: 419-30.

LUCEY, R. F., and M. B. TESAR. 1965. "Frequency and rate of irrigation as factors in forage growth and water absorption." *Agronomy Journal,* 57: 519-23.

LUDWIG, L. J., T. SAEKI, and L. T. EVANS. 1965. "Photosynthesis in artificial communities of cotton plants in relation to leaf area." *Australian Journal of Biological Science,* 18: 1103-1118.

LUMLEY, J. L., and H. A. PANOFSKY. 1964. *The structure of atmospheric turbulence.* New York: John Wiley.

LYDOLPH, P. E. 1964. "The Russian sukhovey." *Annals,* Association of American Geographers, 54: 291-309.

LYSENKO, T. D. 1925. *The theoretical basis of vernalization.* Moscow: Selskosgys.

MAGNESS, J. R., E. S. DEGMAN, and J. R. FURR. 1935. "Soil moisture and irrigation investigations in eastern apple orchards." USDA Technical Bulletin 491.

MAJOR, J. 1963. "A climatic index to vascular plant activity." *Ecology,* 44: 485-98.

MAKKINK, G. F. 1953. "Een nieuw lysimeterstation." ["A new lysimeter station"]. *Water,* The Hague. 13: 159-63.

MAKKINK, G. F. 1955. "Toetsing van de berekening van de evapotranspiratie volgens Penman" ["Testing the method of computing evapotranspiration by Penman"]. *Landbouwk Tijdschrift [Agricultural Journal]*, 67: 267.

MAKKINK, G. F. 1957. "Ekzameno de la formula de Penman." ["Examining the Penman formula"]. *Netherlands Journal of Agricultural Science*, 5: 290-305.

MAKKINK, G. F., and H. D. J. VAN HEEMST. 1956. "The actual evapotranspiration as a function of the potential evapotranspiration and soil moisture tension." *Netherlands Journal of Agricultural Science*, 4: 67-72.

MANTELL, A., and E. GOLDIN. 1964. "The influence of irrigation frequency and intensity on the yield and quality of peanuts (*Arachis hypogaea*)." *Israel Journal of Agricultural Research*, 14: 203-210.

MARANI, A., and M. HORWITZ. 1963. "Growth and yield of cotton as affected by the time of a single irrigation." *Agronomy Journal*, 55: 219-22.

MASON, S. C. 1925. "The inhibitive effect of direct sunlight on the growth of the date palm." *Journal of Agricultural Research*, 31: 455-68.

MASSON, H. 1954. "La rosée et les possibilités de son utilisation" [Dew and possibilities of its use]. Dakar. *Institut des Hautes Études. L'Ecole Supérieure des Sciences, Annales*, 1: 1-44.

MATEER, C. L. 1955. "A preliminary estimate of the average insolation in Canada." *Canadian Journal of Agricultural Science*, 35: 579-94.

MATHER, J. R. 1954. "The measurement of potential evapotranspiration." *Publications in Climatology*, Laboratory of Climatology, 7(1).

MATHER, J. R. 1959. "Determination of evapotranspiration by empirical methods." *Transactions*, American Society of Agricultural Engeneers, 2: 35-38, 43.

MATUSHIMA, S., S. YAMAGUCHI, and T. OKABE. 1955. "Crop-scientific studies on the yield forecast of lowland rice, carbon assimilation of rice plant under near-natural conditions." *Proceedings*, Crop Science Society, Japan, 23: 192-97.

MAXIMOV, N. A. 1929. *The plant in relation to water*. (Translated by R. H. Yapp.) London: Allen and Unwin.

MAY, P., and A. J. ANTCLIFF. 1963. "The effect of shading on fruitfulness and yield in the sultana." *Journal of Horticultural Science*, 38: 85-94.

McCULLOCH, J. S. G. 1965. "Tables for the rapid computation of the Penman estimate of evaporation." *East African Agricultural and Forestry Journal*, 30: 286-95.

McCULLOCH, J. S. G., H. C. PEREIRA, O. KERFOOT, and N. A. GOODCHILD. 1965. "Effect of shade trees on tea yield." *Agricultural Meteorology*, 2: 385-99.

McCULLOCH, J. S. G., H. C. PEREIRA, O. KERFOOT, and N. A. GOODCHILD. 1966. "Shade tree effects in tea gardens." *World Crops*, 17: 26-27.

McGEHEE, R. M. 1963. "Weather: complex cause of aridity." In C. Hodge (Ed.), *Aridity and man*. Washington, D.C.: American Association for the Advancement of Science.

McILROY, I. C. 1957. "The measurement of natural evaporation." *Journal of the Australian Institute of Agricultural Science*, 23: 4-17.

McILROY, I. C., and D. E. ANGUS. 1963. "The Aspendale multiple weighed lysimeter installation." Commonwealth Scientific and Industrial Research

Organization, Australia, Division of Meteorological Physics Technical Paper 14.

McIlroy, I. C., and D. E. Angus. 1964. "Grass, water, and soil evaporation at Aspendale." *Agricultural Meteorology*, 1: 201-24.

McIlroy, I. C., and C. J. Sumner. 1961. "A sensitive high-capacity balance for continuous automatic weighing in the field." *Journal of Agricultural Engineering Research*, 6: 252-58.

Meyer, B. S., and D. B. Anderson. 1952. *Plant physiology*. New York: Van Nostrand.

Mihara, Y. 1961. "The microclimate of paddy rice culture and the artificial improvement of the temperature factor." Paper presented at the 10th Pacific Science Congress, Honolulu.

Millar, B. D. 1964. "Effect of local advection on evaporation rate and plant water status." *Australian Journal of Agricultural Research*, 15: 85-90.

Miller, D. H. 1965. "The heat and water budget of the earth's surface." *Advances in Geophysics*. 11: 175-302.

Miller, E. E., C. A. Shadbolt, and L. Holm. 1956. "Use of an optical planimeter for measuring leaf area." *Plant Physiology*, 31: 484-86.

Milthorpe, F. L. 1961. "The income and loss of water in arid and semiarid zones." In *Plant-water relationships in arid and semiarid conditions*. UNESCO.

Mitchell, K. J., and J. R. Kerr. 1966. "Differences in rate of use of soil moisture by stands of perennial ryegrass and white clover." *Agronomy Journal*, 58: 5-8.

Mitchell, P. K. 1966. "Field measurements of potential evapotranspiration in Malta." *Agricultural Meteorology*, 3: 247-55.

Mohr, H. 1964. "The control of plant growth and development by light." *Biological Review*, 39: 87-112.

Molga, M. 1962. *Agricultural meteorology*. Warsaw.

Molisch, H. 1897. *Untersuchungen über das Erfrieren der Pflanzen*. Jena: Fischer.

Möller, F. 1951. "Long-wave radiation." In *Compendium of meteorology*. Boston: American Meteorological Society.

Monsi, M., and T. Saeki. 1953. "Über der Lichtfaktor in den Pflanzengesellschaften und seine Bedeutung für die Stoffproduktion." *Japanese Journal of Botany*, 14: 22-52.

Monteith, J. L. 1957. "Dew." *Quarterly Journal*, Royal Meteorological Society, 83: 322-41.

Monteith, J. L. 1958. "The heat balance of soil beneath crops." In *Climatology and microclimatology*. UNESCO.

Monteith, J. L. 1959. "The reflection of short-wave radiation by vegetation." *Quarterly Journal*, Royal Meteorological Society, 85: 386-92.

Monteith, J. L. 1963. "Dew: facts and fallacies." In A. J. Rutter and F. H. Whitehead (Eds.), *The Water Relations of Plants*. London: Blackwell Scientific Publications.

Monteith, J. L. 1965a. "Light distribution and photosynthesis in field crops." *Annals of Botany*, 29: 17-37.

Monteith, J. L. 1965b. "Radiation and crops." *Experimental Agriculture*, 1: 241-51.

Monteith, J. L. 1966. "The photosynthesis and transpiration of crops." *Experimental Agriculture*, 2: 1-14.

Monteith, J. L., and G. Szeicz. 1961. "The radiation balance of bare soil and vegetation." *Quarterly Journal*, Royal Meteorological Society, 87: 159-70.

Mooney, H. A., and W. D. Billings. 1961. "Comparative physiological ecology of arctic and alpine populations of *Oxyria digyna*." *Ecological Monographs*, 31: 1-29.

Moore, J. N., A. A. Kattan, and J. W. Fleming. 1958. "Effects of supplemental irrigation, spacing, and fertility on yield and quality of processing tomatoes." *Proceedings*, American Society for Horticultural Science, 17: 356-68.

Morris, L. G. 1959. "A recording weighing machine for the measurement of evapotranspiration and dewfall." *Journal of Agricultural Engineering Research*, 4: 161-73.

Moss, D. M., R. B. Musgrave, and E. R. Lemon. 1961. "Photosynthesis under field conditions: III. some effects of light, carbon dioxide, temperature and soil moisture on photosynthesis, respiration and transpiration of corn." *Crop Science*, 1: 83-87.

Mukammal, E. I. 1961. "Evaporation pans and atmometers." In *Proceedings*, Hydrology Symposium No. 2, Evaporation. Toronto: Department of Northern Affairs and National Resources.

Mukammal, E. I., and J. P. Bruce. 1960. "Evaporation measurements by pan and atmometer." International Union of Geodesy and Geophysics. Association of Scientific Hydrology, No. 53: 408-420.

Munn, R. E. 1961. "Energy budget and mass transfer theories of evaporation." In *Proceedings*, Hydrology Symposium No. 2, Evaporation. Toronto: Department of Northern Affairs and National Resources.

Munn, R. E. 1966. *Descriptive micrometeorology*. New York: Academic Press.

Munro, J. M., and R. A. Wood. 1964. "Water requirements of irrigated maize in Nyasaland." *Empire Journal of Experimental Agriculture*, 32: 141-52.

Murata, Y., and J. Iyama. 1960. "Studies on photosynthesis in upland field crops: I. diurnal changes in photosynthesis of eight summer crops growing in the field." *Proceedings*, Crop Science of Japan, 29: 151-54.

Musick, J. T., and D. W. Grimes. 1961. "Water management and consumptive use of irrigated grain sorghum in western Kansas." Kansas Agricultural Experiment Station Technical Bulletin 113.

Nakagawa, Y. 1963. "Studies on the microclimate in the cultivated land and on the plant temperature." *Bulletin*, National Institute of Agricultural Science (Japan). Series A, 10: 127-65.

Nagel, J. F. 1956. "Fog precipitation on Table Mountain." *Quarterly Journal*, Royal Meteorological Society, 82: 452-60.

Nash, A. J. 1963. "A method for evaluating the effects of topography on the soil water balance." *Forest Science*, 9: 413-22.

NEBIKER, W. A. 1957. "Evapotranspiration studies at Knob Lake, June-September 1956." Montreal: McGill University. McGill Sub-Arctic Research Paper No. 3.

NELSON, L. B. and R. E. UHLAND. 1955. "Factors that influence loss of fall applied fertilizers and their probable importance in different sections of the U. S." *Proceedings,* Soil Science Society of America, 19: 492-96.

NELSON, R. M. 1959. "Drought estimation in southern forest fire control." U. S. Forest Service Southeast Forest Experiment Station Paper 99.

NEUMANN, J. 1953. "On a relationship between evaporation and evapotranspiration." *Bulletin,* American Meteorological Society, 34: 454-57.

NEWTON, O. H., and J. A. RILEY. 1964. "Dew in the Mississippi delta in the fall." *Monthly Weather Review,* 92: 369-73.

NICHIPROVICH, A. A. 1962. "Properties of plant crops as an optical system." *Soviet Plant Physiology,* 8: 428-35.

NIEUWOLT, S. 1965. "Evaporation and water balances in Malaya." *Journal of Tropical Geography,* 20: 34-53.

NJOKU, E. 1959. "An analysis of plant growth in some West African species: I. growth in full sunlight." *Journal of West African Science Association,* 5: 37-56.

NOFFSINGER, T. L. 1961. "Leaf and air temperature under Hawaii conditions." *Pacific Science,* 40: 304-306.

NOFFSINGER, T. L. 1962. "World population and maximum crop yield." In *Biometeorology.* New York: Pergamon Press.

NORDENSON, T. L., and D. R. BAKER. 1962. "Comparative evaluation of evaporation instruments." *Journal of Geophysical Research,* 67: 671-79.

NORUM, E. M., and C. L. LARSON. 1960. "Evaporation losses from standard non-recording rain gages." *Transactions,* American Society of Agricultural Engineers, 3: 82-83, 86.

NUTMAN, F. J. 1937. "Studies of the physiology of *Coffea arabica.*" *Annals of Botany* [N.S.], 1: 353-67, 681-93.

NUTTONSON, M. Y. 1947. "Agroclimatology and crop ecology of Palestine and Transjordan and climatic analogues in the United States." *Geographical Review,* 37: 436-56.

O'BANNON, J. H., and H. W. REYNOLDS. 1965. "Water consumption and growth of root-knot-nematode-infected and uninfected cotton plants." *Soil Science,* 99: 251-55.

OGATA, G., L. A. RICHARDS, and W. R. GARDNER. 1960. "Transpiration of alfalfa determined from soil water content changes." *Soil Science,* 89: 179-82.

OGUNTOYINBO, J. S. 1966. "Evapotranspiration and sugarcane yields in Barbados." *Journal of Tropical Geography,* 22: 38-48.

ORMROD, D. P. 1961. "Photosynthesis rates of young rice plants as affected by light intensity and temperature." *Agronomy Journal,* 53: 93-95.

ORVIG, S. 1961. "Net radiation flux over subarctic surfaces." *Journal of Meteorology,* 18: 199-203.

OWEN, P. C. 1957. "Rapid estimation of the areas of the leaves of crop plants." *Nature,* 180: 611.

OWEN, P. C. 1958. "Photosynthesis and respiration rates of leaves of *Nicotiana glutinosa* infected with tobacco mosaic virus and of *N. tabacum* infected with potato virus X." *Annals of Applied Biology*, 46: 198-204.

OWEN, P. C., and D. J. WATSON. 1956. "Effect on crop growth of rain after prolonged drought." *Nature*, 177: 847.

PALTRIDGE, T. B., and H. K. C. MAIR. 1936. "Studies of selected pasture grasses." Australia Council for Scientific and Industrial Research Bulletin 102.

PARKER, N. W. 1946. "Environment factors and their control in plant environments." *Soil Science*, 62: 109-119.

PASQUILL, F. 1950. "Some further consideration of the measurement and indirect evaluation of natural evaporation." *Quarterly Journal*, Royal Meteorological Society, 76: 287-301.

PATTEN, H. E. 1909. "Heat transference in soils" USDA Bureau of Soils Bulletin 59.

PEARSON, C. H. O., T. G. CLEASBY, and G. D. THOMPSON. 1961. "Attempts to confirm irrigation control factors based on meteorological data in the cane belt of South Africa. In *Proceedings*, South African Sugar Technologists' Association. April, 1961.

PELTON, W. L. 1964. "Evaporation from atmometers and pans." *Canadian Journal of Plant Science*, 44: 397-404.

PENDLETON, J. W., D. B. PETERS, and J. W. PEEK. 1966. "Role of reflected light in the corn ecosystem." *Agronomy Journal*, 58: 73-74.

PENMAN, H. L. 1941. "Laboratory experiments on evaporation from fallow soil." *Journal of Agricultural Science*, 31: 454-65.

PENMAN, H. L. 1948. "Natural evaporation from open water, bare soil, and grass." *Proceedings*, Royal Society, Series A, 193: 120-45.

PENMAN, H. L. 1949. "The dependence of transpiration on weather and soil conditions." *Journal of Soil Science*, 1: 74-89.

PENMAN, H. L. 1952. "The physical bases of irrigation control." In *Report of the Thirteenth International Horticultural Congress*, London.

PENMAN, H. L. 1956a. "Estimating evaporation." *Transactions*, American Geophysical Union,.. 37: 43-50.

PENMAN, H. L. 1956b. "Evaporation: an introductory survey." *Netherlands Journal of Agricultural Science*, 4: 9-29.

PENMAN, H. L. 1962. "Weather and crops." *Quarterly Journal*, Royal Meteorological Society, 88: 209-219.

PENMAN, H. L., and I. F. LONG. 1960. "Weather in wheat—an essay in micrometeorology." *Quarterly Journal*, Royal Meteorological Society, 86: 16-50.

PENMAN, H. L., and R. K. SCHOFIELD. 1951. "Some physical aspects of assimilation and transpiration." *Symposia*, Society of Experimental Biology, 5: 115-29.

PEREIRA, H. C. 1957. "Field measurements of water use for irrigation control in Kenya coffee." *Journal of Agricultural Science*, 49: 459-66.

PEREIRA, H. C., and P. H. HOSEGOOD. 1962. "Consumptive water use of softwood plantations and bamboo forest." *Journal of Soil Science*, 13: 299-313.

PETERS, D. B. 1960a. "Relative magnitude of evaporation and transpiration." *Agronomy Journal*, 52: 536-38.

PETERS, D. B. 1960b. "Growth and water absorption by corn as influenced by soil moisture tension, moisture content, and relative humidity." *Proceedings*, Soil Science Society of America, 24: 523-26.

PETERS, D. B., and L. C. JOHNSON. 1960. "Soil moisture use by soybean." *Agronomy Journal*, 52: 687-89.

PETERS, D. B., and M. B. RUSSELL. 1959. "Relative water losses by evaporation and transpiration in field corn." *Proceedings*, Soil Science Society of America, 23: 170-73.

PHILIP, J. R. 1957. "Evaporation, and moisture and heat fields in the soil." *Journal of Meteorology*, 14: 354-66.

PICHA, K. G., and J. VILLANUEVA. 1962. "Nocturnal radiation measurements, Atlanta, Georgia." *Solar Energy*, 6: 151-54.

PIERCE, L. T. 1958. "Estimating seasonal and short-term fluctuations in evapotranspiration from meadow crops." In American Meteorological Society Bulletin 39.

PIETERS, G. A. 1960. "On the relation between the maximum rate of photosynthesis and the thickness of the mesophyll in sun and shade leaves of *Acer Pseudoplatanus L.*" *Mededelingen van de Landbouwhogeschool te Wageningen*, 60(17): 1-6.

POOL, R. J. 1914. "A study of the vegetation of the sandhills of Nebraska." Minnesota Botanical Studies, 4: 185-312.

POPOV, O. V. 1952. "Soveshchanie po voprosam metodiki nabliudenii nad ispareniem s pochvy" ["Conference on the methods of observation of soil evaporation"]. *Meteorologiia i Gidrologiia* [*Meteorology and Hydrology*]. 7: 51-53.

PORTMAN, D. J., and F. DIAS. 1959. "Influence of wind and angle of incident radiation on the performance of a Beckman and Whitley total hemipherical radiometer." Final Report, U.S. Department of Commerce, Weather Bureau, UMRI Project 2715.

POSEY, J. W., and P. F. CLAPP. 1964. "Global distribution of normal surface albedo." *Geofisca Internacional*, 4: 33-48.

PRESCOTT, J. A. 1948. "The response of the growth of pastures to temperature." *Australian Journal of Science*, 2: 24-25.

PRIESTLEY, C. H. B. 1959a. "Heat conduction and temperature profiles in air and soil." *Journal of the Australian Institute of Agricultural Science*, 25: 94-107.

PRIESTLEY, C. H. B. 1959b. *Turbulent transfer in the lower atmosphere.* Chicago: University of Chicago Press.

PRIESTLEY, C. H. B. 1966. "The limitation of temperature by evaporation in hot climates." *Agricultural Meteorology*, 3: 241-46.

PRUITT, W. O. 1960a. "Relation of consumptive use of water to climate." *Transactions*, American Society of Agricultural Engineers, 3: 9-13, 17.

PRUITT, W. O. 1960b. "Correlation of climatological data with water requirement of crops." Department of Irrigation, University of California, Davis.

PRUITT, W. O. 1962a. "Correlation of climatological data with water requirements of crops." Water Resources Center Contract Research. Annual Report 3. Davis, California.

PRUITT, W. O. 1962b. "Evapotranspiration of perennial ryegrass in the Sacramento Valley as measured by lysimeters." Paper presented at the ARS-SCS Workshop on consumptive use, U.S. Water Conservation Laboratory, Tempe, Arizona, March 6-8, 1962.

PRUITT, W. O. 1963. "Application of several energy balance and aerodynamic evaporation equations under a wide range of stability." In *Final report, Investigation of energy and mass transfer near the ground including influences of the soil-plant atmosphere system.* University of California. Davis, California.

PRUITT, W. O., and D. E. ANGUS. 1960. "Large weighing lysimeters for measuring evapotranspiration." *Transactions,* American Society of Agricultural Engineers, 3: 13-15, 18.

PRUITT, W. O., and D. E. ANGUS. 1961. "Comparison of evapotranspiration with solar and net radiation and evaporation from water surfaces. In *Investigation of energy and mass transfers near the ground.* Davis: University of California Press.

PRUITT, W. O., and M. C. JENSEN. 1955. "Determining when to irrigate." *Agricultural Engineering,* 36: 389-93.

RABINOWITCH, E. I. 1948. "Photosynthesis." *Scientific American,* 183: 3-13.

RABINOWITCH, E. I. 1956. *Photosynthesis and related processes.* New York: Interscience.

RAMDAS, L. A. 1935. "Frost hazard in India." *Current Science,* 3: 325-33.

RAMDAS, L. A., and S. YEGNANARAYANAN. 1956. "Solar energy in India." In *Wind and solar energy.* UNESCO.

RANEY, W. A. 1959. "Evaluation of evapotranspiration in field plots." *Transactions,* American Society of Agricultural Engineers, 2: 41.

RASCHKE, K. 1956. "Über die physikalischen Beziehungen zwischen Wärmeübergangszahl, Strahlungsaustausch, Temperatur und Transpiration eines Blattes." *Planta,* 48: 200-238.

RASUMOV, V. I. 1930. "Über die photoperiodische Nachwirkung in Zusammenhang mit der Wirkung verschiedener Aussaatterminen auf die Pflanzen." *Planta,* 10: 345-77.

RENSE, W. A. 1961. "Solar radiation in the extreme ultraviolet region of the spectrum and its effect on the earth's upper atmosphere." *Annals,* New York Academy of Science, 95: 33-38.

RICHARDS, L. A., and H. T. STUMPF. 1966. "Graphical recorder for a pan evaporimeter." *Water Resources Research,* 2: 209-212.

RICKARD, D. S. 1960. "The occurrence of agricultural drought at Ashburton, New Zealand." *New Zealand Journal of Agriculture Research,* 3: 431-41.

RIDER, N. E. 1954a. "Evaporation from an oat field." *Quarterly Journal,* Royal Meteorological Society, 80: 198-211.

RIDER, N. E. 1954b. "Eddy diffusion of momentum, water vapor, and heat near the ground." *Philosophical Transactions,* Royal Society Series A, 246: 481-501.

RIJKOORT, P. J. 1954. *A nomogram for the determination of the pan evaporation adapted from the Penman Method.* Amsterdam: Royal Netherlands Meteorological Institute.

RILEY, J. A. 1957. "Soil temperatures as related to corn yield in central Iowa." *Monthly Weather Review,* 85: 393-400.

ROBINS, J. S., and C. E. DOMINGO. 1953. "Some effects of severe soil moisture deficits at specific growth stages of corn." *Agronomy Journal*, 45: 618-21.

ROBINS, J. S., and H. R. HAISE. 1961. "Determination of consumptive use of water by irrigated crops in the western United States." *Proceedings*, Soil Science Society of America, 25: 150-54.

ROBINSON, G. D. 1950. "Notes on the measurement and estimation of atmospheric radiation—2." *Quarterly Jorrnal*, Royal Meteorological Society, 76: 37-51.

ROHWER, C. 1931. "Evaporation from free water surface." USDA Technical Bulletin 271.

ROSENBERG, N. J. 1966. "Microclimate, air mixing, and physiological regulation of transpiration as influenced by wind shelter in an irrigated bean field." *Agricultural Meteorology*, 3: 197-224.

RUDOLF, P. O., and S. R. GEVORKIANTZ. 1935. *Possibilities of shelterbelt planting in the plains region: shelterbelt experience in other lands.* U.S. Forest Service.

RUNGE, E. C. A., and R. T. ODELL. 1958. "The relation between precipitation, temperature, and the yield of corn on the Agronomy South Farm, Urbana, Illinois." *Agronomy Journal*, 50: 448-54.

RUNGE, E. C. A., and R. T. ODELL. 1960. "The relation between precipitation, temperature, and the yield of soybeans on the Agronomy South Farm, Urbana, Illinois." *Agronomy Journal*, 52: 245-47.

SAEKI, T. 1960. "Interrelationships between leaf amount, light distribution, and total photosynthesis in a plant community." *Botanical Magazine*, Tokyo, 73: 404-408.

SALE, P. J. M. 1966. "The response of summer lettuce to irrigation at different stages of growth." *Journal of Horticultural Science*, 41: 43-52.

SALIM, M. H., and G. W. TODD. 1965. "Transpiration patterns of wheat, barley and oat seedlings under varying conditions of soil moisture." *Agronomy Journal*, 57: 593-96.

SALTER, P. J. 1959. "The effect of different irrigation treatments on the growth and yield of early summer cauliflower." *Journal of Horticultural Science*, 34: 23-31.

SALTER, P. J. 1960. "The growth and development of early summer cauliflowers in relation to environmental factors." *Journal of Horticultural Science*, 35: 21-33.

SALTER, P. J. 1961. "The irrigation of early summer cauliflower in relation to stage of growth, plant spacing, and nitrogen level." *Journal of Horticultural Science*, 36: 241-53.

SALTER, P. J. 1962. "Some responses of peas to irrigation at different growth stages." *Journal of Horticultural Science*, 37: 141-49.

SALTER, P. J. 1963. "The effect of wet or dry soil conditions at different growth stages on the components of yield of a pea crop." *Journal of Horticultural Science*, 38: 321-34.

SAMISH, R. M., and P. SPIEGEL. 1961. "The use of irrigation in growing olives for oil production." *Israel Journal of Agricultural Research*, 11: 87-95.

SANDERSON, M. 1950. "Measuring potential evapotranspiration at Norman Wells, 1949." *Geographical Review*, 40: 636-45.

SCHNEIDER, G. W., and N. F. CHILDERS. 1941. "Influence of soil moisture on photosynthesis, respiration, and transpiration of apple leaves." *Plant Physiology,* 16: 565-83.

SCHOLTE-UBING, D. W. 1959. "Over stralingsmetingen, de warmtebalans en de verdamping van grass" ["Studies on solar and net radiation and on evapotranspiration from grass"]. *Mededelingen van de Landbouwhogeschool [Bulletin of Agricultural University],* Wageningen, 59: 1-93.

SCHOLTE-UBING, D. W. 1961a. "Short-wave and net radiation under glass as compared with radiation in the open." *Netherlands Journal of Agricultural Science,* 9: 163-67.

SCHOLTE-UBING, D. W. 1961b. "Solar and net radiation, available energy, and its influence on evapotranspiration from grass." *Netherlands Jounral of Agricultural Science,* 9: 81-93.

SCHULTZ, H. B., and R. R. PARKS. 1957. "Frost protection by sprinklers." *California Agriculture,* 2: 8-9.

SCHUURMAN, J. J., and G. F. MAKKINK. 1955. "The influence of sprinkler upon grass root development in a field experiment at IJsselstein (Utrecht)." *Landbouwkundig Tijdschrift.* 67: 283-96.

SCHWAB, G. O., W. D. SHRADER, P. R. NIXON, and R. H. SHAW. 1958. "Research on irrigation of corn and soybeans at Conesville and Ankeny, Iowa, 1951-1955." In Research Bulletin 458, Iowa State College, Agricultural and Home Economics Experiment Station.

SCRASE, F. J. 1930. "Some characteristics of eddy motion in the atmosphere." London Meteorological Office, Geophysical Memoir 52.

SEKIHARA, K., and M. KANO. 1957. "On the distribution and variation of solar radiation in Japan." *Papers in Meteorology and Geophysics,* 8: 144-49.

SELLERS, W. D. 1965. *Physical Climatology.* Chicago: University of Chicago Press.

SEYBOLD, A. 1929. "Die pflanzliche Transpiration." *Ergebnisse der Biologie,* 5: 29-165.

SEYBOLD, A. 1931. "Weitere Beiträge zur Kenntnis der Transpiration-analyze." *Planta,* 13: 18-28; 14: 386-410.

SEYBOLD, A. 1932. "Weitere Beiträge zur Kenntnis der Transpiration-analyze." *Planta,* 16: 518-25.

SEYBOLD, A. 1933. "Zur Klarung des Begriffes Transpirationswiderstand." *Planta,* 21: 353-67.

SHAW, C. F. 1926. "The effect of a paper mulch on soil temperature." *Hilgardia,* 1: 341-64.

SHAW, R. H. 1954. "Leaf and air temperatures under freezing conditions." *Plant Physiology,* 29: 102-104.

SHAW, R. H. 1956. "A comparison of solar radiation and net radiation." In Bulletin 37, American Meteorological Society.

SHAW, R. H. 1959. "Transpiration and evapotranspiration as related to meteorological factors." Final Report, USWB Contract No. Cwb-9295, Iowa State College.

SHMUELI, E. 1953. "Irrigation studies in the Jordan Valley: I. physiological activity of the banana in relation to soil moisture." In Bulletin 3, Research Council of Israel.

Shreve, F. 1924. "Soil temperature as influenced by altitude and slope exposure." *Ecology*, 5: 128-36.

Sibbons, J. L. H. 1962. "A contribution to the study of potential evapotranspiration." *Geografiska Annaler*, 44: 279-92.

Singh, B. N., and K. N. Lal. 1935. "Investigation of the effect of age on assimilation of leaves." *Annals of Botany*, 49: 291-306.

Singh, R., and R. B. Alderfer. 1966. "Effects of soil moisture stress at different periods of growth of some vegetable crops." *Soil Science*, 101: 69-80.

Slatyer, R. O. 1956. "Absorption of water from atmospheres of different humidity and its transport through plants." *Australian Journal of Biological Science*, 9: 552-58.

Slatyer, R. O. 1958. "Availability of water to plants." In *Climatology and microclimatology*. UNESCO.

Slatyer, R. O., and J. F. Bierhuizen. 1964. "The influence of several transpiration suppressants on transpiration, photosynthesis, and water-use efficiency of cotton leaves." *Australian Journal of Biological Science*, 17: 131-46.

Smith, A. 1931. "Effect of paper mulches on soil temperature, soil moisture, and yield of certain crops." *Hilgardia*, 6: 160-201.

Smith, A. M. 1909. "On the internal temperature of leaves in tropical insolation." *Annals*, Royal Botanical Garden, Peradeniya, 4: 229-98.

Smith, G. E. 1953. "Less water required per bushel of corn with adequate fertility." Missouri Farmers Association Bulletin No. 583.

Smith, K. 1964. "A long period assessment of the Penman and Thornthwaite potential evapotranspiration formulae." *Journal of Hydrology*, 2: 277-90.

Sonmor, L. G. 1963. "Seasonal consumptive use of water by crops grown in southern Alberta and its relationship to evaporation." *Canadian Journal of Soil Science*, 43: 287-97.

Spector, W. S. 1956. *Handbook of biological data*. Philadelphia: Saunders.

Spiegel, P. 1955. "The water requirement of the olive tree, critical periods of moisture stress, and the effect of irrigation upon the oil content of its fruit." In Report 2, Fourteenth International Horticultural Congress. Scheveningen, The Netherlands.

Spoehr, H. A. 1956. *Essays on Science*. Stanford: Stanford University Press.

Sreenivasan, P. S. 1949. "A study of cotton grown under constant soil moisture." *Proceedings*, Indian Academy of Science, 30: 134-54, 249-58.

Stalfelt, M. G. 1932. "Der Einfluss des Windes auf die Kutifuläre und Stomatare Transpiration svensk." *Botanisk Tidsskrift*, 26: 45-69.

Stanhill, G. 1958. "Effects of soil moisture on the yield and quality of turnips: II. response at different growth stages." *Journal of Horticultural Science*, 33: 264-74.

Stanhill, G. 1960. "The relationship between climate and the transpiration and growth of pastures. In *Proceedings*, Eighth International Grassland Congress.

Stanhill, G. 1961. "A comparison of methods calculating potential evapotranspiration from climatic data." *Israel Journal of Agricultural Research*, 11: 159-71.

STANHILL, G. 1962a. "The control of field irrigation practice from measurements of evaporation." *Israel Journal of Agricultural Research*, 12: 51-62.

STANHILL, G. 1962b. "The use of Piche evaporimeter in the calculation of evaporation." *Quarterly Journal*, Royal Meteorological Society, 88: 80-82.

STANHILL, G. 1962c. "The effect of environment factors on the growth of alfalfa in the field." *Netherlands Journal of Agricultural Science*, 10: 247-53.

STANHILL, G. 1964. "Potential evapotranspiration at Caesarea." *Israel Journal of Agricultural Research*, 14: 129-35.

STANHILL, G., G. J. HOFSTEDE, and J. D. KALMA. 1966. "Radiation balance of natural and agricultural vegetation." *Quarterly Journal*, Royal Meteorological Society, 92: 128-40.

STAPLE, W. J., and J. J. LEHANE. 1954. "Weather conditions influencing wheat yields in tanks and field plots." *Canadian Journal of Agricultural Science*, 34: 552-64.

STAPLE, W. J., and J. J. LEHANE. 1955. "The influence of field shelterbelts on wind velocity, evaporation, soil moisture, and crop yield." *Canadian Journal of Agricultural Science*, 35: 440-53.

STEELMANN NIELSEN, E. 1953. "Carbon dioxide concentration, respiration during photosynthesis, and maximum quantum yield of photosynthesis." *Physiologia Plantarum*, 6: 316-32.

STERN, W. R. 1965. "Evapotranspiration of safflower at three densities of sowing." *Australian Journal of Agricultural Research*, 16: 961-71.

STERN, W. R., and C. M. DONALD. 1963. "The influence of leaf area and radiation on the growth of clover in swards." *Australian Journal of Agricultural Research*, 14: 20-38.

STOCKER, O., G. LEYER, and G. H. VIEWEG. 1954. "Wasserhaushalt und Assimilation." *Kuratoriums für Kulturbauwesen*, 3: 45-77.

STOECKELER, J. H. 1965. "The design of shelterbelts in relation to crop yield improvement." *World Crops*, 17: 27-32.

STOLL, A. M., and J. D. HARDY. 1955. "Thermal radiation measurements in summer and winter, Alaskan climates." *Transactions*, American Geophysical Union, 36: 213-25.

STONE, E. C. 1957a. "Dew as an ecological factor: I. a review of the literature." *Ecology*, 38: 407-413.

STONE, E. C. 1957b. "Dew as an ecological factor: II. the effect of artificial dew on the survival of *Pinus ponderosa* and associated species." *Ecology*, 38: 414-22.

STOUGHTON, R. H. 1955. "Light and plant growth." *Journal of Royal Horticultural Science*, 80: 454-66.

STOUGHTON, R. H., and D. VINCE. 1954. "Possible applications of photoperiodism in plants." *World Crops*, 6: 311-13.

STOUGHTON, R. H., and D. VINCE. 1957. "Use of artificial light in horticulture." *Agricultural Review*, 3: 8-15.

STRUCHTEMEYER, R. A. 1961. "Efficiency of the use of water by potatoes." *American Potato Journal*, 38: 22-24.

SUMNER, C. J. 1963. "Unattended long-period evaporation recorder." *Quarterly Journal*, Royal Meteorological Society, 89: 414-17.

Suomi, V. E. 1953. The heat budget over a corn field. Unpublished doctoral dissertation, University of Chicago.

Suomi, V. E. 1958. "An economical net radiometer." *Tellus*, 10: 160-63.

Suomi, V. E., and C. B. Tanner. 1958. "Evapotranspiration estimates from heat-budget measurements over a field crop." *Transactions*, American Geophysical Union, 39: 298-304.

Sutton, O. G. 1953. *Micrometeorology*. New York: McGraw-Hill.

Suzuki, S., and H. Fukuda. 1958. "A method of calculating potential evapotranspiration from pan evaporation data." *Journal of Agricultural Meteorology*, Tokyo, 13: 81-85.

Swinbank, W. C. 1951. "The measurement of vertical transfer of heat and water vapor by eddies in the lower atmosphere, with some results." *Journal of Meteorology*, 8: 135-45.

Swinbank, W. C. 1958. "Turbulent transfer in the lower atmosphere." In *Climatology and microclimatology*. UNESCO.

Swinbank, W. C. 1963. "Long-wave radiation from clear skies." *Quarterly Journal*, Royal Meteorological Society, 89: 339-48.

Takeda, T. 1961. "Studies on the photosynthesis and production of dry matter in the community of rice plants." *Japanese Journal of Botany*, 17: 403-437.

Tamm, E., and G. Krysch. 1961. "Zum Verlauf des CO_2-Gehaltes der Luft" *Zeitschrift für Acker-und Pflanzenbau*, 112: 253-377.

Tanner, C. B. 1957. "Factors affecting evaporation from plants and soils." *Journal of Soil Water Conservation*, 12: 221-27.

Tanner, C. B. 1960a. "Energy balance approach to evapotranspiration from crops." *Proceedings*, Soil Science Society of America, 24: 1-9.

Tanner, C. B. 1960b. "A simple aero-heat budget method for determining daily evapotranspiration." *Transactions*, Seventh International Congress of Soil Science. Madison, Wisconsin.

Tanner, C. B. 1963a. "Energy relations in plant community." In L. T. Evans (Ed.), *Environment control of plant growth*. New York: Academic Press.

Tanner, C. B. 1963b. "Basic instrumentation and measurements for plant environment and micrometeorology." University of Wisconsin College of Agriculture, Soils Bulletin 6.

Tanner, C. B., and W. L. Pelton. 1960. "Potential evapotranspiration estimates by the approximate energy balance method of Penman." *Journal of Geophysical Research*, 65: 3391-3413.

Tanner, C. B., A. E. Peterson, and J. R. Love. 1960. "Radiation energy exchange in a corn field." *Agronomy Journal*, 52: 373-79.

Taylor, R. J., and A. J. Dyer. 1958. "An instrument for measuring evaporation from natural surfaces." *Nature*, 181: 408-409.

Taylor, R. J., and E. K. Webb. 1955. "A mechanical computer for micrometeorological analysis." Division of Meteorology and Physics, Commonwealth Scientific and Industrial Research Organization, Australia, Technical Paper 6.

Taylor, S. A. 1961. "Water relations of field crops." In *Plant-Water Relations*. UNESCO.

Taylor, S. A., and B. Rognerund. 1959. "Water management for potato production." *Utah Farm and Home Science*, 20: 82-84.

THIMANN, K. V. 1951. "Biological utilization of solar energy." *Proceedings*, American Academy of Arts and Sciences, 79: 323-26.

THIMANN, K. V. 1956. "Solar energy utilization by higher plants." In *Proceedings*, World Symposium on Applied Solar Energy, Phoenix, Arizona.

THODAY, D. 1910. "Experimental researchers on vegetable assimilation and respiration: VI. some experiments on assimilation in the open air." *Proceedings*, Royal Society, London, Series B, 82: 421-50.

THOMAS, M. D., and G. R. HILL. 1937. "The continuous measurements of photosynthesis, respiration and transpiration of alfalfa, and wheat growing under field conditions." *Plant Physiology*, 12: 285-307.

THOMAS, M. D., and G. R. HILL. 1949. "Photosynthesis under field conditions." In J. Franck and W. E. Loomis (Eds.), *Photosynthesis in plants*. Ames: Iowa State College Press.

THOMPSON, B. W. 1965. *The Climate of Africa*. Nairobi and New York: Oxford University Press.

THOMPSON, G. D., C. H. O. PEARSON, and T. G. CLEASBY. 1963. "The estimation of the water requirements of sugar cane in Natal." In *Proceedings*, South African Sugar Technologists' Association, April, 1963.

THORNE, G. N. 1959. "Photosynthesis in the lamina and sheath of barley leaves." *Annals of Botany* [N.S.], 23: 365-70.

THORNE, G. N. 1960. "Variations with age in net assimilation rate and other growth attributes of sugar beet, potato, and barley in a controlled environment." *Annals of Botany* [N.S.], 24: 356-71.

THORNTHWAITE, C. W. 1948. "An approach toward a rational classification of climate." *Geographical Review*, 38: 55-94.

THORNTHWAITE, C. W. 1954. "A re-examination of the concept and measurement of potential evapotranspiration." *Publications in Climatology*, Johns Hopkins University, 7(1): 200-209.

THORNTHWAITE, C. W., and M. H. HALSTEAD. 1942. "Note on the variation of wind with height in the layer near the ground." *Transactions*, American Geophysical Union, 23: 249-55.

THORNTHWAITE, C. W., and B. HOLZMAN. 1939. "The determination of evaporation from land and water surface." *Monthly Weather Review*, 67: 4-11.

THORNTHWAITE, C. W., and B. HOLZMAN. 1942. "Measurement of evaporation from land and water surface." USDA Technical Bulletin 817.

THORNTHWAITE, C. W., and J. R. MATHER. 1955a. *The water balance*. Centerton, N.J.: Laboratory of Climatology.

THORNTHWAITE, C. W., and J. R. MATHER. 1955b. "The water budget and its use in irrigation." In *U.S. Department of Agriculture Yearbook*.

THORNTHWAITE, C. W., and J. R. MATHER. 1957. "Instruction and tables for computing potential evapotranspiration and the water balance." Centerton, N.J.: Laboratory of Climatology Publications, 3: 185-311.

TILL, M. R. 1965. "Methods of timing irrigations with particular reference to horticultural crops." *Journal of the Australian Institute of Agricultural Science*, 31: 196-204.

TOMLINSON, B. R. 1953. "Comparison of two methods of estimating consumptive use of water." *Agricultural Engineering*, 34: 459-64.

TRANSEAU, E. N. 1926. "The accumulation of energy by plants." *Ohio Journal of Science*, 26: 1-10.

TRICKETT, E. S., L. J. MOULSLEY, and R. I. EDWARDS. 1957. "Measurement of solar and artificial radiation with particular reference to agriculture and horticulture." *Journal of Agricultural Engineering Research*, 2: 86-110.

TURC, L. 1954. "*Le bilan d'eau des sols: relations entre les precipitations, l'evaporation et l'ecoulement.*" *Sols Africains (Paris)*, 3: 138-72.

TURLEY, R. H., G. R. WEBSTER, and R. B. CARSON. 1963. "The effect of irrigation on yield, seasonal distribution and chemical composition of three pasture mixtures." *Canadian Journal of Plant Science*, 43: 575-82.

TURRELL, F. M. 1944. "Correlation between internal surface and transpiration rate in mesomorphic and xeromorphic leaves grown under artificial light." *Botanical Gazette*, 105: 413-26.

TURRELL, F. M., and S. W. AUSTIN. 1965. "Comparative nocturnal thermal budgets of large and small trees." *Ecology*, 46: 25-34.

UCHIJIMA, Z. 1961. "On characteristics of heat balance of water layer under paddy plant cover." In Bulletin 8, National Institute of Agricultural Science, Tokyo.

ULRICH, A. 1951. "Sugar beets and climate." *California Agriculture*, 5: 3, 12.

U.S. BUREAU OF RECLAMATION. 1951. "Irrigation land use." U.S. Department of the Interior, Manual 5.

UPCHURCH, R. P., M. L. PETERSON, and R. M. HAGAN. 1955. "Effect of soil moisture content on the rate of photosynthesis and respiration in ladino clover (*Trifolium repens* L.). *Plant Physiology*, 30: 297-303.

URIU, K. 1964. "Effect of post-harvest soil moisture depletion on subsequent yield of apricots." *Proceedings*, American Society for Horticultural Science, 84: 93-97.

VAN BAVEL, C. H. M. 1953a. "Chemical composition of tobacco leaves as affected by soil moisture conditions." *Agronomy Journal*, 45: 611-14.

VAN BAVEL, C. H. M. 1953b. "A drought criterion and its application in evaluating drought incidence and hazard." *Agronomy Journal*, 45: 167-72.

VAN BAVEL, C. H. M. 1956. "Estimating soil moisture conditions and time for irrigation with the evapotranspiration method." U.S. Department of Agriculture, Agricultural Research Service, Soil and Water Conservation Research Series 41(11).

VAN BAVEL, C. H. M., and L. J. FRITSCHEN. 1962. "Energy balance of bare surfaces in an arid climate." In International Symposium on the Methodology of Plant Eco-Physiology, Montpellier, France.

VAN BAVEL, C. H. M., L. J. FRITSCHEN, and W. E. REEVES. 1963. "Transpiration by sudangrass as an externally controlled process." *Science*, 141: 269-70.

VAN BAVEL, C. H. M., and L. E. MEYERS. 1962. "An automatic weighing lysimeter." *Agricultural Engineering*, 43: 580-83, 586-88.

VAN DER BIJL, W. 1957. *The evapotranspiration problem*. Manhattan: Kansas State College.

VAN DER PAAUW, F. 1949. "Water relations of oats with special attention to the influence of periods of drought." *Plant and Soil*, 1: 303-341.

van Dobben, W. H. 1962. "Influence of temperature and light conditions on dry-matter distribution, development rate, and yield in arable crops." *Netherlands Journal of Agricultural Science*, 10: 377-89.

van Hylckama, T. E. A. 1956. "The water balance of the earth." *Laboratory of Climatology Publications*, 9(2): 57-117.

van Sluis, E. J. H. 1952. "Artificial light in horticulture." *World Crops*, 4: 161-63, 166.

van Wijk, W. R., and D. A. de Vries. 1954. "Evapotranspiration." *Netherlands Journal of Agricultural Science*, 2: 105-119.

Veihmeyer, F. J., and A. H. Hendrickson. 1955. "Does transpiration decrease as the soil moisture decreases?" *Transactions*, American Geophysical Union, 36: 425-48.

Veihmeyer, F. J., and A. H. Hendrickson. 1956. "Responses of fruit trees and vines to soil moisture." *Proceedings*, American Society for Horticultural Science, 55: 11-15.

Veihmeyer, F. J., W. O. Pruitt, and W. D. McMillan. 1960. "Soil moisture as a factor in evapotranspiration equations." Presented at the 1960 annual meeting of American Society of Agricultural Engineers.

Ventikeshwaran, S. P., P. Jagannathan, and S. S. Ramakishran. 1959. "Some experiments with U.S.A. Standard evaporimeter." *Indian Journal of Meteorology and Geophysics*, 10: 25-36.

Ventskevich, G. Z. 1961. *Agrometeorology*. (Translated from Russian.) Israel Program for Scientific Translation, Jerusalem.

Verduin, J., and W. E. Loomis. 1944. "Absorption of carbon dioxide by maize." *Plant Physiology*, 19: 278-93.

Verle, E. K., and G. V. Svinukhov. 1960. "Radiation balance in the Primorskiy Kray region." Leningrad. Nauchno-issledovatel'skii institut gidrometeorologicheskogo Priborostroeniia [People's Research Institute of Hydrometeorologiacl Instruments.], 6: 30-34.

Viets, F. G. 1962. "Fertilizers and the efficient use of water." *Advances in Agronomy*, 14: 223-64.

Vincente-Chandler, J., S. Silva, and J. Figarella. 1959. "The effect of nitrogen fertilization and frequency of cutting on the yield and composition of three tropical grasses." *Agronomy Journal*, 51: 202-206.

Vittum, M. T., R. B. Alderfer, B. E. Janes, C. W. Reynolds, and R. A. Struchtemeyer. 1963. "Crop response to irrigation in the Northeast." New York State Agricultural Experiment Station Bulletin 800.

Voight, G. K. 1962. "The role of carbon dioxide in soil." In T. T. Kozlowski (Ed.), *Tree Growth*. New York: Ronald Press.

von Mohl, H. 1848. "Über das Erfrieren der Zweigspizten mancher Holzgewächse." *Botanische Zeitung*, 8: 6-9.

Wadleigh, C. H., H. G. Gauch, and O. C. Magistad. 1946. "Growth and rubber accumulation in guayule as conditioned by soil salinity and irrigation regime." USDA Technical Bulletin 925.

Wadsworth, R. M. 1959. "An optimum wind speed for plant growth." *Annals of Botany* [N.S.], 23: 195-99.

Waggoner, P. E., P. M. Miller, and H. D. de Roo. 1960. "Plastic mulching principles and benefits." Connecticut Agricultural Experiment Station Bulletin 634.

WAGGONER, P. E., D. M. MOSS, and J. D. HESKETH. 1963. "Radiation in the plant environment and photosynthesis." *Agronomy Journal*, 55: 36-38.

WAGGONER, P. E., A. B. PACK, and W. E. REIFSNYDER. 1959. "The climate of shade." Connecticut Agricultural Experiment Station Bulletin 626.

WAGGONER, P. E., and W. E. REIFSNYDER. 1961. "Difference between net radiation and water use caused by radiation from the soil surface." *Soil Science*, 91: 246-50.

WAGGONER, P. E., and I. ZELITCH. 1965. "Transpiration and the stomata of leaves." *Science*, 150: 1413-20.

WALLEN, C. C. 1966. "Global radiation and potential evapotranspiration in Sweden." *Tellus*, 18: 786-800.

WALLIN, J. R., and D. N. POLHEMUS. 1954. "A dew recorder." *Science*, 119: 294-95.

WANG. JEN-YU. 1962a. "Crop forecast without weather prediction." *Crops and Soils*, 14: 7-9.

WANG. JEN-YU. 1962b. "The influence of seasonal temperature ranges on pea production." *Proceedings*, American Society for Horticultural Science. 80: 436-48.

WANG. JEN-YU, and S. C. WANG. 1962. "A simple graphical approach to Penman's method for evaporation estimates." *Journal of Applied Meteorology*. 1: 582-88.

WARBURG, O. 1919. "The velocity of the photochemical decomposition of carbon dioxide in the living cell." *Biochemische Zeitschrift*. 100: 230-70.

WARD, R. C. 1963. "Observations of potential evapotranspiration on the Thames floodplain 1959-1960." *Journal of Hydrology*, 1: 183-94.

WARREN WILSON, J. 1959. "Notes on wind and its effect in arctic-alpine vegetation." *Journal of Ecology*, 47: 415-27.

WARREN WILSON, J. 1960a. "Observations on net assimilation rates in arctic environments." *Annals of Botany* [N.S.], 24: 372-81.

WARREN WILSON, J. 1960b. "Influence of spatial arrangement of foliage area on light interception and pasture growth." In *Proceedings*, Eighth International Grassland Congress.

WARREN WILSON, J., and R. M. WADSWORTH. 1958. "The effect of wind speed on assimilation rate—a reassessment." *Annals of Botany* [N.S.], 22: 285-90.

WARTENA, L. 1959. "The climate and the evaporation from a lake in central Iraq." *Mededelingen van de Landbouwhogeschool*, Wageningen, 59(9): 1-59.

WARTENA, L., and E. C. VELDMAN. 1961. "Estimation of basic irrigation requirement." *Netherlands Journal of Agricultural Science*, 9: 293-98.

WARMING, E. 1909. *Oecological Plant Geography (Translated by P. Groom and I. B. Balfour)*. Oxford: Clarendon Press.

WASSINK, E. C. 1948. "De lichtfactor in de photosynthese en zijn relatid tot andere milieufactoren" ["Light as a factor in photosynthesis and its relation to other environmental factors"]. *Mededelingen Dir Tuinbouw* [*Bulletin of Horticulture*], 11: 503-513.

WATSON, D. J. 1947. "Comparative physiological studies on the growth of field crops: I. variation in net assimilation rate and leaf area between species and varieties, and within and between years." *Annals of Botany* [N.S.] 11: 41-76.

Watson, D. J. 1963. "Climate, weather and plant yield." In L. T. Evans (Ed.), *Environmental control of Plant Growth*. New York: Academic Press.

Watson, D. J., and S. A. W. French. 1962. "An attempt to increase yield by controlling leaf area index." *Annals of Applied Biology*, 50: 1-10.

Watson, D. J., and K. J. Witts. 1959. "The net assimilation rates of wild and cultivated beets." *Annals of Botany* [N.S.], 23: 431-39.

Weaver, H. A., and R. W. Pearson. 1956. "Influence of nitrogen fertilization and plant population density on evapotranspiration by sudangrass." *Soil Science*, 81: 443-51.

Weaver, H. A., and J. C. Stephens. 1963. "Relation of evaporation to potential evapotranspiration." *Transactions*, American Society of Agricultural Engineering, 6: 55-56.

Weaver, J. E., and W. J. Himmel. 1930. "Relation of increased water content and decreased aeration to root development in hydrophytes." *Plant Physiology*, 5: 69-92.

Webb, E. K. 1960. "On estimating evaporation with fluctuating Bowen ratio." *Journal of Geophysical Research*, 65: 3415-17.

Weger, B. 1953. "Effect of overhead irrigation on the vine and the wine in the hilly portion of the Bolzano Valley." *Rivista di Viticoltura i di Enologia*, 6: 176-78.

Weihing, R. N. 1963. "Growth of ryegrass as influenced by temperature and solar radiation." *Agronomy Journal*, 55: 519-21.

Wellensiek, S. J. 1957. "The plant and its environment." In J. P. Hudson (Ed.), *Control of the Plant Environement*. London: Butterworth Scientific Publications.

Wells, S. A., and S. Dubetz. 1966. "Reaction of barley varieties to soil water stress." *Canadian Journal of Plant Science*, 46: 507-512.

Went, F. W. 1950. "The response of plant to climate." *Science*, 112: 489-94.

Went, F. W. 1955. "Fog, mist, dew and other sources of water." In *USDA Yearbook of Agriculture*.

Went, F. W. 1956. "The role of environment in plant growth." *American Scientist*, 44: 378-98.

Went, F. W. 1957a. "Climate and agriculture." In *Plant life*. New York: Simon and Schuster.

Went, F. W. 1957b. *Experimental Control of Plant Growth*. Waltham, Mass.: Chronica Botanica.

Went, F. W. 1958. "The physiology of photosynthesis in higher plants." *Preslia*, 30: 225-49.

West, E. S., and O. Perkman. 1953. "Effect of soil moisture on transpiration." *Australian Journal of Agricultural Research*, 4: 326-33.

Westlake, D. F. 1963. "Comparisons of plant productivity." *Biological Review*, 38: 385-425.

Wetzel K. 1924. "Die Wasseraufnahme der höheren Pflanzen gemässigter Klimate durch oberirdische Organe." *Flora*, 117: 221-69.

Wheaton, R. Z., and E. P. Kidder. 1965. "The effect of frequency of application on frost protection by sprinkling." *Quarterly Journal*, Michigan State University Agricultural Experiment Station, 47: 439-45.

WHITEHEAD, F. H. 1957. "Wind as a factor in plant growth." In J. P. Hudson (Ed.), *Control of the plant environment*. London: Butterworth Scientific Publications.
WHITEHEAD, F. H. 1963. "The effects of exposure on growth and development." In A. J. Rutter and F. H. Whitehead (Eds.), *Water relations of plants*. Blackwell Scientific Publications.
WHITTINGHAM, C. P. 1955. "Energy transformation in photosynthesis and the relation of photosynthesis to respiration." *Biological Review*, 30: 40-64.
WIGGANS, S. C. 1936. "The effect of seasonal temperatures on maturity of oats planted at different dates." *Agronomy Journal*, 48: 21-25.
WILLIAMS, R. F. 1954. "Estimation of leaf area for agronomic and plant physiological studies." *Australian Journal of Agricultural Research*, 5: 235-46.
WILLIAMS, R. F., L. T. EVANS, and L. J. LUDWIG. 1964. "Estimation of leaf area for clover and lucerne." *Australian Journal of Agricultural Research*, 15: 231-33.
WILSON, H. W. 1924. "Studies on the transpiration of some Australian plants." *Victoria Proceedings*, Royal Society, 36: 175-237.
WINNEBERGER, J. H. 1958. "Transpiration as a requirement for growth of land plants." *Physiologia Plantarum*, 11: 56-61.
WINTER, E. J. 1960. "The irrigation of potatoes." *Agriculture*, London, 66: 549-51.
WINTER, E. J. 1963. "A new type of lysimeter." *Journal of Horticultural Science*, 38: 160-68.
WITHROW, R. B. (Ed.). 1959. *Photoperiodism and related phenomena in plants and animals*. Washington, D.C.: American Association for Advancement of Science. Publication 55.
WITTWER, S. H., and W. ROBB. 1964. "Carbon dioxide enrichment of greenhouse atmospheres for food crop production." *Economic Botany*, 18: 34-56.
WOLF, F. A. 1962. *Aromatic or Oriental Tobaccos*. Durham, N.C.: Duke University Press.
WORLD METEOROLOGICAL ORGANIZATION. 1956. *First Report of Working Group on Climatic Atlases*. Centerton, N.J. Laboratory of Climatology.
WOODRUFF, N. P., R. A. READ, and W. S. CHEPIL. 1959. "Influence of a field windbreak on summer wind movement and air temperature." In Bulletin 100, Kansas Agricultural Experiment Station.
XIAO. WEN-JUN. 1959. "The distribution of total annual and seasonal insolation in China." *Acta Meteorologica Sinica*, 30: 186-90.
YAKUWA, R. 1946. "Über die Bodentemperaturen in den verschiedenen Bodenarten in Hokkaido." *Geophysical Magazine*, Tokyo, 14: 1-12.
YAMADA, N., Y. MURATA, A. OSADA, and J. IYAMA. 1955. "Photosynthesis of rice plant." *Proceedings*, Crop Science Society, Japan, 23: 214-22.
YAMAMOTO, G. 1936. "The measurement of the amount of dew at Hukuoka, Japan." *Geophysical Magazine*, Tokyo, 10: 265-68.
YAMAMOTO, G. 1952. "On a radiation chart." *Geophysics*. Tohoku University, Science Report, Series 5, 4: 9-23.
YAO, Y. M., and R. H. SHAW. 1964a. "Effect of plant population and planting

pattern of corn on water use and yield." *Agronomy Journal*, 56: 147-52.

YAO, Y. M., and R. H. SHAW. 1964b. "Effect of plant population and planting pattern of corn on the distribution of net radiation." *Agronomy Journal*, 56: 165-69.

YOCUM, C. S., L. H. ALLEN, and E. R. LEMON. 1964. "Photosynthesis under field conditions: VI. solar radiation balance and photosynthetic efficiency." *Agronomy Journal*, 56: 249-53.

YOUNG, A. A. 1947. "Some recent evaporation investigations." *Transactions*, American Geophysical Union, 28: 279-84.

YOUNG, C. P. 1963. "A computer programme for the calculation of mean rates of evaporation using Penman's formula." *Meteorological Magazine*, 92: 84-89.

ZAHNER, R., and A. R. STAGE. 1966. "A procedure for calculating daily moisture stress and its utility in regressions of tree growth on weather." *Ecology*, 47: 64-74.

ZELITCH, I. 1961. "Biochemical control of stomatal opening in leaves." *Proceedings*, National Academy of Science, 47: 1423-33.

ZELITCH, I. 1964. "Reduction of transpiration of leaves through stomatal closure induced by alkenylsuccinic acids." *Science*, 143: 692-93.

ZELITCH, I., and P. W. WAGGONER. 1962. "Effect of chemical control stomata on transpiration and photosynthesis." *Proceedings*, National Academy of Science, 48: 1101-1108.

ZUBRISKI, J. C., and E. B. NORUM. 1955. "What effect do fertilizers have on soil moisture utilization by wheat?" North Dakota Agricultural Experiment Station, *Bimonthly Bulletin*, 17: 126-27.

GLOSSARY

Adiabatic process—A thermodynamic change of state of a system in which there is no transfer of heat or mass across the boundaries of the system.
Advection—The movement of air, water, and other fluids in a horizontal direction.
Albedo—The ratio of the amount of visible light reflected by a body to the amount incident upon it.
Atmometer—The general name of an instrument that measures the evaporation rate of water into the atmosphere.
Austausch coefficient—*See* exchange coefficient.
Back radiation—The downward flux of atmospheric long-wave radiation.
Bar—A unit of pressure equal to one million dynes per square centimeter.
Black body—A hypothetical "body" that absorbs all of the electromagnetic radiation striking it; that is, one that neither reflects nor transmits any of the incident radiation.
Bowen's ratio—At a water surface, the ratio of the energy flux upward as sensible heat to the energy flux used in evaporation.
Climatic analogues—Areas that are enough alike with respect to some of the major weather characteristics affecting the production of crops.
Chloroplast—The body in the cell cytoplasm that contains the green chlorophyll pigment.
Clothesline effect—Horizontal heat transfer from a warm upwind area to a relatively cooler crop field.
Compensation point—The point at which the respiration rate equals the photosynthetic rate.
Consumptive use—The sum of volume of water used by the vegetation growth of a given area in transpiration or building of plant tissue and that evaporated from adjacent soil.
Convection—In general, mass motions within a fluid resulting in transport and mixing of the properties of that fluid. Distinction is made between: (1) free convection (or gravitational convection), motion caused only by density differences within the fluid, and (2) forced convection, motion induced by mechanical forces such as deflection by a large-scale surface irregularities, turbulent flow caused by friction at the boundary of a fluid, or motion caused by any applied external force.

Cuticle—An outer skin or pellicle.

Degree day—A measure of the departure of the mean daily temperature from a given standard.

Dew—Water condensed onto plants and other objects near the ground.

Dew point—The temperature at which air, being cooled, becomes saturated with water vapor.

Diffuse radiation—Radiation reaching the Earth's surface after having been scattered from the direct solar beam by molecules or suspensoids in the atmosphere.

Dry adiabatic lapse rate—The rate of decrease of temperature with height of a parcel of dry air lifted adiabatically.

Eddy diffusivity—The exchange coefficient for the diffusion of a conservative property by eddies in a turbulent flow.

Effective outgoing radiation—The difference between the outgoing long-wave radiation of the Earth's surface and the downward long-wave radiation from the atmosphere.

Einstein—A unit for measuring the energy output of light that is equal to one mole of quanta (6.02×10^{23} quanta).

Elsasser's radiation chart—A radiation chart developed by W. M. Elsasser for the graphical solution of the radiative transfer problems.

Emissivity—The ratio of the emittance of a given surface to the emittance of an ideal black body.

Eppley pyrheliometer—An instrument for measuring direct and diffuse radiation.

Evapotranspiration—The combined loss of water from a given area by evaporation from the soil surface and by transpiration from plants.

Exchange coefficient (also called austausch coefficient)—Coefficients of eddy flux in turbulent flow.

Extinction coefficient—A measure of the space of diminution, or extinction, of any transmitted light.

Field capacity—The percentage of water remaining in a soil two or three days after having been saturated and after free drainage has practically ceased.

Foot-candle—A unit of illuminance equal to one lumen per square foot. This is the illuminance provided by a light source of one standard candle at a distance of one foot, hence the name.

Freezing point—The temperature at which a liquid solidifies.

Global radiation—The total of direct solar radiation and diffuse sky radiation received by a unit horizontal surface.

Greenhouse effect—The effect that short-wave radiation can pass easily through the atmosphere to the surface of the Earth, while a proportion of the resulting heat is retained in the atmosphere, since the outgoing long-wave reradiation cannot penetrate the atmosphere easily, especially when there is a cloud cover.

Heat capacity (*sometimes called volumetric specific heat*)—The ratio of the heat absorbed by a system to the corresponding temperature rise.

Hygrometer—An instrument that measures the water vapor content of the atmosphere.

Hysteresis—Lag in one of two associated processes or phenomena.

Isopleth—A line of equal value of a given quantity.

Kinetic energy—The energy that a body possesses as a consequence of its motion.

Laminar flow—A flow in which the fluid moves smoothly in streamlines in parallel layers or sheets; a non-turbulent flow.

Langley—A unit of radiation; equal to one gram-calorie per square centimeter.

Latent heat—The amount of heat energy expended in changing the state or phase of a body, without raising its temperature.

Leaf area index—The leaf area subtended per unit area of land.

Livingston atmometer—A clay atmometer in the form of a sphere.

Lysimeter—A device for measuring gains (precipitation, condensation, and irrigation) and losses (evapotranspiration) by a column of soil.

Mesoclimate—The climate of small areas of the Earth's surface that may not be representative of the general climate of the district. The mesoclimate is intermediate in scale between the macroclimate and the microclimate.

Microclimate—The fine climatic structure of the air space that extends from the very surface of the earth to a height at which the effects of the immediate character of the underlying surface no longer can be distinguished from the gneral local climate (mesoclimate or macroclimate).

Moisture tension—The equivalent pressure that must be applied to the soil water to bring it to hydraulic equilibrium, through a porous permeable wall or membrane, with a pool of water of the same composition.

Mole—A unit of mass numerically equal to the molecular weight of the substance. The gram-mole or gram-molecule is the mass in grams numerically equal to the molecular weight.

Momentum—The property of a particle that is given by the product of its mass with its velocity.

Net assimilation rate—See Net photosynthesis.

Net photosynthesis (also called net assimilation rate)—The excess of dry matter gain by photosynthesis over loss by respiration.

Net radiation—The difference between upward and downward radiation flux, or, alternatively, the difference between the net short-wave and the net long-wave radiation.

Neutral stability—The state of an unsaturated column of air in the atmosphere when its environmental lapse rate of temperature is equal to the dry adiabatic lapse rate.

Oasis effect—Vertical energy transfer from the air to the crop.

Optical air mass—A measure of length of the path through the atmosphere to sea level traversed by light rays from a celestial body, expressed as a multiple of the path length for a light source at the zenith.

Organic soil—A soil that contains a high percentage ($>$ 15 or 20 per cent) of organic matter throughout the solum.

Peat soil—An organic soil containing more than 50 per cent organic matter.

Permanent wilting point—The amount of water in the soil below which plants will be unable to obtain further supplies, and will therefore wilt.

pF—The logarithm of the soil moisture tension expressed in centimeters height of a column of water.

Photoperiod—Duration of daily exposure to light.

Photoperiodism—Response of an organism to the relative duration of day and night.

Photosynthesis (also called assimilation)—The formation of carbohydrates from water and the carbon dioxide of the air in the chlorophyll-containing tissues of plants exposed to light.

Piche evaporimeter—A porous paper wick atmometer.

Potential energy—The energy that a body possesses as a consequence of its position in the field of gravity.

Potential evapotranspiration—The amount of moisture that, if available, would be removed from a given land area by evapotranspiration.

Prygeometer—An instrument that measures the effective terrestrial radiation.

Psychrometric constant—The ratio of the specific heat of air to the latent heat of evaporation of water.

Quantum—An elemental unit of energy with no electric charge and very little mass. In its interactions with matter, light acts as though it were composed of quanta.

Rayleigh scattering—Scattering by particles whose radii are considerably smaller than the wavelength of the incident radiation.

Reflection coefficient—The ratio of the amount of solar radiation reflected by a body to the amount incident upon it.

Relative humidity—The ratio of the air's vapor pressure to the saturation vapor pressure.

Respiration—The process by which a plant absorbs oxygen from the air and gives out carbon dioxide.

Richardson number—A nondimensional index of stability arising in the study of shearing flows of a stratified fluid.

Roughness parameter—A measure of the roughness of a surface over which a fluid is flowing.

Runoff—The portion of the precipitation on an area that is discharged from the area through stream channels.

Saturation deficit—The amount of water-vapor required to bring nonsaturated air at a given temperature and pressure to the point of saturation.

Saturation light intensity—The light intensity beyond which additional light will not produce an increase in photosynthesis.

Saturation vapor pressure—The vapor pressure of a parcel of saturated air at a given temperature.

Sensor—The component of an instrument that converts an input signal into a quantity that is measured by another part of the instrument.

Shearing stress—The force of retardation per unit horizontal area in the direction of the wind.

Soil texture—The relative proportions of the various soil separates in a soil as described by the classes of soil texture shown in Figure 1.

Solar constant—The rate at which solar radiation is received outside the Earth's atmosphere on a surface normal to the incident radiation, and at the Earth's mean distance from the sun.

Solar spectrum—The part of the electromagnetic spectrum occupied by the wavelengths of solar radiation.

Solar radiation—The total electromagnetic radiation emitted by the sun.

Glossary

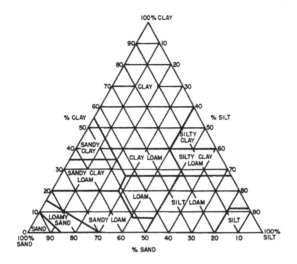

Figure 1—Graph showing the percentages of sand, silt, and clay in the soil classes.

Specific heat—The amount of heat in calories required to raise the temperature of 1 gram of a substance by 1° C.

Specific humidity—The ratio of the weight of the water-vapor in a "parcel" of the atmosphere to the total weight of the air (including the water vapor) stated in gram of water-vapor per kilogram of air.

Stable—An atmosphere in which temperature decrease with height is less than the dry adiabatic lapse rate.

Stefan-Boltzman constant—A universal constant of proportionality between the radiant emittance of a black body and the fourth power of the body's absolute temperature.

Stoma (plu. Stomata)—Minute openings, with guard cells, in epidermis of plants, especially on under-surface of leaves.

Sunspot—A dark area in the photosphere of the sun caused by a lowered surface temperature.

Superadiabatic lapse rate—An environmental lapse rate greater than the dry-adiabatic lapse rate.

Temperature inversion—An increase of air temperature with height, so that warmer air overlies colder, contrary to the normal lapse rate.

Tensiometer—A device for measuring the tension of water in soil *in situ*; a porous, permeable ceramic cup connected through a tube to a manometer.

Terrestrial radiation—The long-wave radiation emitted from the Earth's surface.

Thermal conductivity—The quantity of heat flowing per unit of time through unit area of a plate of unit thickness when a unit temperature difference is maintained between the two opposing faces.

Thermal diffusivity—The quotient of thermal conductivity and heat capacity; hence, a measure of the facility with which a substance will undergo temperature change.

Thermocouple—A device that uses the voltage developed by the junction of two dissimilar metals to measure temperature differences.

Thermoperiodicity—Effects of temperature difference between light and dark periods upon plants.

Thermopile—A transducer for converting thermal energy directly into electrical energy. It is composed of pairs of thermocouples which are connected either in series or in parallel.

Topoclimatology—The study of local climate as affected by slope, aspect, and surface characteristics.

Translocation—Transportation of products of metabolism from one part of a plant to another.

Turbulence—An irregular eddying flow, in contrast to a smooth laminar flow.

Unstable—An atmosphere in which temperature decrease with height is greater than the dry adiabatic lapse rate.

Vernalization—The exposure of plants to low temperature to induce or accelerate the development of the ability to form flowers.

Wet-bulb temperature—The temperature recorded on a thermometer that has its bulb surrounded by a moist muslin bag, thus lowering the temperature by loss of latent heat through evaporation. With the aid of a dry-bulb thermometer and a set of tables, relative humidity can be ascertained.

Wig-wag—An instrument for measuring solar radiation.

Zero plane displacement—An empirically determined constant introduced into the logarithmic wind profile to extend its applicability to tall vegetation.

INDEX

Abd El Rahman, A. A., 91, 129, 247
Abdel-Aziz, M. H., 170, 171, 247
Absorption of solar radiation, 6, 7
Absorptivity, 98, 99
Acer Pseudoplatanus L., 28
Actual evapotranspiration, factors affecting, 133-40; relationship between, and crop yield, 211-15
Adams, J. E., 247
Adiabatic lapse rate, 111, 292
Advection, 117, 131, 140-43, 163, 164, 170, 172, 174, 179, 188, 291
Advective frost, 100, 102, 107
Aerodynamic approach, 117, 145, 157-62, 165, 166, 201, 228
Africa violet, 81
Ageratum, 81
Ahmad, M. S., 155, 247
Aikman, J. M., 94
Aitken, J., 225, 247
Air temperature, 117, 131, 132
   effect of shelterbelt on, 239, 240
   effect on photoperiodism 71, 73
   effect on photosynthesis, 28, 62
   effect on plant growth, 75-81
   effect on respiration, 32, 59
   the use of, for estimating evapotranspiration, 149-56
Albedo, 11, 13, 14, 129, 133, 188, 244, 291
Alberda, T., 58, 220, 247, 259
Alderfer, R. B., 216, 223, 281, 286
Aldrich W. W., 224, 247
Alessi, J., 222, 255
Alfalfa, 20, 25, 27, 32-35, 41, 42, 58, 72, 115, 128, 129, 153, 186, 187, 190, 192, 193, 211, 241, 242; *see also* Lucerne
Alisov, B. P., 241, 247
Allard, H. A., 70, 71, 247, 262
Allen, L. H., 11, 14, 42-44, 247, 248, 290
Allison, F. E., 212, 222, 248
Allmaras, R. R., 96, 248
Allmendinger, D. F., 119, 248
Almond, 104
Alter, J. C., 95, 248
Althea, 72
Alway, F. J., 198, 248
Andersen, P. C., 242, 248
Anderson, C. H., 248
Anderson, D. B., 119, 273
Anderson, D. T., 54, 248
Anderson, E. R., 16, 248
Anderson, M. C., 248
Angot's value, 8, 166
Ångström, A., 8, 9, 11, 16, 248
Angus, D. E., 17, 102-105, 122, 147, 182, 186-89, 229, 248, 263, 272, 273, 278
Anita, N., 223, 249
Ansari, A. Q., 86, 249
Anstey, T. H., 78, 249
Antcliff, A. J., 66, 272
Apple, 25, 78, 119-21, 211
Apricot, 77, 192, 193, 216
Arkley, R. J., 125-27, 232, 249
Arthur, I. P., 249
Arthur, J. M., 66, 249
Artichoke, 72, 190, 192, 193
Artificial light, 65-70, 74
Arvidsson, I., 227, 229, 249
Ash, 87
Ashbell, C. W., 211, 255
Ashcroft, G. L., 171, 247
Ashley, D. A., 47, 186, 211, 249, 259
Ashton, F. M., 58, 119-21, 249, 255
Aslyng, H. C., 17, 187, 211, 249
Asparagus, 190
Aspect, 94, 95
Aster, 71
Atmometer, 178, 190-93, 291
Atmospheric window, 15
Aubertin, G. M., 177, 249
Austausch coefficient, 117, 291, 292
Austin, S. W., 285
Azalea, 72

Back radiation, 15, 16, 22, 291
Bahrani, B., 20, 249
Baier, W., 229, 250
Bainbridge, R., 250
Baker, D. G., 150, 170, 250
Baker, D. N., 41, 42, 63, 64, 231, 250
Baker, D. R., 183, 275
Baker, F. S., 87, 250
Bald, J. G., 47, 250
Balfour, I. B., 287
Balsam, 72, 119
Bamboo, 187
Banana, 78, 190, 211
Barber, D. A., 210, 250
Bark, L. D., 20, 98, 264
Barley, 33, 35, 53, 54, 61, 62, 72, 75, 101, 127, 186, 211, 216, 220, 223, 234, 235
Barnes, A. C., 62, 250
Barry, R. G., 14, 250
Bartels, L. F., 195, 250
Bartlett, R. J., 210, 250
Batanouny, K. H., 129, 247
Batchelor, L. D., 198, 250
Bates, C. G., 240, 241, 250
Baver, L. D., 156, 250
Bean, 25, 72, 82, 101, 135, 136, 153, 155, 190, 211
Beck, G. E., 20, 265
Beech, 87
Beer's law, 39, 41
Beet, 58, 72, 101, 128, 155, 242; *see also* Sugar beet
Begg, J. E., 43, 250
Begonia, 72
Belcher, B. A., 71, 256
Bellani atmometer, 190-92
Bendelow, V. M., 223, 251
Benedict, H. W., 65, 251
Bennett, O. L., 47, 186, 211, 249, 259
Benoit, G. R., 205, 206, 270
Berliand, T. G., 11, 251
Bernard, E. A., 143, 251
Berry, G., 167, 251
Best, R., 23, 60, 251
Bielorai, H., 122, 249
Bierhuizen, J. F., 43, 91, 123, 124, 135, 136, 200, 202, 210, 223, 247, 250, 251, 281
Bilham, E. G., 184, 251
Billings, W. D., 28, 274
Birch, J. W., 258
Black body, 14, 291
Black, J. N., 8, 9, 11, 34, 49, 50, 251
Blackwell, J. H., 89, 251
Blanc, M. L., 88, 269
Blaney-Criddle formula, 151-53, 178, 180
Blaney, H. F., 151, 152, 186, 252
Blaser, R. E., 35, 254
Bleksley, A. E. H., 11, 252
Blueberry, 106
Blumenstock, D. I., 203, 252
Böhning, R. H., 25, 26, 252
Bolas, B. D., 28, 252
Bonnen, C. A., 211, 252
Bonner, J., 24, 27, 29, 30, 59, 252
Bonython, C. W., 8, 9, 181, 251, 252
Bookkeeping method, 196, 197
Bordeau, P., 120, 252
Borthwick, H. A., 74, 252
Bourget, S. J., 211, 252
Bouyoucos, G. J., 95, 252
Bowen, I. S., 165, 252
Bowen ratio, 165, 244, 291
Bowers, S. A., 11, 12, 20, 98, 252, 264
Bowman, D. H., 178, 195, 252
Boyko, B. H., 228, 252
Brackett, F. S., 29, 266
Bradley, G. A., 211, 252
Brassica napus, 238
Brassica rapa, 238
Breazeale, E. L., 231, 232, 253
Breazeale, J. F., 231, 253
Brierley, W. G., 232, 253
Briggs, L. J., 125-27, 198, 233, 253
Brilliant, B., 119, 253

Brix, H., 253
Broccoli, 190, 216
Brodie, H. W., 7, 58, 121, 253, 255
Brooks, C. E. P., 200, 253
Brooks, C. F., 199, 253
Brooks, F. A., 5, 14, 15, 99, 102, 104, 105, 108, 253, 263
Brooks, R. M., 77, 253
Broomsedge, 72
Brougham, R. W., 37, 39, 55, 56, 253, 254
Brouwer, R., 133, 210, 254
Brown, D. M., 77, 254
Brown, K. W., 11, 14, 247
Brown, M. A., 233, 254
Brown, P. L., 185, 254
Brown, R. H., 35, 254
Bruce, J. P., 131, 191, 274
Brunt, D., 15, 254
Brunt formula, 15, 16, 102
Brushing, 107
Brutsaert, W., 178, 254
Bryophyllum, 72
Bryson, R. A., 14, 269
Buckwheat, 66, 101
Budyko, M. I., 14, 17, 224, 254
Buettner, K., 7, 88, 255, 262
Bullrush, 118
Bulrush millet, 42, 43
Burman, R. D., 150, 255
Burnside, C. A., 25, 26, 252
Burov, D. I., 140, 255
Burr, G. O., 28, 58, 121, 255
Burr, W. W., 198, 255
Burrows, W. C., 96, 248
Businger, J. A., 106, 170, 255
Butler, P. F., 150, 187, 255
Butler, W. L., 255

Cabbage, 71, 101, 190, 216
Caborn, J. M., 238, 240, 255
Cactus, 72
Calder, K. L., 110, 255
Calvert, A., 255
Campbell, R. B., 211, 215, 256
Cannell, G. H., 210, 255
Cantaloupe, 190, 224
Carbon black, 98, 99
Carbon dioxide, absorption of radiation by, 6, 15
  effect of wind on the supply of, 234, 238
  effect on photosynthesis, 23, 24, 27-32
  source of, 32
Carder, A. C., 255
Cardinal temperature, 75
Carlson, C. W., 222, 255
Carnation, 32
Carrot, 58, 71, 101, 155
Carruthers, N., 200, 253
Carson, R. B., 223, 285
Carter, A. R., 241, 266
Castor plants, 101
Cauliflower, 63, 190, 210, 216
Cavrielsen, E. K., 256
Celery, 71, 210
Celosia, 48, 49
Cerastium astrovirens, 234, 235
Chambers, R. E., 14, 250
Chandraratha, N. F., 74, 256
Chang, Jen-hu, 12, 14, 17, 88, 117, 151, 166, 169, 170, 186, 200, 208, 211, 215, 256
Chapas, L. C., 169, 187, 256
Chapman, H. W., 25, 29, 256
Chapman, L. J., 31, 256
Chappuis band, 6
Chatterjee, B. N., 28, 256
Chaverra, H., 73, 257
Chepil, W. S., 240, 289
Cherry, 78, 216
Chicory, 58, 72
Childers, N. F., 25, 119-21, 265, 280
Chili pepper, 80
China aster, 81
Chisholm, E. de C., 211, 263
Chrelashvili, M. N., 119, 256
Christidis, B. G., 56, 256
Chrysanthemum, 32, 68, 72, 73
Church, T. W., 125, 257
Cineraria, 72
Citrus, 77, 104, 107, 153, 190, 242
Clapp, P. F., 13, 277
Clark Greenhouse, 3
Cleasby, T. G., 186, 187, 209, 276, 284
Climatic analogues, 1, 291
Clothesline effect, 141, 142, 183, 291

Closs, R. L., 137, 138, 256
Cloudiness, effect on outgoing radiation, 15; relationship between radiation and, 9
Clover, 34, 37, 39, 40, 47, 49, 51, 72, 119, 141, 144, 151, 155, 170, 182, 186, 242
Cocoa, 60, 66
Coffee, 27, 66, 187
Coleman, E. A., 146, 256
Coleman, O. H., 71, 256
Coleman, R. E., 58, 121, 255
Coleus, 25
Collins, B. G., 256
Collins, P. E., 239, 257
Collins-George, N., 2, 257
Committee on Plant Irradiation, 5, 257
Compensation point, 33, 49, 291
Conrad, V. E., 200, 257
Consumptive use, 129, 153, 182, 192, 200, 291
Control pan ratio, 217-19
Convection, 112, 159, 291
Cooper, R. B., 35, 254
Copeland, E. B., 233, 257
Coriander, 101
Corn, 11, 14, 20, 25, 28, 29, 31, 39-44, 47, 57, 58, 60, 63, 66, 72, 73, 77, 82, 96, 97, 99, 101, 109, 110, 112, 113, 115, 120, 121, 137, 146, 153, 155, 163, 164, 175, 177, 186, 189, 190, 200, 216, 218, 220, 223, 229, 232, 241
Cornflower, 71
Cosmos, 71
Cotton, 14, 20, 25, 47, 56, 72, 99, 101, 125, 153, 171, 186, 192, 193, 211, 216, 230
Cottonwood, 153
Counterradiation; *see* Back radiation
Covey, W., 109, 110, 141, 264
Cowpea, 74
Cox, H. J., 105, 257
Craddock, J. M., 225, 257
Cranberry, 105, 108
Crawford, T. V., 259
Criddle, W. D., 152, 252
Crider, F. J., 216, 257
Crop quality, effect of irrigation on, 222-24; effect of radiation on, 67, 68
Crowder, L. V., 73, 257
Crowe, P. R., 150, 257
Cucumber, 28-30, 32, 68, 72, 101, 229, 230, 242
Culbertson, M. F., 15, 260
Cunningham, R. K., 66, 257
Curry, J. R., 125, 257
Cyprus, 187

Dagg, M., 153, 179, 257
Dahlia, 66
Dale, R. F., 2, 257
Dalton equation, 157, 158
Daniel, F., 253
Das, U. K., 80, 257
Date palm, 66
Davey, B. G., 2, 257
Davidson, J. L., 47, 48, 257
Davidson, J. M., 147, 268
Davies, J. A., 257
Davis, J. R., 106, 178, 220, 257, 258
Davis, R. M., 224, 261
Day, G. L., 11, 258
Day-length indifferent plants; *see* Day-neutral plants
Day-neutral plants, 70, 72
Deacon, E. L., 16, 111, 112, 146, 160, 227, 229, 258
Deacon's wind equation, 111, 112
Decker, W. L., 20, 90, 91, 150, 164, 170, 203, 258, 260, 263
Deficit, 203-206
Degman, E. S., 211, 271
Degree-day, 77, 78, 292
De Lis, B. R., 216, 258
Deneke, H., 234, 258
Denmead, O. T., 43-45, 136, 137, 179, 216, 258
Dermine, P., 187, 256
De Roo, H. C., 98, 286
De Vos, N. M., 200, 202, 223, 251
De Vries, D. A., 11, 88, 143, 150, 258, 286
Dew, 94, 225-30, 245, 292
De Wit, C. T., 58-62, 77, 123-25, 128, 209, 220, 258, 259
Dias, F., 16, 277
Diffuse sky radiation, 7, 12, 42, 94, 245, 292
Dill, 71
Diurnal temperature range, 1, 34, 77, 224; *see also* Thermoperiodicity
Domingo, C. E., 216, 279
Donald, C. M., 47, 49, 51, 259, 282
Dore, J., 73, 259

298

Doss, B. D., 47, 186, 211, 249, 259
Dow, B. K., 211, 252
Downs, R. J., 255
Dreibelbis, F. R., 147, 175, 209, 218, 221, 228, 259, 265, 268
Drew, D. H., 216, 259
Drinkwater, W. O., 216, 267
Drosometer, 225
Drought, 204-206
Drummond, A. J., 7, 11, 259
Dry matter production, as affected by leaf area index, 49-51; as affected by radiation, 49-51; efficiency of water use in, 125, 232
Dryopteris, 25
Dubetz, S., 54, 216, 248, 288
Duffie, J. A., 253
Dunkie, R. V., 16, 263
Duration of sunshine, 8, 9
Duvdevani, S., 225-27, 229, 230, 259
Dyer, A. J., 161, 259, 283

Eagleman, J. R., 203, 260
Earhart Plant Research Laboratory, 3
Earley, E. B., 66, 260
Eddy correlation technique, 160-62, 202
Eddy diffusivities, 158-60, 165, 292
Edwards, R. I., 285
Effective outgoing radiation, 15, 292; see also Outgoing radiation
Effective pan ratio, 217-21
Effective rainfall, 208, 218
Efimova, N. A., 11, 251
Ehrler, W. L., 260
Eik, K., 260
Einstein, 24, 292
Ekern, P. C., 14, 17, 20, 130, 179, 186, 260
Elam, A. B., 205, 206, 270
Elmus americana, 86
El Nadi, A. H., 143, 260
Elsasser, W. M., 15, 260
Emmisivity, 14, 245, 292
Empirical formulae for estimating evapotranspiration, 145, 149-56, 178, 201
Energy budget, 163-77
  basic equation, 163-65
  diurnal variation of, 171
  variation throughout the crop cycle, 171-74
Enger, I., 82, 260
England, C. B., 150, 260
English daisy, 81
English holly, 71
English ivy, 119
Ertel, H., 110, 260
European yellow lupine, 101
Evans, G. C., 250
Evans, L. T., 47, 258, 262, 267, 270, 271, 283, 288, 289
Evaporation, from bare soil, 140
  difference between, and transpiration, 133
  effect of shelterbelt on, 240, 241
  relative magnitude of, and transpiration, 174-77
Evaporation pan, 178-91; design and installation, 179-85; ratios between evapotranspiration and pan evaporation, 185-90, 202
Evaporimeter, 178-93
Evapotranspiration, 1, 117, 129-44, 195, 195, 196, 201-204, 213-15, 228, 245, 292
  actual; see Actual evapotranspiration
  definition, 129-31
  fraction of radiation used in, 142-43, 170-73
  meteorological factors determining, 131-33
  potential; see Potential evapotranspiration
Evapotron, 161, 162
Everson, J. N., 99, 260
Extinction coefficient, 39, 43, 47-49, 292

Facultative long-day plants, 71
Fedrov, C. F., 147, 260
Ferguson, W. S., 260
Fern *Nephrolephis*, 25
Fescue, 72
Fibras, F., 233, 260
Field capacity, 134, 136, 137, 198, 202, 292
Figarella, J., 58, 286
Finkelstein, J., 179, 187, 260
Finn, B. J., 211, 252
Finnel, H. H., 238, 261
Fir, 115
Fischer, R. A., 130, 261
Fitzgerald, P. D., 150, 261
Fitzpatrick, E. A., 14, 261
Flax, 61, 63, 82, 101, 121, 122, 155, 211
Fleming, J. W., 211, 216, 268, 274

Flocker, W. J., 224, 261
Flooding, 104, 105
Fog, 225, 230, 231, 245
Foley, J. C., 105, 261
Foot-candle, 25, 292
Forsgate, J., 135, 263
Fortanier, E. J., 80, 231, 261
Fournier d'Albe, E. M., 242, 261
Fox, R. L., 119, 268
Fox, W. E., 167, 268
Foxglove, 72
Foxtail, 72
Franck, J., 284
Frear, D. E. H., 46, 261
Free-water evaporation, 124, 128, 157, 166; see also Evaporation
French, S. A. W., 47, 267, 288
Frith, H. J., 261
Fritschen, L. J., 16, 17, 43-45, 129, 140, 159, 186, 189, 190, 258, 261, 285
Fritz, S., 9, 11, 261
Frost, injury, 100; protection, 100-108; weather, 100-102
Fuchsia, 72
Fujito, M., 11, 268
Fukuda, H., 178, 283
Fuller, H. J., 32, 261
Fulton, J. M., 211, 262
Funk, J. P., 16, 17, 262
Fuquay, D., 7, 262
Furr, J. R., 211, 271

Gaastra, P., 24, 27, 29, 30, 51, 53, 58, 60, 61, 262
Gabites, J. F., 11, 262
Gal'tsov, A. P., 142, 262
Gard, L. E., 222, 262
Gardenia, 72
Gardner, W. R., 129, 275
Garner, W. W., 70, 71, 262
Garnier, B. J., 151, 262
Gates, D. M., 14, 28, 84, 123, 263
Gauch, H. G., 224, 286
Gay, L. W., 268
Geiger, R., 21, 22, 245, 263
Geranium, 72
Gerber, J. F., 170, 263
Gevorkiantz, S. R., 241, 279
Gier, J. T., 16, 263
Gilbert, M. J., 170
Ginseng, 66
Glaser, A. H., 125, 140, 142, 171, 186, 270
Gleason, L. S., 31, 256
Global radiation, 17, 245, 296; distribution, 9-11; measurement, 7-8
Glover, J., 9, 135, 200, 263
Gloyne, R. W., 239, 263
Goldenseal, 66
Goldin, E., 211, 272
Goodall, G. E., 101, 104, 105, 263
Goodchild, N. A., 66, 67, 272
Goss, J. R., 15, 263
Goto, Y., 210, 263
Grable, A. R., 171, 186, 263
Graham, W. G., 14, 143, 170, 263
Grainger, J., 211, 263
Grape, 192, 193, 203
Grass, 11-14, 16, 17, 19, 20, 36, 40, 47, 58, 65, 72, 101, 113, 115, 117, 121, 129, 135, 143, 153-55, 162, 178, 179, 186, 187, 190, 209, 211, 227, 228
Gray, H. E., 186, 264
Greenhouse effect, 98, 292
Gregory, F. G., 33, 61, 62, 264
Grimes, D. W., 220, 274
Groom, P., 287
Grossman, R., 211, 223, 267
Guayule, 224
Guerrini, V. H., 151, 264
Gunness, C. I., 105, 264
Gwynne, M. D., 200, 263

Haas, H. J., 177, 264
Hagan, H. R., 98, 264
Hagan, R. M., 119, 209, 210, 260, 261, 264, 285
Hageman, R. H., 66
Hagood, M. A., 220, 258
Haines, F. M., 231, 264
Haise, H. R., 171, 173, 174, 179, 186, 263, 267, 279
Halkais, N. A., 192, 264
Hallstead, A. L., 185, 254
Halstead, M. H., 109, 110, 141, 160, 170, 264, 284
Hand, D. W., 264
Hanks, R. J., 11, 12, 20, 98, 171, 186, 222, 247, 252, 263, 264

Hansen, E., 67, 186, 264
Hansen, V. E., 189, 199, 264, 267
Hanway, J. J., 260
Harbeck, G. E., 160, 264
Hardy, J. D., 84, 282
Harris, D. G., 170, 264
Harrold, L. L., 147, 175, 209, 218, 221, 228, 259, 265
Hart, S. A., 107, 265
Hartley band, 6
Hartt, C. E., 25, 58, 121, 234, 255, 265
Hasselkus, E. R., 20, 265
Hastie, A., 211, 263
Haude, W., 156, 265
Hawaiian Sugar Planters' Association, 35, 221, 265
Hearn, A. B., 179, 265
Heat capacity, 88, 90, 92, 93, 98, 104, 292
Heat flux, to the air, 163-65; to the soil, 163, 164
Heater, 102-104
Heating coefficient, 20
Heeney, H. B., 192, 265
Heinicke, A. J., 25, 234
Heinicke, D. R., 7, 265
Helianthus annuus, 119, 236
Hemp, 101
Hemp mallow, 101
Hendricks, S. B., 252
Hendrickson, A. H., 134, 135, 192, 203, 211, 223, 264, 265, 286
Hershfield, D. M., 208, 265
Hesketh, J. D., 28, 29, 234, 265, 287
Hesse, W., 233, 265
Hildreth, A. C., 216, 266
Hill, G. R., 25, 27, 32, 33, 284
Himmel, W. J., 118, 288
Hobbs, E. H., 192, 269
Hodge, C., 272
Hoffman, M. B., 234, 265
Hofmann, G., 226, 266
Hofstede, G. J., 20, 282
Hogg, W. H., 241, 266
Højendahl, K., 226, 266
Holm, L., 46, 273
Holmes, R. M., 137-39, 191, 203, 204, 266
Holt, R. F., 177, 266
Holzman, B., 159, 228, 284
Hoover, W. H., 29, 266
Hopen, H. J., 32, 266
Hopkins, J. W., 47, 266
Hord, H. H. V., 78, 266
Horwitz, M., 216, 272
Hosegood, P. H., 187, 276
Hounam, C. E., 9, 266
House, G. J., 160, 170, 266
Howe, O. W., 216, 220, 266
Hudson, J. P., 143, 144, 171, 260, 267, 288
Hughes, R., 267
Hughes, W. F., 211, 252
Humidity, effect on evapotranspiration, 131-33, 137-39; effect on plant growth, 231, 232
Humphries, E. C., 47, 267
Humus, 88, 89
Hurd, R. G., 46, 267
Huxley, P. A., 66, 267
Hydrangea, 72
Hydrophytes, 118, 233

Iceland poppy, 81
Iizuka, H., 239, 267
Iljin, W. S., 119, 267
Infrared radiation, 5, 6, 11, 14, 42
Infrared radiometer, 83, 84
Inoue, E., 115, 116, 157, 267
Intermediate plants, 71, 72
Irrigation, effect on crop quality, 22-24; experiment, 215, 217, 218; guide to cultural practice, 209-211
Israelsen, O. W., 199, 267
Iyama, J., 25, 63, 274, 289

Jackson, E. A., 153, 187, 267
Jacob, W. C., 211, 223, 267
Jacobs, W. C., 82, 269
Jagannathan, P., 181, 286
Janes, B. E., 216, 223, 267, 286
Jenkins, H. V., 46, 267
Jennings, E. G., 225, 267
Jensen, M., 242, 267
Jensen, M. C., 153, 278
Jensen, M. E., 173, 174, 178, 187, 267
Jerusalem sage, 119
Jessep, C. T., 191, 267
Johnson, A. F., 156, 271

Johnson, E. S., 29, 266
Johnson, F. A., 4, 267
Johnson, L. C., 175, 277
Jones, B. A., 222, 262
Jones, L. G., 210, 264
Jones, S. T., 211, 267
Jordan, R. C., 270

Kalanchoe, 72
Kale, 13, 14, 40
Kalma, J. D., 20, 282
Kano, M., 11, 280
Karsten, H., 89, 267
Kasanaga, H., 36, 37, 267
Katschon, R., 73, 268
Kattan, A. A., 211, 216, 268, 274
Kawabata, V., 11, 268
Keegan, H. J., 14, 263
Keen, B. A., 89, 268
Kelley, C. F., 104, 253
Kelley, O. J., 222, 268
Kemp, C. D., 47, 268
Kennedy, W. K., 186, 264
Kenworthy, A. L., 119, 248
Kerfoot, O., 66, 67, 272
Kerr, J. R., 144, 273
Khalil, M. S. H., 82, 268
Kidder, E. P., 106, 107, 288
Kihlman, A. O., 87, 268
Kimball, H. H., 25
Kimmey, J. W., 230, 268
Kinbacker, E. J., 123, 268
King, F. H., 90, 268
King, K. M., 14, 143, 147, 151, 170, 171, 178, 195, 252, 263, 268
Klute, A., 211, 223, 267
Kmoch, H. G., 119, 268
Knoerr, K. R., 268
Koehler, F. E., 119, 268
Kohler, M. A., 167, 268
Kohn, G. D., 130, 261
Kohnke, H., 147, 268
Kok-saghyz, 101
Kolasew, F. E., 241, 269
Konis, E., 84, 269
Kortschak, H. P., 58, 121, 255
Korven, H. C., 192, 269
Kozlowski, T. T., 286
Kramer, P. J., 118, 120, 269
Krause, H., 229, 269
Kristensen, K. J., 211, 249
Krogman, K. K., 187, 192, 269
Krysch, G., 31, 283
Kuiper, P. J. C., 91, 247
Kung, E. C., 14, 113, 115, 269
Kurtyka, J. C., 199, 269

Lachenbruch, A. H., 88, 269
Lake, J. V., 83, 269
Lal, K. N., 35, 47, 269, 281
Lamb, J., 66, 257
Lamoureux, W. W., 167, 269
Landsberg, H. E., 11, 82, 88, 200, 269
Langley, 4, 26, 293
Larkspur, 72
Larsen, A., 61, 63, 121, 122, 269
Larson, C. L., 199, 275
Larson, W. E., 96, 248
Larsson, P., 14, 269
Lawrence, E. N., 239, 270
Laycock, D. H., 66, 188, 270
Leaf age, effect of transmissibility, 36; effect on photosynthesis, 35
Leaf area index, basic concept, 46, 47
 dry matter as a function of, 49-51
 guide to cultural practice, 53-56
 measurement of, 46, 47
 net photosynthesis as a function of, 47-49
 optimum, 49, 51-53, 56, 209
 variation throughout the crop cycle, 51-53
Leaf arrangement, 36-39, 210
Leaf temperature, 83-86, 102, 123, 165, 228, 245
Leaf transmissiblity, 36, 37, 39, 40
Lee, D. H. K., 59, 270
Lehane, J. J., 213, 214, 216, 241, 242, 270, 282
Lehenbauer, P. A., 76, 270
Lehmann, P., 225, 228, 270
Lemon, E. R., 29, 31, 42-44, 60, 63, 112, 113, 115, 120, 125, 140, 142, 171, 186, 248, 250, 270, 274, 290
Lenschow, D. H., 14, 269
Lentil, 101, 242
Leonard, A. S., 102, 104, 105, 263

Lespedeza, 72
Letey, J., 175, 270
Lettau, H., 113, 115, 117, 270
Lettuce, 32, 68, 72, 91, 190, 210, 216, 241
Levine, G., 186, 211, 223, 264, 267
Levitt, J., 100, 270
Lewis, M. R., 224, 247
Leyer, G., 123, 282
Lichen, 130
Light distribution within the canopy, 39-42
Light utilization, efficiency of, 25-28, 37-39, 57, 58
Ligon, J. T., 205, 206, 270
Linacre, E. T., 84, 85, 270
Lindstrom, R. S., 32, 270
Lingle, J. C., 224, 261
List, R. J., 9, 270
Liu, B. Y. H., 270
Livingston atmometer, 190, 293
Lloyd, M. G., 226, 271
Logarithmic wind equation, 109-111, 113
Lomas, J., 186, 271
Long, I. F., 115, 225, 227, 271, 276
Long-day plants, 70-74
Lönnqvist, O., 16, 271
Loomis, R. S., 224, 271
Loomis, W. E., 25, 29, 31, 84, 86, 121, 249, 256, 271, 284, 286
Lougee, C. R., 150, 271
Loustalot, A. J., 119, 271
Love, J. R., 44, 175, 283
Lovett, W. J., 211, 223, 271
Lowry, R. L., 156, 271
Lowry, W. P., 134, 271
Lucerne, 14, 47, 144, 155, 171; see also Alfalfa
Lucey, R. F., 211, 271
Ludwig, L. J., 47, 271, 289
Lumley, J. L., 112, 271
Lupine, 101, 242
Lutwick, L. E., 187, 269
Lydolph, P. E., 233, 271
Lysenko, T. D., 75, 271
Lysimeter, 145-48

Magee, A. C., 211, 252
Magistad, O. C., 224, 286
Magness, J. R., 211, 216, 266, 271
Maher, F. J., 161, 259
Mair, H. K. C., 118, 276
Maize; see Corn
Major, J., 271
Makkink formula, 153-54
Makkink, G. F., 57, 137, 147, 150, 153, 180, 271, 272, 284
Mandioca, 190
Mangrove, 118
Mantell, A., 211, 272
Marani, A., 216, 272
Marigold, 238
Mason, S. C., 66, 272
Masson, H., 227, 272
Mateer, C. L., 9, 272
Mather, J. R., 132, 135, 136, 141, 150, 181, 199, 203, 272, 284
Matushima, S., 25, 272
Maximov, N. A., 129, 272
May, P., 66, 272
McArthur, W. C., 211, 252
McCulloch, J. S. G., 9, 66, 67, 167, 263, 272
McDole, G. R., 198, 248
MacDonald, T. H., 9, 261
McGehee, R. M., 230, 272
McGeorge, W. T., 230, 232, 253
McGuinness, J. L., 175, 265
McKibben, G. E., 222, 262
McIlroy, I. C., 146, 147, 161, 182, 186, 187, 272, 273
McMillan, W. D., 134, 286
Melons, 75, 101, 190, 242
Mesophytes, 118, 229
Meyer, B. S., 119, 273
Meyers, L. E., 147, 285
Mickelson, R. H., 222, 255
Mid-day depression, 28, 123, 147
Mihara, Y., 175, 176, 273
Millar, B. D., 141, 142, 273
Miller, D. H., 244, 273
Miller, E. E., 46, 273
Miller, P. M., 98, 286
Miller, R. J., 66, 224, 260, 261
Miller, S. R., 192, 265
Millet, 101
Milthorpe, F. L., 125, 227, 273
Mint, 190

Misra, D. K., 43, 250
Mitchell, J. W., 216, 266
Mitchell, K. J., 144, 273
Mitchell, P. K., 273
Modulated water balance method, 203, 204, 215
Mohr, H., 273
Moisture sensitive periods, 215, 216
Molecular diffusivity, 157, 159
Molga, M., 28, 29, 273
Molisch, H., 100, 273
Möller, F., 15, 273
Momentum, 109, 140, 158-61, 244, 293
Monsi, M., 36, 37, 267, 273
Monteith, J. L., 4, 11-14, 17, 19, 20, 40, 41, 102, 125, 164, 225-27, 267, 273, 274
Mooney, H. A., 28, 274
Moore, J. N., 211, 274
Morin, K. V., 151, 252
Morning glory, 72
Morris, L. G., 147, 274
Moss, D. M., 29, 63, 120, 274, 287
Moulsley, L. J., 285
Muck, 32, 89, 93
Mukammal, E. I., 131, 183, 191, 274
Mulch, 98, 99
Munn, R. E., 161, 244, 274
Munro, J. M., 274
Murata, Y., 25, 63, 274, 289
Murwin, H. F., 211, 262
Musgrave, R. B., 28, 29, 41, 42, 63, 120, 250, 265, 274
Musick, J. T., 220, 274
Mustard plants, 138, 139, 154

Nakagawa, Y., 19, 274
Nakayama, F. S., 260
Nagel, J. F., 230, 274
Nageli, W., 238, 239
Nash, A. J., 274
Nebiker, W. A., 130, 275
Nelson, L. B., 208, 275
Nelson, R. M., 198, 275
Net assimilation rate; see Net photosynthesis
Net photosynthesis, 32-35, 37, 38, 46, 47, 53, 57, 63, 76, 79, 123, 293
  as affected by leaf age, 35, 49
  as affected by virus, 34, 35
  as a function of extinction coefficient, 47-49
  as a function of leaf area, 47-49
Net radiation, 16-20, 22, 42-45, 125, 137, 142, 143, 163, 164, 166-68, 170-72, 174-76, 179, 222, 245, 293
Net radiometer, 16, 102
Neumann, J., 133, 275
Neutron scattering meter, 194, 195
Newton, O. H., 230, 275
Nichiprovich, A. A., 39, 275
Nielsen, B. F., 17, 249
Nieuwolt, S., 275
Nixon, P. R., 223, 280
Njoku, E., 73, 74, 275
Noffsinger, T. L., 59, 86, 275
Nordenson, T. J., 167, 183, 268, 275
Norum, E. B., 222, 290
Norum, E. M., 199, 275
Nutman, F. J., 25, 275
Nuttonson, M. Y., 1, 275
Nyctotemperature, 79

Oak, 87, 120
Oasis effect, 141-43, 293
Oats, 72, 75, 78, 101, 117, 123, 126, 128, 216, 220
O'Bannon, J. H., 130, 275
Odell, R. T., 216, 279
Ogata, G., 129, 275
Oguntoyinbo, J. S., 275
Oil palm, 60
Okabe, T., 25, 272
Olive, 216, 224
Onion, 72, 190, 216
Orange, 107, 115
Orchard, 72
Ormrod, D. P., 33, 274
Orvig, S., 19, 274
Osada, A., 25, 289
Outgoing radiation, 14-16, 98, 106, 224, 227, 228, 245
Overholser, E. L., 119, 248
Owen, P. C., 35, 47, 200, 274, 276
Oxalis, 25
Oxygen atom, 6, 7
Oxyria digyna, 238
Ozone, 6, 7, 15

Pack, A. B., 287
Palti, J., 230, 259
Paltridge, T. B., 118, 276
Pan evaporation, ratio between evapotranspiration and, 185-90
Panofsky, H. A., 112, 271
Pansy, 72
Papaya, 86
Parker, N. W., 75, 276
Parks, R. R., 106, 280
Partridge, J. R., 150, 255
Pasquill equation, 158, 160
Pasquill, F., 158, 276
Pasture, 47, 56, 73, 76, 153, 187, 192, 220, 222, 223
Patten, H. E., 89, 273
Pea, 72, 77, 80, 82, 101, 128, 190, 216
Peach, 78, 192, 193, 211, 216
Peanut, 80, 101, 211, 232
Pear, 78, 211
Pearson, C. H. O., 186, 187, 209, 276, 284
Pearson, R. W., 222, 288
Peat, 93, 94
Pecan, 119
Peck, A. J., 88, 258
Peek, J. W., 99, 276
Pelton, W. L., 113, 117, 164, 276, 283
Pendleton, J. W., 99, 276
Penman equation, 166-70, 178, 180, 191, 201
Penman, H. L., 31, 115, 125, 128, 129, 131, 133, 140, 166, 170, 179, 209, 211, 276
Pepper, 72
Percolation, 195, 206
Pereira, H. C., 66, 67, 187, 272, 276
Perkman, O., 211, 288
Permanent wilting point, 120, 134, 137, 198, 202, 204, 210, 293
Peters, D. B., 99, 175, 177, 232, 249, 265, 270, 276, 277
Peterson, A. E., 44, 174, 283
Peterson, M. L., 119, 210, 264, 285
Petunia, 72, 81
Philip, J. R., 47, 48, 140, 257, 277
Philodendron, 25
Phlox, 72
Photochemical tubes, 7
Photoelectric method, 46
Photometer, 11
Photoperiod, 68-74, 293
Photoperiodic after-effect; see Photoperiodic induction
Photoperiodic induction, 71
Photoperiodism, 6, 23, 70-74, 294
  classification, 70-72
  as a factor in plant distribution, 73
  historical background, 70
  practical applications of, 74
  relation to temperature, 71, 73
  response of tropical plants to, 73, 74
Photosynthesis, 6, 23-35, 163, 164, 209, 210, 231, 233, 294
  basic process, 23, 24
  effect of water on, 118-21
  efficiency of, 24-28
  in relation to carbon dioxide, 28-32
  in relation to light, 24-28
  in relation to temperature, 28
Phototemperature, 79
Phototron, 3, 79
Picha, K. G., 16, 277
Piche evaporimeter, 167, 168, 180, 191, 240, 241, 294
Pierce, L. T., 135, 136, 203, 277
Pieters, G. A., 28, 277
Pine, 73, 187, 215, 217, 230
Pineapple, 14, 17, 20, 47, 86, 98, 130, 186, 188
Pogosian, Kh. P., 224, 254
Poinsettia, 72
Polar region, 14, 28, 87, 94, 130
Polhemus, D. N., 226, 287
Pollak, L. W., 200, 257
Ponce, I., 216, 258
Pool, R. J., 94, 277
Popov, O. V., 147, 277
Poppy, 101
Populus acuminata, 84
Portman, D. J., 16, 277
Posey, J. W., 13, 277
Potato, 14, 25, 28, 29, 35, 47, 53, 54, 58, 72-74, 80, 95, 96, 101, 155, 190, 211, 216, 223, 242
Potential evapotranspiration, 143, 144, 196, 220, 221; definition, 129-31; methods of determining or estimating, 145-93

Potential maximum evapotranspiration, 131, 143, 144, 170, 171, 183, 201
Potential photosynthesis, 58-61, 77, 243
Pratt, A. J., 211, 252
Precipitation, 184, 185, 195-97, 199-202, 204, 206-208, 224, 225, 243, 245
Prescott, J. A., 8, 9, 76, 150, 187, 251, 255, 277
Priestley, C. H. B., 16, 84, 146, 160, 161, 227, 229, 258, 277
Principle of similarity, 158-60, 165
Pruitt, W. O., 17, 131, 134, 147, 151, 153, 160, 161, 169-72, 178, 182, 187-89, 191, 277, 278, 286
Prune, 78, 192, 193
Psychrometric constant, 165, 166, 294
Pyrgeometer, 15, 294
Pyrheliometer, 7, 11, 292

Quantum, 24, 57, 294
Quercus macrocarpa, 84, 86

Rabinowitch, E. I., 24, 32, 278
Rackham, O., 250
Radiation, solar, depletion by the atmosphere, 6, 7
  distribution, 9
  distribution within plant community, 36-45
  effect on plant growth, 23
  fraction of, used in evapotranspiration, 170-72
  in greenhouse, 20-22
  interception by tall vegetation, 143
  relation between, and net radiation, 17
  in relation to cloudiness, 9
  in relation to duration of sunshine, 8, 9
  see also Global radiation
Radiation frost, 100, 102, 107
Radiation utilization, during successive stages of crop development, 60-62; efficiency of, by field crops, 40, 57-69
Radish, 72, 210, 216
Rainfall; *see* Precipitation
Ramakishran, S. S., 181, 286
Ramdas, L. A., 11, 83, 278
Raming, R. E., 119, 268
Ramirez, R., 73, 257
Raney, W. A., 141, 212, 222, 248, 278
Raschke, K., 133, 278
Raspberry, 232
Rasumov, V. I., 74, 278
Rauwolfia vomitoria Afzel, 73
Read, D. W. L., 248
Read, R. A., 240, 289
Redwood, 230
Reed, H. S., 198, 250
Rees, A. R., 46, 169, 256, 267
Reeves, W. E., 17, 129, 285
Reflection coefficient, 11-14, 16, 20, 99, 166, 174, 181, 224, 245, 294
Reflectivity; *see* Reflection coefficient
Reichert, G. L., 66, 260
Reichert, I., 230, 259
Reifsnyder, W. E., 287
Rense, W. A., 4, 278
Respiration, 32-34, 57-59, 119, 210, 220, 294
Reynold stress, 158
Reynolds, C. W., 216, 223, 286
Reynolds, H. W., 130, 275
Rhoades, D. G., 104, 253
Rhoades, H. F., 216, 220, 266
Rice, 19, 25, 33, 39, 47, 53, 54, 60, 72-74, 101, 115, 118, 153, 175, 176, 183, 187, 190, 210
Richards, L. A., 129, 185, 275, 278
Richardson number, 111, 112, 294
Rickard, D. S., 150, 204, 261, 278
Rider, N. E., 115-17, 159, 160, 170, 266, 278
Ries, S. K., 32, 266
Rijkoort, P. J., 167, 278
Riley, J. A., 96, 230, 275, 278
Robb, W., 32, 289
Robertson, G. W., 137, 139, 191, 203, 204, 250, 266
Robins, J. S., 179, 216, 279
Robinson, F. E., 211, 215, 256
Robinson, G. D., 15, 279
Rognerund, B., 216, 283
Rohwer, C., 158, 279
Roller, E. M., 212, 222, 248
Root/shoot ratio, 119, 236
Rose, 32
Rosenberg, N. J., 279
Roughness, 109, 110, 113-17, 129, 141, 143, 157, 161, 178, 188, 191, 236-38, 244, 245, 294
Rudolph, P. O., 241, 279
Runge, E. C. A., 216, 279
Runoff, 195, 206, 294

Russell, G. C., 54, 248
Russell, M. B., 175, 211, 223, 267, 277
Rutherford, W. M, 192, 265
Rutter, A. J., 273
Rye, 14, 72, 75
Ryegrass, 37, 39, 72, 78, 134, 144, 171, 189

Sacaton, 153
Saeki, T., 37, 39, 48, 49, 271, 273, 279
Safflower, 101, 130
Saintpaulia, 25
Sale, P. J. M., 216, 279
Salim, M. H., 129, 279
Salter, P. J., 63, 210, 216, 279
Salvia, 72
Samish, R. M., 224, 279
Sanderson, M., 150, 279
Sanding, 108
Satterwhite, J. E., 125, 140, 142, 171, 186, 270
Saturation light intensity, 25-28, 47, 210, 294
Schanderl, H., 225, 228, 270
Schleter, J. C., 14, 263
Schneider, G. W., 119-21, 280
Schofield, R. K., 128, 133, 276
Scholte-Ubing, D. W., 17, 20, 137, 170, 280
Schultz, H. B., 104, 106, 253, 280
Schuurman, J. J., 57, 280
Schwab, G. O., 223, 280
Scrase, F. J., 110, 280
Sedum, 72
Seif, R. D., 66, 260
Sekihara, K., 11, 280
Sellers, W. D., 117, 244, 280
Senecio nebrodensis, 234, 235
Sesame, 101
Seybold, A., 233, 280
Shadbolt, C. A., 46, 273
Shade experiment, 65-69
Shade species, 25, 26, 28, 33
Shantz, H. L., 125-27, 198, 233, 253
Shaw, C. F., 98, 280
Shaw, R. H., 19, 43-45, 85, 86, 136, 137, 175, 179, 186, 189, 190, 216, 223, 258, 261, 280, 289, 290
Shearing stress, 109, 110, 158, 294
Shelterbelts, 238-42
Shmueli, E., 211, 280
Short-day plants, 70-74
Shrader, W. D., 223, 280
Shreve, F., 94, 281
Sibbons, J. L. H., 151, 281
Silva, S., 58, 286
Singh, B. N., 35, 281
Singh, R., 216, 281
Slatyer, R. O., 43, 231, 250, 281
Slope, 94, 95
Smith, A., 98, 281
Smith, A. M., 84, 281
Smith, G. E., 222, 281
Smith, K., 178, 281
Smithsonian Institution, 4
Snapbean, 211, 216, 223
Snapdragon, 32, 72
Sneddon, J. L., 211, 263
Soil moisture depletion curve, 133-40, 203, 204, 245
Soil moisture measurements, 194, 195
Soil moisture storage capacity, 88, 93, 195-200, 204-208, 217, 245
Soil temperature, 87-99, 164, 245
 annual range, 89, 90
 diurnal range, 89, 90, 92-95
 effect of aspect on, 94
 effect of slope on, 94
 effect of, on crop yield, 95-97
 effect of texture on, 90-94
 effect of tilth on, 95
 lag of, 89, 90
 methods of measurement, 87, 88
 methods of modifying, 97-99
 physical laws governing the change of, 89, 90
 records of, 88
 significance of, 87
Soil texture, classification, 294; effect on moisture depletion, 136, 138, 139; effect on soil temperature, 90-94
Solar constant, 4, 292
Solar spectrum, 4-6, 294
Sonmor, L. G., 281
Sorghum, 72, 75, 101, 128, 153, 171, 173, 186, 220
Soybean, 25, 28, 71, 72, 101, 175, 203, 216, 223
Specific heat, 88, 158, 166
Spector, W. S., 71, 281

Spell, D. P., 78, 266
Spiegel, P., 216, 224, 279, 281
Spinach, 14, 72, 73, 210
Spoehr, H. A., 57, 281
Sprinkling, 104-106
Squash, 107, 229
Sreenivasan, P. S., 211, 281
Stage, A. R., 215, 217, 290
Stalfelt, M. G., 233, 281
Stanhill, G., 20, 34, 41, 42, 58, 153, 154, 167-69, 179, 183, 184, 186, 191, 216, 220, 222, 281, 282
Staple, W. J., 213, 214, 216, 239, 241, 242, 270, 282
Steelmann Nielsen, E., 282
Stefan-Boltzman constant, 14, 15, 166, 292
Stephens, J. C., 178, 288
Stern, W. R., 14, 43, 49, 51, 130, 250, 261, 282
Stewart, W. D., 66, 249
Stock, 72, 81
Stocker, O., 123, 282
Stoeckler, J. H., 242, 282
Stoll, A. M., 84, 282
Stoma, 119, 120-23, 130, 133, 178, 188, 236, 295
Stone, E. C., 229, 230, 282
Stoughton, R. H., 66, 68, 72, 282
Strawberry, 29, 67, 71, 72, 80, 98, 99, 242
Strict long-day plants, 71
Strokhina, L. A., 17, 254
Struchtemeyer, R. A., 216, 223, 282, 286
Stumpf, H. T., 185, 278
Subba Rao, M. S., 47, 269
Sugar beet, 14, 25, 27, 29, 33, 35, 39, 47, 49, 51, 53, 54, 56, 58, 60, 61, 72, 80, 101, 128, 153, 190, 192, 193, 200, 211, 224, 227
Sugar cane, 12, 14, 17, 25-29, 35, 47, 56, 58, 60, 62-65, 72, 73, 80, 117, 119-21, 143, 146, 151, 160, 170, 186, 187, 190, 203, 209, 211, 214, 215, 220, 234
Sukhovey, 233
Sultana, 66
Sumner, C. J., 185, 273, 282
Sun species, 25, 26, 28, 33
Sunflower, 25, 28, 66, 72, 101, 119
Sunshine duration, 8, 9, 245
Suomi, V. E., 16, 146, 147, 163, 268, 283
Surplus, 196, 197, 205-208
Sutton, O. G., 244, 283
Suzuki, S., 178, 283
Svinukhov, G. V., 11, 286
Swedes, 58
Swinbank, W. C., 16, 146, 160, 227, 229, 258, 283
Szeicz, G., 12-14, 17, 19, 20, 274

Tai, K., 210, 263
Takahashi, D., 58, 121, 255
Takeda, T., 39, 40, 52, 53, 283
Tall vegetation, 158, 183, 188
 evapotranspiration rate of, 143, 144
 net radiation of, 20
 radiation intercepted by, 143
 wind profile over, 112, 113
Tamarisk, 153
Tamm, E., 31, 283
Tanimoto, T., 58, 121, 255
Tanner, C. B., 44, 110, 113, 117, 133, 140, 145, 147, 163-65, 175, 222, 264, 268, 283
Taylor, R. J., 161, 256, 283
Taylor, S. A., 20, 171, 211, 216, 247, 249, 283
Tea, 66, 67, 188, 242
Temperature, air: *see* Air temperature
 inversion, 83, 101-105, 111, 112, 114, 159, 165, 295
 leaf: *see* Leaf temperature
 length of record, 82, 83
 soil: *see* Soil temperature
Tensiometer, 194, 295
Tephrosia, 72
Terrestrial radiation; *see* Outgoing radiation
Tesar, M. B., 211, 271
Theobroma cacao L., 73
Thermal conductivity, 88-91, 93, 98, 104, 295
Thermal diffusivity, 89, 90, 295
Thermal properties of soil, 88, 89
Thermometer, mercury, 87; resistance, 87; thermocouple, 84, 87, 296
Thermoperiodicity, 79-81, 296
Thimann, K. V., 29, 284
Thoday, D., 119, 284
Thomas, M. D., 25, 27, 32, 33, 284
Thompson, B. W., 11, 284
Thompson, G. D., 186, 187, 209, 276, 284
Thorne, G. N., 35, 46, 284

Thornthwaite, C. W., 14, 135, 136, 141, 146, 149, 159, 160, 178, 198, 199, 203, 228, 284
Thornthwaite formula, 149-51, 169, 170, 179
Thornthwaite-Holzman equation, 159, 160, 165
Till, M. R., 284
Timothy, 39, 72, 187
Tizio, R., 216, 258
Tobacco, 25, 34, 54, 66, 70, 72, 80, 101, 211, 223, 224, 242
Todd, G. W., 129, 279
Tomato, 15, 27-29, 32, 47, 54, 72, 79, 80, 85, 101, 107, 190, 192, 193, 211, 232, 242
Tomlinson, B. R., 153, 284
Transeau, E. N., 58, 285
Translocation, 79, 80, 122, 234, 296
Transmissibility, leaf; see Leaf transmissibility
Transpiration, 121-23
 difference between evaporation and, 133
 effect of wind on, 233, 234
 relationship between, and dry matter production, 123-28, 176, 209
 relative magnitude of evaporation and, 174-77
Transpiration/photosynthesis ratio, 123-25
Transplantation, 54, 56
Trickett, E. S., 285
Tropical plants, 28, 73, 100
Tropics, 13, 17, 19, 26, 59, 60, 66, 73, 74, 80, 82, 95, 125, 164, 170
Trumbull, R. S., 198, 248
Tugwell, C. P., 160, 170, 266
Tule, 186
Turbulence, 111, 112, 157, 159, 161, 296
Turbulent diffusivity, 157
Turc, L., 154, 285
Turc's formula, 154-56
Turley, R. H., 223, 285
Turnip, 58, 72, 101, 216
Turrell, F. M., 129, 285
Tyler, K. B., 211, 255

Uchijima, Z., 285
Uhland, R. E., 208, 275
Ultraviolet radiation, 4-7
Ulrich, A., 80, 285
U.S. Bureau of Reclamation, 199, 285
Upchurch, R. P., 119, 210, 264, 285
Uriu, K., 216, 285

Vaadia, Y., 209, 264
Van Bavel, C. H. M., 17, 129, 140, 147, 150, 159, 170, 204, 205, 206, 211, 223, 260, 261, 263, 264, 285
Van der Bijl, W., 151, 285
Van der Paauw, F., 216, 285
Van Dobben, W. H., 73, 82, 286
Van Doren, C. A., 177, 266
Van Heemst, H. D. J., 137, 272
Van Hylckama, T. E. A., 198, 286
Van Sluis, E. J. H., 74, 286
Van Wijk, W. R., 150, 286
Van't Hoff's law, 76, 77
Vapor pressure deficit, 131, 132, 156, 171, 172, 231
Veihmeyer, F. J., 134, 135, 192, 203, 211, 223, 264, 265, 286
Veldman, E. C., 203, 287
Ventikeshwaran, S. P., 181, 286
Ventskevich, G. Z., 100, 286
Verduin, J., 25, 121, 286
Verle, E. K., 11, 286
Vernalization, 75, 296
Vetch, 72, 101
Vetchling, 101
Viets, F. G., 212, 286
Vieweg, G. H., 123, 282
Villanueva, J., 16, 277
Vince, D., 68, 71, 282
Vincente-Chandler, J., 58, 286
Vineyards, 105
Violet, 72
Visible light, 5, 6, 11, 14, 24, 42, 43, 57, 68
Vittum, M. T., 216, 223, 286
Voight, G. K., 32, 286
Volumetric heat capacity, 88
Von Karman's constant, 109, 159
Von Mohl, H., 87, 286
Vowinckel, E., 11, 259

Wadleigh, C. H., 224, 286
Wadsworth, R. M., 234, 238, 286, 287
Waggoner, P. E., 79, 98, 122, 286, 287, 288
Waldo, G. F.,
Wallen, C. C.,

Wallin, J. R., 226, 287
Walnut, 192, 193
Wang, Jen-yu, 77, 96, 97, 169, 287
Wang, S. C., 169, 287
Warburg, O., 24, 287
Ward, R. C., 178, 287
Warren Wilson, J., 28, 37, 38, 234, 238, 287
Wartena, L., 183, 203, 287
Warming, E., 118, 287
Wassink, E. C., 57, 58, 287
Water, effect on photosynthesis, 118-21; general effect on plant growth, 118-28; relationship between, and yield, 209-224
Water absorption by leaves, 229
Water balance, 1, 194-209, 215, 218, 225, 243
Water use, 22; efficiency of, in dry matter production, 125, 218-22
Water vapor, radiation absorption by, 6, 14
Watson, D. J., 28, 39, 46, 53, 54, 200, 211, 276, 287, 288
Wave number, 4, 5
Weaver, H. A., 178, 211, 222, 259, 288
Weaver, J. B., 99, 260
Weaver, J. E., 118, 288
Webb, E. K., 161, 283, 288
Webster, G. R., 223, 285
Weger, B., 203, 288
Weidner, V. R., 14, 263
Weihing, R. N., 78, 288
Wellensiek, S. J., 79, 288
Wells, S. A., 216, 288
Went, F. W., 3, 27, 79-81, 229, 230, 288
West, E. S., 211, 288
Westlake, D. F., 288
Wetzel, K., 229, 288
Wheat, 14, 25, 29, 47, 53, 54, 58, 59, 72, 75, 101, 115, 128, 130, 155, 160, 177, 211, 213, 214, 216, 218, 220, 227, 241, 242
Wheaton, R. Z., 106, 107, 288
Whillhite, F. M., 171, 186, 263
Whitehead, F. H., 234-37, 273, 289
Whittingham, C. P., 289
Wiggans, S. C., 78, 289
Wig-wag, 7, 296
Wilcox, J. C., 192, 269
Williams, R. F., 47, 289
Willis, W. O., 177, 264
Willow, 153
Wilson, H. W., 233, 289
Wind, effect on carbon dioxide intake, 234; effect on mechanical breakage, 234; effect on transpiration, 233, 234
Wind frost; see Advection frost
Wind machine, 98, 104
Wind profile, 109-117, 160; logarithmic equation, 109-111; over tall vegetation, 112, 113
Windbreaks; see Shelterbelts
Winneberger, J. H., 122, 289
Winter, E. J., 216, 289
Withrow, R. B., 289
Witts, K. J., 39, 288
Wittwer, S. H., 32, 289
Wolf, F. A., 224, 289
Wood, R. A., 66, 179, 188, 265, 270, 274
Woodruff, N. P., 240, 289
Work, R. A., 224, 247
Worker, G. F., 224, 271
World Meteorological Organization, 88, 182, 245, 289

Xerophytes, 118, 129, 199, 229, 236
Xiao, Wen-jun, 11, 289

Yakuwa, R., 92, 93, 289
Yamada, N., 25, 289
Yamaguchi, S., 25, 272
Yamamoto, G., 15, 226, 289
Yao, Y. M., 289, 290
Yapp, R. H., 272
Yefimova, N. A., 17, 254
Yegnanarayanan, S., 11, 278
Yocum, C. S., 42-44, 248, 290
Young, A. A., 181, 290
Young, C. P., 167, 290

Zahner, R., 215, 217, 290
Zelitch, I., 122, 287, 290
Zero plane displacement, 113, 117, 159, 160, 296
Zink, F. W., 107, 265